Biogeography of Freshwater Algae

Developments in Hydrobiology 118

Series editor
H. J. Dumont

Biogeography of Freshwater Algae

Proceedings of the Workshop on Biogeography of Freshwater Algae, held during the Fifth International Phycological Congress, Qingdao, China, June 1994

Edited by

Jørgen Kristiansen

Reprinted from Hydrobiologia, vol. 336 (1996)

Kluwer Academic Publishers
Dordrecht / Boston / London

Library of Congress Cataloging-in-Publication Data

Biogeography of freshwater algae / edited by Jørgen Kristiansen.
p. cm. -- (Developments in hydrobiology ; 118)
Based on a workshop held during the Fifth International Phycological Congress in Qingdao, China, 25 June - 1 July 1994.
ISBN 0-7923-4253-4 (HB : alk. paper)
1. Freshwater algae--Geographical distribution--Congresses.
2. Freshwater algae--Ecology--Congresses. I. Kristiansen, Jørgen.
II. Series.
QK570.25.B56 1996
589.392'9--dc20 96-40994

ISBN 0-7923-4253-4

Published by Kluwer Academic Publishers,
P.O. Box 17, 3300 AA Dordrecht, The Netherlands

Kluwer Academic Publishers incorporates
the publishing programmes of
D. Reidel, Martinus Nijhoff, Dr W. Junk and MTP Press.

Sold and distributed in the U.S.A. and Canada
by Kluwer Academic Publishers,
101 Philip Drive, Norwell, MA 02061, U.S.A.

In all other countries, sold and distributed
by Kluwer Academic Publishers Group,
P.O. Box 322, 3300 AH Dordrecht, The Netherlands

Printed on acid-free paper

Printed in Belgium

Contents

Hydrobiologia **336:** vi, 1996.
J. Kristiansen (ed.), Biogeography of Freshwater Algae.

Introduction

This volume on "Biogeography of Freshwater Algae" is based on a workshop held during the Fifth International Phycological Congress in Qingdao, China, 25 June–1 July 1994.

During the workshop several of the participants had expressed the view that the proceedings should be published. It was felt that such a publication would fill a serious gap in the phycological literature, as there was no recent survey of freshwater algal biogeography.

However, the contributions presented by the workshop participants only gave some aspects of the whole complex. Therefore, it was decided to give a more comprehensive coverage of the subject by inviting several other specialists on various groups of algae or in different fields to contribute.

I thank both categories of authors for their contributions and for their cooperation. It has been tried in this volume to give a survey of the state of the art in the biogeographical studies of freshwater algae. As many as possible of the different aspects and approaches have been included, using classical and modern methods, and from taxonomic, ecologic, or more theoretical viewpoints.

As far as I can see, this is the first attempt to publish such a volume. It is certainly incomplete, reflecting the status of this research field, and certain areas could not be covered. Still, I hope it will be of interest and that it will help to promote studies of distribution and dispersal of freshwater algae.

THE EDITOR

Hydrobiologia **336**: 1–17, 1996.
J. Kristiansen (ed.), Biogeography of Freshwater Algae.

2. Genome rearrangement and speciation in freshwater algae

Terunobu Ichimura
Institute of Algological Research, Hokkaido University, Muroran, Hokkaido, Japan 051

Key words: biogeography, biological species, chromosomal rearrangement, life cycle, isolating mechanisms

Abstract

Speciation problems are reviewed in the context of biogeography of fresh-water algae. Currently accepted species concept in phycology is based on morphological characters, and according to this concept, most freshwater algal species are considered cosmopolitan. This implies whether they have a highly efficient means of dispersal or their morphological characters are very static through a long evolutionary time. Recent studies of reproductive isolation show that some biological species of fresh-water algae are not so static or may not have such a high power of dispersal means, though some are indeed very static in morphological characters. The life cycle of most freshwater algae is composed of a vegetative cycle of growth and reproduction and a sexual cycle of gametic fusion and meiosis in the zygote, which forms a dormant spore-like structure. Since any freshwater habitat is ephemeral in terms of evolutionary time scale, each species has a capacity of forming germlings from a dormant cell in order to recycle its life history. The genome of freshwater algae, therefore, contains various coadapted gene systems, at least two, for the vegetative and for the sexual cycle. Homothallism and heterothallism are two contrasting mating systems that represent two opposing ways of life to harmonize antagonism between the vegetative stage of growth and reproduction and the sexual and dormant stage. Geographic and ecological distribution, polyploidy, and sex determination are discussed in conjunction with sexual and postzygotic isolating mechanisms.

Introduction

The problem of speciation has long been debated since Darwin and Wallace's discovery of the natural selection theory. During the founding age of the modern synthetic theories of evolution, many active biologists, such as Dobzhansky (1951), Mayr (1942) and Wright (1931), explored the empirical and theoretical problems of modes and mechanisms for speciation. Meanwhile, under the strong influence of inductive experimental biology, it may appear to be eclipsed somehow from the disciplines of major biological sciences. Nevertheless, recent developments in knowledge about and technologies analyzing diverging populations of organisms resurrect the speciation problems as a major issue in evolutionary biology (Coyne, 1992; Otte et al. 1989). Although such a trend appears to be still poor in phycology, there are several lines of information that are most pertinent and valuable for seeking a materialized and fundamental theory of speciation for all eukaryotic organisms including algae. In this essay, I attempt to review such information obtained in some groups of freshwater algae.

Taxonomic species and biological species

The species concept for most phycologists is based on the morphologic characters and hence the term "species" means morphospecies. On the other hand, for many students of evolutionary studies, the term means the biological species. According to Mayr (1982, p. 273), 'a species is a reproductive community of populations (reproductively isolated from others) that occupies a specific niche in nature.'

As argued by Mishler et al. (1982) and McCourt et al. (1990), however, it seems unproductive and inconvenient to restrict the usage of the term "species" to mean exclusively one or another. Therefore, I use it for both the meanings of morphospecies and biological species, but whenever necessary, choose either one to specify the nature of species. Other terms, such as 'syngen' (Sonneborn, 1957) and 'mating group' (Ichimura, 1981; 1985), which are almost equivalent to biological species, may be used to follow the progress in this field on particular organisms.

Geographic and ecological distribution and reproductive isolation

According to the morphological species concept, most species of freshwater algae are considered to be cosmopolitan. In other words, they may be found all over the world, if studied extensively by specialists. This implies either they have a highly efficient means of dispersal or their morphological characters are very static through a long evolutionary time. Recent phycologists, however, seem to pay attention to intraspecific variation in physiological and biochemical traits (Manhart et al., 1992; Wood et al., 1992). On the other hand, careful observations of natural populations sometimes may reveal the presence of reproductive isolation within a morphospecies or between infraspecific taxa (Geitler, 1975; Mann, 1989). Unfortunately we have not yet obtained adequate information on such beings as to how they become diversified.

Allopatry and sympatry

The problem of speciation contains two basic questions: "(1) how do new populations within a species develop?; (2) how do such populations become reproductively isolated from other populations of the parental species?" (Mayr, 1976, p. 146). Mayr is the champion of allopatric speciation who has consistently held the view that the differentiation of populations must be preceded by geographical or other means of physical isolations (Mayr, 1942; 1959; 1963). When he has discussed the cases of presumptive sympatric differentiation of ecological races of phytophagous insects that obligatorily depend on particular species or races of host plants, Mayr (1959) has interpreted such cases as being isolated by microgeographical conditions because of their low vagility. According to him, sympatric speciation in land snails on oceanic islands also can be understood in the same way. White (1978) has defined vagility in evolutionary biology as "the mean distance between the point at which an individual comes into being (by fertilization) and the point at which it meets a mate (or strictly speaking, gives rise to a new zygote)." Therefore, mere dispersal of propagules or individuals has little relation to the change of population genetic parameters, unless such immigrants experience sexual cycles with resident population. In the phycological literature, we have very limited records of the occurrence of sexual reproduction of freshwater algae in nature (Hoham, 1975; Hoham et al., 1977). Even in the Zygnematophyceae, in which relatively large numbers of observations on conjugation and zygospore formation have been recorded (Gerrath, 1993; Homfeld, 1929; Hoshaw et al., 1988), we have as yet few analytical works to estimate the frequency of sexual cycles in nature (Dell'Uomo et al., 1985; Ohtani, 1985). We may, however, be able to obtain indirect estimation of the frequency from analyses of size variation by frequent samplings of vegetative cells in diatoms (Mann, 1988) or from quantitative assays of dried soil (mud) samples in which the resistant zygospores of freshwater algae have been deposited (Yamagishi et al., 1974).

Allopatry and sympatry are, however, debatable concepts (Mayr, 1976; White, 1978). If most cosmopolitan freshwater algae had such high powered dispersal means as tempting Provasoli et al. (1959) to imagine constant ubiquitous inoculation of propagules, allopatry and sympatry would be meaningless concepts for our discussion of speciation in freshwater algae. We know, indeed, many common morphospecies of freshwater algae like *Chlorella* and *Stichoccous* that have often been found in temporal waters everywhere in the world. From the presently available information that will be shown below, such freshwater algae may not be the majority but the minority. Perhaps, only certain limited groups of species and genera, compared to the large number and diversity of freshwater algal taxa, seem to have such a strong colonizing ability for a wide variety of habitat (Round, 1981, p. 47–51). In general, however, it is admittedly true that like ferns and mosses, most freshwater algae have more chances of passive transport than most higher plants and animals. Neverthelss, facts and figures for their dispersal and colonizing abilities seem to be very obscure. For instance, the wide distribution patterns of some chrysophycean species make a keen scientist wonder about the existence of certain unknown transport mechanisms other than that known in many green algae (Brown, 1971; Coesel et al., 1988; Maguire, 1963; Proctor, 1959, 1962, 1965; Proctor et al., 1965; Stewart et al., 1966), since they seem to be too delicate to be transported for a long distance by such a means (Kristiansen 1981).

Recent molecular studies indicate that some morphospecies of freshwater algae seem to encompass larger or comparable genetic variation in mitochondrial and plastid genomes (Coleman et al., 1994; Kowallik, 1989) than do angiosperms at the generic level (cf. Soltis et al., 1992). This may suggest their antiquity of origin and the static nature of their morphological characters. If so, small wonder that most of them are cosmopolitan, even if they have only small or moderate abilities of dispersal. But another question must be asked as to why they can be so static. It may be plausible that a paleontological hypothesis, for animals, that more complex forms change more rapidly (Schopf et al., 1975) may well be extended down to the primitive freshwater algae.

Nevertheless, we still must inquire about the underlying genetic mechanisms by which the morphological identities have been maintained among world-wide populations of the same species (Van Valen, 1982). Dobzhansky (1951) has introduced the biological species concept based on the idea that species is a common gene pool. Collaborating with Dobzhansky for the development of the biological species concept, Mayr (1963) has argued that gene flows hold the integrity of species. The effects of gene flow between natural populations have been extensively investigated both theoretically and empirically (Endler, 1977). Ehrlich et al. (1969) have argued "that selection is both the primary cohesive and disruptive force in evolution." Their argument is based on the following three points: "(i) gene flow in nature is much more restricted than commonly thought; (ii) populations that have been completely isolated for long periods often show little differentiation; and (iii) populations freely exchanging genes but under different selective regimes may show marked differentiation." Hutchinson (1968) also has taken a similar ecological approach on the species problem by extending his niche concept (Hutchinson, 1957). Such ecological views including the niche theory explaining the organic diversity in a given geographic area, however, inevitably embrace theoretical possibilities of sympatric speciation (speciation without physical isolation of populations). Accordingly, primitive microorganisms like freshwater algae might be likely to speciate more easily and more often than higher plants and animals because of their relatively large possible vacant niche (Hutchinson, 1959, 1961) and the tremendously long time since their appearance (Knoll, 1994; Tappan, 1980). But, there should be certain genetic constraints for sympatric speciation (Mayr, 1976, p. 144–175). One important genetic problem is how the mutations critical to the establishment of isolating mechanism are fixed in either of the two ecologically diverging populations. Theoretical predictions suggest that inbreeding by assortative mating may allow sympatric speciation (Felsenstein, 1981; Maynard Smith, 1966). Recently, Rice et al. (1993) have reviewed the experimental aspects on the problem. Another problem is the genetic homeostasis or the cohesiveness of coadapted gene systems that holds the integrity of species (Carson, 1975; Mayr, 1963, 1975). Although unfortunately most of these arguments are focused and based on research in animals, especially *Drosophila*, there is no reason to consider that such basic problems are not related to speciation of freshwater algae.

Natural populations studied by culture techniques

After the advent of the culture techniques to induce algal sexuality by Starr (1955), many clonal isolates of some freshwater algae from wide geographic areas were studied and the problem of reproductive isolation within a single morphospecies has arisen. Sexual isolation has been reported in *Astrephomene gubernaculifera* Pocock (Brooks, 1966), *Chlamydomonas eugametos* Moewus, *C. moewusii* Gerloff (Wiese, 1974; Wiese et al., 1977), *Closterium* spp. (Coesel, 1988; Ichimura, 1981, 1985; Ichimura et al., 1987; Lippert, 1967), *Cosmarium* spp. (Starr, 1959), *Gonium pectorale* Müller (Stein, 1958), *Pandorina morum* Bory (Coleman, 1959), *Eudorina* spp. (Goldstein, 1964), and *Volvulina steinii* Playfair (Carefoot, 1966), while no sexual isolation was observed in *Platydorina caudata* Kofoid (Harris et al, 1969). On the other hand, interspecific hybrid zygote formation has been reported in *Chlamydomonas* (Bell et al., 1983; Cain, 1979; Gowans, 1963; Hoshaw, 1965; Strehlow, 1929), *Pleurotaenium* (Ling et al., 1974, 1976) and *Spirogyra* (Allen, 1985; Hoshaw et al., 1985; Wang et al., 1986). Before considering the meanings of these findings, we should examine some details of the following three cases of cosmopolitan species that are, by contrast, very significant for considering the evolution of morphological characters and reproductive isolating mechanisms.

Gonium pectorale *Müller*

Although about six morphospecies have been described in the isogamous colonial green algal genus *Gonium* (Nozaki et al., 1994), extensive biosystematic studies are focused on only *G. pectorale*, which is known to be distributed world-wide (Coleman et al., 1994; Sako et al., 1991; Stein, 1958, 1965, 1966a, 1966b; Stein et al., 1976). At present, we have very limited biogeographic information of the other species (Nozaki, 1993; Pocock, 1955; Pringsheim, 1959; Watanabe, 1977). In *G. viridistellatum* Watanabe, recently, Nozaki (1989) has reported sexual and postzygotic reproductive isolation in one of the reciprocal crosses between strains from Kanagawa and Okinawa Prefectures, Japan, two hitherto known localities of this species. Further studies based on more localities will give light on the speciation problem of this and morphologically related species *G. octonarium* Pocock, *G. quadratum* Pringsheim, *G. multicoccum*, Pocock and perhaps, *G. pectorale*.

Most clonal isolates of *G. pectorale* show heterothallic sexuality and hence are easily tested for crossability between different geographic strains. Stein (1958) has first reported sexual isolation between certain strains from the United States, but her further investigations to cover more wide geographic areas turned out to be rather unsatisfactory as to concluding whether or not the populations studied might be divided into two or more sexually isolated groups (Stein, 1965). Degrees of sexual isolation or compatibility were variable among the strains studied. As suggested from the great variety of original habitats of those isolates, degrees of sexual compatibility seemed to be largely affected by temperatures, nutrients, and pHs of the mating conditions (Stein, 1966a, 1966b). Stein et al. (1976), however, have studied sexual compatibility among 31 clonal cultures isolated from soil of a single pond at Lemon Cove, California and recognized two distinct groups of clones between which no zygote formation was observed. Puzzlingly, both the two groups are sexually compatible with three additional clones. In addition, they also recognized complete sexual isolation between the 31 Lemon Cove clones and an additional 9 clones collected from the same pond or nearby in different year. Although mating intensity was variable among the strains of *G. pectorale*, germination in all zygote-producing crosses has been observed (Coleman et al., 1994; Stein, 1965). In short, Stein seemed to have a doubt whether all populations of *G. pectorale* might belong to a single biological species, because she could not find clear evidence to divide into two or more discrete entities from the extensive investigations on sexual isolation.

Recently, however, Coleman et al. (1994) have suggested that although similarly inconclusive from the basis of sexual compatibility, populations of the United States, Nepal, Tibet and Japan could be all interrelated genetically. Using four strains from Japan, India, and the east (Maryland) and west (British Columbia) coasts of North America, they examined the sequence of the internal transcribed spacer (ITS) of the nuclear ribosomal DNA and found a remarkable uniformity both in length (615–621 bp) and in sequence. Less than 5% of the ITS positions differ in pairwise comparisons, and fewer than 7% of the bases vary among all the clones, including sites of insertion/deletion. Coleman et al. (1994) have suggested that *G. pectorale* could be a good species both morphologically and biologically, having relative uniformity in morphology and chromosome number and some limited variation in ITS sequence with more variation in mitochondrial DNA sequence.

The Pandorina morum *species complex*

In the genus *Pandorina*, which is characterized by a spherical vegetative colony and by iso- or anisogamy with male gametes not forming sperm packets as in other volvocacean genera, seven morphospecies have been described, but only three of them were studied in culture (Nozaki, 1991). Only *P. morum*, however, is known to be distributed world-wide, often found together with *G. pectorale* in same ponds and pools, but the two other cultured species are known only from one or two localities. Most clonal isolates of *P. morum* also show heterothallic sexuality. In contrast to the case of *G. pectorale* described above, however, sexual isolation is so clearly demonstrated as to divide all natural populations studied into more than 20 discrete groups or syngens as they are called by Coleman (Coleman, 1959, 1977; Coleman et al., 1994). Clones belonging to the same syngen will cross and form zygotes, while no mixture of clones belonging to different syngens will display even the initial stages of gametogensis. In spite of the uniformity in gross morphology throughout the whole life cycle of *P. morum*, Coleman et al. (1977) have reported variable chromosome numbers from two to 12 among the known syngens, perhaps each having a fixed karyotype of that syngen. There seem to be some differences in zygospore arrangements and time of cell division in culture between some syngens. Known distribution ranges of the majority of *P. morum* syngens are within 300 km, while two of them are possibly world-wide in distribution (Coleman, 1977). Using five or six strains covering the distribution ranges, Coleman et al. (1994) have compared the ITS sequence of the nuclear ribosomal DNA of the two cosmopolitan syngens with each other and with *Gonium pectorale* mentioned above. Although one was more variable than the other, both the two cosmopolitan syngens of *P. morum* showed the degrees of variation comparable to that of the cosmopolitan morphospecies *Gonium pectorale*.

Coleman et al. (1994) have recognized some degrees of sexual and postzygotic isolation within a cosmopolitan syngen. Especially in syngen II, which is the most variable in the ITS sequence, Coleman (1977) has reported the appearance of morphologically abnormal F_1 hybrids and less than 50% F_1 survival in crosses between different geographic strains.

The Closterium ehrenbergii *species complex*

The morphospecies *Closterium ehrenbergii* Meneghini ex. Ralfs is a common desmid species and found worldwide, mostly in slightly eutrophic and alkaline stagnant waters (Ruzicka, 1977). Many populations sampled from paddy fields, ponds and lakes in Japan, Taiwan, Nepal and Australia have been studied biosystematically, and they were divided into more than a dozen mating groups, primarily based on sexual isolation (Ichimura, 1981; Ichimura et al., 1987). Morphologic and cytological studies, as well as those on the reproductive barriers of sexual and postzygotic isolation, clearly suggest that each mating group of this alga is a biological species (Ichimura, 1985; Ichimura et al., 1990b). It has been shown that populations of *C. ehrenbergii* are rich in variation, especially in temporal waters distributed from subtropical to warm temperate climatic zones, and that they are, in many cases, composed of two or more mating groups. It is noteworthy from a view-point of evolutionary ecology that populations of two mating groups that are known to be completely isolated at the initial stages of the mating reaction often coexist, intermingled with each other in one and the same habitat, whereas those that are incompletely sexually isolated are not so. Different mating groups, between which partial mating interactions are known to occur in the laboratory, are usually found in different habitats, even if they are distributed in the same restricted geographic area (Ichimura, 1981; Ichimura et al., 1987). At any event, in most cases, it was possible, with little difficulty, to identify complementary mating types for clonal isolates from such habitats. In other words, one may find only a few exceptional isolates which would not mate with any known mating-type strains of the stock cultures. We may well consider, therefore, that seasonal and/or occasional occurrences of drought and/or extremely high and low temperatures in such habitats would select, at least once a year, against any defective mutation that affects the sexual process of zygote formation and germination.

Francke and Coesel (1985) and Coesel (1989) have, however, reported the occurrence of populations of *C. ehrenbergii* apparently lacking sexuality in peaty sites in Holland and France (inland Normandy) which seldom or never dry up. They have, on the other hand, found sexual populations of the same morphospecies in shallow water bodies on sandy or clayey soil in France (coastal Normandy) which regularly dry out in dry summers. Coesel and his colleagues have reported that the coastal Normandy populations are sexually compatible with populations of this alga from the southern part of England (Coesel, 1988). Recently, it has become known that populations of *C. ehrenbergii* sampled from a stream and a lake in Denmark and also in Japan, exclusively from cold clear water streams and springs, are sexually compatible with those from England and France studied by Coesel and his colleagues (Ichimura et al., 1995). It has been confirmed that all these *C. ehrenbergii* populations belong to the same mating group, P, on the basis of sexual compatibility and viability of F_1 progeny (T. Ichimura, F. Kasai & P.F.M. Coesel, unpublished).

Although both the two complementary mating types of Group P were found in each locality in northern Europe (Coesel, 1988, 1989; Francke et al., 1985), no population of Group P has so far been found to have both *plus* and *minus* mating types coexisting in a single locality in Japan. Instead, populations of either *plus* or *minus* mating type alone have been found in every one of the localities surveyed (Figure 1). As to the distribution of mating types of Group P, there seem to be four geographic zones in Japan: Zone I, north-eastern Hokkaido; Zone II, south-western Hokkaido to Aomori; Zone III, Akita to Fukushima and Zone IV, Niigata to Shizuoka. We may find only *plus* mating type in Zones I and III and only *minus* mating type in Zones II and IV. From this, we cannot expect to find zygospores of Group P at natural habitats in Japan. All Japanese populations of Group P seem to propagate only by asexual reproduction, being maintained without sexual reproduction. Although we know nothing about how such an interesting distribution pattern has occurred, it clearly shows that Group P across the distribution range in Japan is divided into patchworks of parapatric populations. If studied by an appropriate means, such as by molecular genetic techniques, we may find patchwork distribution patterns of local popu-

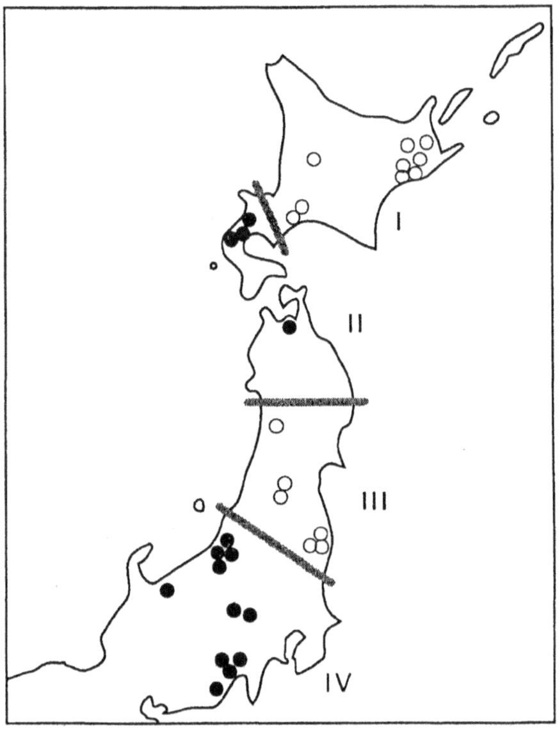

Figure 1. Geographic distribution in northern Japan of natural populations of Group P of the *Closterium ehrenbergii* species complex. Based on unpublished data of Ichimura & Kasai. Zone I, north-eastern Hokkaido (I); zone II, south-western Hokkaido to Aomori (II); zone III, Akita to Fukushima (III); zone IV; Niigata to Shizuoka (IV). Population of exclusively mating-type *plus* individuals (open circle); *minus* individuals (closed circle).

lations or races with different genetic backgrounds for the whole range of worldwide Group P distribution. At present it is known, ecologically, that the Japanese populations are distributed exclusively in running waters, while the northern European populations are found in both running and stagnant waters.

As shown by the three paradigms of *Gonium pectorale, Pandorina morum* and *Closterium ehrenbergii*, each biological species of freshwater algae has a distribution range that may occupy a rather restricted geographic area or a worldwide vast area, and whether small or large in the geographic area, because of the patchiness of habitat for freshwater algae, each is deemed to be subdivided into local populations with distinct genetic backgrounds that have been selected for the local environment. The evolutionary outcomes of such conditions, however, may not be uniform in variation of both morphologic and reproductive characters among the three morphospecies, as follows. *G. pectorale* seems to retain unaltered both the specific morphology and mating affinity, while adapting to ecologically variable habitats both from acidic to alkaline waters and from lower to warmer temperatures. *P. morum*, which seems to exploit the same range of habitat variation as that of *G. pectorale*, retains the specific morphology unaltered, but it has extensively altered genetic systems of mating behaviors to differentiate into many syngens. *C. ehrenbergii* has extensively altered both morphologic and mating characteristics to differentiate into many mating groups, most of which if statistically compared, seem to be morphologically distinguishable (Kasai, 1986; Ichimura et al., 1990b).

Polyploidy, aneuploidy and sex determination

Knowledge of nuclear cytology in algae remains to be much improved, compared to that in higher plants and animals. Nevertheless, it is worth considering some pertinent experimental data in the light of speciation by polyploidy. Some groups of freshwater algae have higher numbers of chromosomes, while others have reasonable numbers. Especially, the Euglenophyceae, the Cryptophyceae, the Dinophyceae and the Zygnematophyceae contain many species with extremely high chromosome numbers of over one hundred (Sarma, 1982). According to Grant (1981), in higher plants, species with the gametic chromosome number of $n = 14$ or more can be classified as polyploid. Although we know only a few examples of polyploid series of related species in algal taxa (Khan et al., 1984), species with such a high chromosome number, and perhaps also many others with chromosomes of more than some tens, which are widely scattered in almost all the cytologically studied orders (Sarma, 1982), are no doubt of polyploid origin. Therefore, it may be of interest for many students of freshwater algae to survey the occurrence and the mechanism of ploidy changes in nature and under experimental conditions.

Natural and artificially-induced diploids in clonal cultures

It seems not so rare events that all of the sister chromatids are incorporated into the same nucleus instead of one half being distributed into each of the two daughter nuclei after the mitotic or meiotic division. Low temperatures, centrifugal forces and some chemicals are known to cause such abnormal chromatid distribution (Berliner et al., 1976; Gerassimoff, 1905; Kallio, 1959; Kasai et al., 1987; Waris, 1958), although many freshwater algae are usually insensitive to common mitotic inhibitors, such as colchicine, used for higher plants and animals.

In the Zygnematophycean genus *Spirogyra*, Hoshaw and his colleagues (Hoshaw et al., 1985, 1987; McCourt et al., 1990; Wang et al., 1986) have presented interesting data suggesting that ploidy changes in clonal populations, which are revealed by culture experiments and confirmed also in field observations, may cause some difficulties in correspondence of breeding groups, morphology, and monophyletic groups. Two or three taxonomic species may appear interchangeable by ploidy change in a clonal population, and according to them, there is some degree of reproductive isolation among the taxa. The ploidy changes are euploidal and the accompanying cell width changes are so discrete that each level of ploidy appear to fit well the currently accepted taxonomic diagnoses. By studying F_1 progeny clones, however, Miller et al. (1974) have argued that cell width is not a valid taxonomic character for *Zygnema circumcarinatum* Czurda and should be questioned also for other algal taxa, although cell dimensions, especially cell width, are much used taxonomically for many microalgae. Reviewing the so-called giant cells in culture and the occurrence of similar forms in nature in some desmid species, Ichimura et al. (1990) have suggested that mere ploidy changes are not sufficient for speciation in haplontic algae.

In *Chlamydomonas reinhardtii* Dangeard, a few zygotes (3–4%) undergo mitosis to form stable diploid vegetative cells, instead of entering into the normal course of zygospore formation, dormancy and meiotic divisions to restore the haploid condition (Ebersold, 1967). Such diploid strains are heterozygous for the mating-type genes and express the mating-type *minus* phenotype. Sex determination in *C. reinhardtii* by the dominant mt^- gene has been confirmed by other workers (Gillham, 1969; Goodenough, 1985; Matagne et al., 1979). Zygotes resulting from haploid × diploid crosses yield a few viable F_1 progeny (13–20%), which are aneuploid segregants of triploid meiosis (Dutcher et al., 1988; Eves et al., 1982, 1984). In *Closterium ehrenbergii*, Kasai et al. (1987) have reported that upon zygospore germination, some rare zygotes yield a pair of giant cells both mating-type *minus*, instead of the normal mating-type *plus* and *minus* pair. Since such a pair of *minus* giant germlings can appear only when each contains both the two daughter nuclei of second division of a meiosis in which crossing-over between the centromere and the mating-type locus has occurred in the first division, they are presumed to be heterozygous diploids for the mating types genes, with mt^- being dominant. This has been supported by genetic analyses using artificially constructed homozygous mt^-/mt^- diploid strains (Kasai et al., 1990). The evolutionary significance of the sex determining mechanism by a single dominant genetic factor has been discussed for desmid speciation by polyploidy (Ichimura et al., 1990b). In *Pandorina morum*, Coleman (1977) has reported a naturally occurring diploid clone ($n = 10$) that is interfertile with other haploid clones ($n = 5$) of the same syngen II of the *P. morum* species complex. Although F_1 viability is considerably reduced (ca. 40%), the viable zygote progeny of the diploid and haploid cross appears to be normal in both morphology and mating activity, but backcrosses show that the mating types do not segregate in a 1:1 ratio. It is intriguing to know, but not yet clarified, whether sex determination in *P. morum* is based on the same mechanism as known in *C. reinhardtii* and *C. ehrenbergii*. At any event, diploidization or polyploid formation itself may not be an instantaneous, direct cause for new species formation in freshwater algae, because it is known that both diploid derivatives and aneuploid offspring are freely interfertile with the original haploids in all three species mentioned above. It may rather provide chances for contributing genome size enlargements on some special occasions that are not yet clarified.

Interspecific hybridization and amphiploid formation

In the *Closterium ehrenbergii* species complex, Kasai (1986) has described that cell sizes of most, if not all, rare F_1 survivors of the intergroup (interspecific) crosses between closely related mating groups are larger than the maximum cell size of the larger mating group parent. Figure 2 shows a typical example of cell-size relationships between an intergroup hybrid and the parental strains. Intergroup crosses in this species complex are known to cause abnormal meiosis accompanied by irregular disjunction of chromatids. Intergroup hybrid progeny, therefore, may have no possibility of survival in most cases in which any functional chromosome had been lost. But, in such rare cases in which all chromosomal complements of both parents had been incorporated into a gone, such a hybrid could survive to form a clonal population. That is the reason that most survivors of intergroup crosses are extremely large in cell size. They are amphiploids or something like amphiploids that have almost all chromosomal complements of both the parental groups (Ichimura et al., 1990b). The predominance of mating-type *minus* among F_1 survivors of intergroup crosses (Ichimura, 1983) can be well explained from such an amphiploid condition and the dominance of mt^- over mt^+ gene (Kasai et al., 1990). Such an intergroup hybrid might have chance to establish a founding population for new species with larger cell dimensions than those of the progenitors, because the sex determining mechanism with its one dominant genetic factor, mt^-, would not make any hindrance to polyploid evolution, which is known to be hindered in most animals largely by the appearance of abnormal sexes when their genome duplicates (Ohno, 1970; White, 1978). Intergroup hybrid individuals, however, seem to be very rare in nature, although we have surveyed many natural populations in Japan, perhaps because of the reproductive isolation by sexual and ecological means that have been elucidated for such closely related mating groups as A and B (Ichimura, 1981; Ichimura et al., 1987). Group P, however, may possibly be the one that has originated as an amphiploid hybrid, considering that it has extremely large cell dimensions and that it inhabits cold (mostly running) waters that are not necessarily preferable habitats for most other relatives (Ichimura et al., 1995). We have not so far succeeded in identifying the progenitor mating groups for Group P.

Goldstein (1964) has reported the appearance of amphiploid F_1 hybrid progeny in interspecific crosses between *Eudorina elegans* Ehrenberg and *E. illinoisensis* (Kofoid) Pascher in which low viability and abnormal morphology of the progeny have been observed. Although both the reciprocal crosses yielded a large excess of male progeny, *E. elegans* male × *E. illinoisensis* female cross (but not the reciprocal) resulted in the appearance of selfing male progeny. Selfed zygotes of such amphiploid (diploid) selfing clones were viable and yielded a 1:1 ratio of female and selfing male, in contrast to inviable zygotes of naturally found haploid selfing clones. Zygotes of crosses between the amphiploids also showed a good F_2 progeny survival, but segregated five types of sexuality; female, male, selfing male, selfing female and homothallic, with female progeny largely exceeding the other types. It is intriguing to analyze the genetic factors determining these five types of sexuality, but crosses of these amphiploid strains with normal wild type haploid strains seemed to result in no viable zygotes (Goldstein, 1964). This has hindered such genetic investigations.

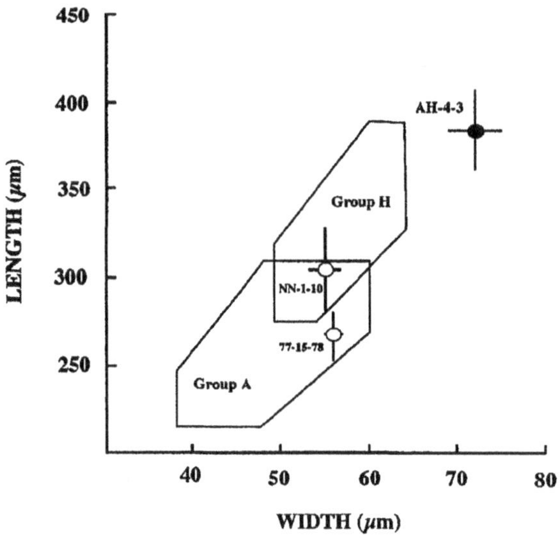

Figure 2. The cell size of an intergroup (interspecific) hybrid clone in the *Closterium ehrenbergii* species complex. The cell size of an amphiploid F_1 hybrid clone (AH–4–3) between a mating-type *plus* strain (77–15–78) of Group A and a mating-type *minus* (NN–1–10) of Group H is exceedingly larger than the full range of the larger parental mating group.

In another colonial volvocacean, *Volvox carteri* Stein f. *nagariensis* Iyengar, a cross between geographic isolates from India and Japan seems to produce rare survivors of an amphiploid nature with a dominant male determining factor (Starr, 1989; Adams et al., 1990). It may, therefore, be likely also in these colonial Volvocaceae, *Pandorina, Eudorina* and *Volvox* that a genetic mechanism with one dominant factor determines the two sex phenotypes as in *Chlamydomonas* and *Closterium*.

The genome of freshwater algal species

The life cycle of most freshwater algae is composed of the vegetative cycle of growth and reproduction and the sexual cycle of gametic fusion and meiosis in the zygote, which forms a dormant spore-like structure in many species. Any freshwater habitat may not be stable in terms of the evolutionary time scale and the external conditions suitable for the vegetative growth and reproduction for any one species may vanish sooner or later for a while. Each species, therefore, should have an ability to form germlings from specially organized cells, after a short or long period of dormancy, in order to recycle the life history (Fryxell, 1983).

The genome components for the life cycle of freshwater algae

From the evolutionary view-point, we can take the genome of a species as a well organized coadapted genetic system (Mayr, 1963, p. 263–296). Ecologically, however, different components of the genome may be subjected in different ways to selection pressures of the environment and biotic community. By selecting for mating efficiency and measuring the correlated response in vegetative growth rate for three sets of lineages of mixed strains of *Chlamydomonas reinhardtii*, Da Silva et al. (1992) have found a negative genetic correlation between sexual competence and vegetative growth rate. This seems to show, in just the opposite way, the same phenomenon often experienced in the maintenance of laboratory stocks of many rapidly growing species, such *Astrephomene* and *Gonium*, that sexual strains tend to become incompetent to respond to the same, once effective sexual induction method after being maintained for a short or long period only by vegetative means. Wiese (1984) has suggested that the dual nature of the life cycle of unicellular algae is fundamentally advantageous in exploring new sexual types during their evolution, compared to the life cycle characters of higher plants and animals. His reasoning is based on the idea that the continuity of existence can be protected by asexual reproduction in unicellular algae, but not in higher plants and animals. This also, however, suggests the possibility of the appearance of forms that have completely lost their sexual competence more often in the former than the latter. Possibly, some of them may be able to maintain the species status, as found among contemporary Euglenoids etc., if they could find a niche that would not demand the operation of a sexual cycle.

Since most freshwater habitats are ephemeral as mentioned above, each freshwater algal species should have a specific genetic system to switch from the vegetative cycle for reproduction to the sexual cycle for dormancy before the habitats become too unfavorable. The environmental cues for the switching vary in quality and quantity according to species, although the most general potent factors may be decrement of nutrients such as nitrogen and phosphate and higher or lower temperatures (Coleman, 1962; Sager et al., 1954). I have found many homothallic species of *Closterium*, such as *C. acerosum* (Schr.) Ehr. ex Ralfs, *C. calosporum* Wittr., *C. dianae* Ehr. ex Ralfs, and *C. navicula* (Bréb.) Lütkem., in small temporary pools and ponds on hills in Nepal and on the Ryukyu Islands in Japan, that tend to dry up within a short time after rain falls. Some of them are so-called precocious strains that after a very short span (a couple of weeks) of asexual reproduction, are soon inclined to form zygospores even in growth media with sufficient nutrients (Ichimura, 1973; Yamagishi et al., 1974). In contrast, heterothallic strains are found more often in more stable habitats in terms of a predictable, long period of submerged conditions. Strains that stringently require low levels of nitrogen source for sexual induction are often found in permanent water bodies of large lakes, or spring-fed streams as in Group P of *Closterium ehrenbergii* mentioned earlier. Homothallism and heterothallism, therefore, are considered to represent two contrasting mating systems that are related to two opposing modes of life style to harmonize antagonism between vegetataive and sexual stages. In general, the vegetative growth period seems to be much longer in heterothallic than in homothallic algae and the sexual cycle tends to go round more often in homothallic than in heterothallic algae. Unfortunately there has been as yet little analytical work on these problems.

There is, however, another problem of mating systems from the view-point of population genetics. When a natural population of algae has long been maintained exclusively by asexual reproduction, it may accumulate deleterious mutations at the gene loci expressed during the sexual stage. Recently Ichimura et al. (1995) have shown, by genetic analysis, a dominant or mating-type *minus* specific asynaptic mutation found in a Group P population of the *Closterium ehrenbergii* species complex. In other mating groups of the same complex that preferentially inhabit small temporary waters in warmer regions, on the other hand, polygenic deleterious factors have been detected even in normal standard strains, by systematic inbreeding experiments, at loci that are expressed only during the sexual cycle, when the nuclear condition is diploid, although no such marked mutation as mentioned above has been detected (T. Ichimura & F. Kasai, unpublished). Hence, we may well assume that every apparently normal vegetative clone may contain some recessive mutations that affect the sexual cycle if they become homozygous. As inhabitants of temporary waters, Group A populations are deemed to go around the sexual cycle at least once a year and any deleterious dominant mutation for the sexual cycle must be eliminated. Nevertheless, because of its mating system of outbreeding to maintain genetic variation, the genome of Group A may allow deleterious recessive mutations to accumulate, concealed by normal alleles at the diploid phase.

To be sure, any mutation that affects the vegetative phase is subjected to direct selection pressure soon after appearance because of its haploid condition.

Genome rearrangements and survivals of meiotic progeny in intercrosses

Intercrosses, if possibly, between species whose genome rearrangements are already considerably diversified from each others will result in mostly inviable F_1 progeny, with rare survivors that have become amphiploid, as in the case of the interspecific crosses between *Eudorina elegans* and *E. illinoisensis* and of the inter-mating-group of the *Closterium ehrenbergii* complex. On the other hand, intercrosses between populations with similar chromosome structures and common genetic regulatory systems for meiotic events will result in viable F_1 progeny with the same ploidy as the parental. Even in such cases, if there are still cryptic or minor chromosomal rearrangements, intercrosses may reduce F_1 viability in various degrees depending upon the nature and scale of the rearrangements. Along with a somewhat complicated historical account of the taxonomy and the origins of the strains, Gowans (1963) has reported that presently available strains of two closely related species of *Chlamydomonas, C. eugametos* and *C. moewusii*, are morphologically indistinguishable and that they show an immunologically close relationship and yield many viable zygotes when crossed. Such interspecific hybrid zygotes are capable of germinating to form four cells each, as in the cases of intra-specific crosses of either parental species. However, only one or sometimes two of the meiotic products (gones) from each zygote give rise to a clonal population of vegetative cells, the other gones die without cell division after several days. Gowans (1963) concluded that such lethality of the interspecific cross is due to differences of more than two genes, but not chromosomal difference, because several genetic markers tested segregate 1:1. The post-zygotic lethality of the interspecific hybrids of these two species were studied further to confirm the Gowans' conclusion by several investigators (Cain, 1979; Lee et al., 1990; Lemieux et al., 1980, 1981, 1984; Mets, 1980). Showing the polygenic nature of the interspecific hybrid lethality by backcrossing hybrids recurrently to the same parental mating-type strain of either species, Lee et al. (1990) demonstrated that the mt^+ gene of either species introduced into the other species genome works as well as does the original one. The genome rearrangements that cause the hybrid lethality between these two species, therefore, are not related to the structure of mating-type gene and its controlling systems. In the morphospecies *C. moewusii*, however, three other heterothallic pairs of strains are known that are completely sexually isolated from each other and also from the one described above. Although not clearly described, all of the *C. moewusii* strains seem to be morphologically indistinguishable (Wiese et al., 1977, 1978). A model for microalgal speciation caused by mutations in gamete recognition systems (sexual agglutinins) has been proposed by Wiese et al. (1977, 1978) and supported by Bell (1992). In this context, the work of Van Winkle-Swift et al. (1991) is very intriguing, although they work on a homothallic species, *Chlamydomonas monoica* Strehlow, because they have suggested that unequal crossing over between highly homologous flagellar agglutinin genes may account for the unusual sexual preferences of a mutant strain, *mtl–3* (*m*ating-*t*ype *l*imited).

Chlamydomonas reinhardtii and *C. smithii* Hoshaw et Ettl are also closely related species that are heterothallic and intercrossable with each other, but they are morphologically distinct (cell shape, chloroplast structure and the presence or absence of marked apical papilla) from each other (Hoshaw, 1965; Hoshaw et al., 1966). In contrast to the case of *C. eugametos* and *C. moewusii*, however, Bell et al. (1983) have shown that the possibility of gene flow is considerable between *C. reinhardtii* and *C. smithii*, as high as between local populations of a single biological species. Recently, more strains that produce zygotes when crossed to either mating type of *C. reinhardtii* have been isolated from various localities in the northern America; one mt^- strain from the bank of a small lake in Plymouth, Minnesota by Gross et al. (1988), one mt^- strain from soil, Pittsburgh, Pennsylvania, and two mt^+ strains from soil, Melbourne, Florida and soil, Malverne Pennsylvania by Spanier et al. (1992) and 19 strains all from arable fields on the outskirts of Farnham, Quebec by Sack et al. (1994). Since the *C. smithii* strains were previously identified as *C. reinhardtii* by G.M. Smith because of the crossability but were distinguished by the morphology (Hoshaw, 1965), morphological characteristics of these strains should be studied in detail. From the view of speciation problems, it is of particular interest to study these presumptive *C. reinhardtii* strains for crossability with the *C. smithii* strains as well and with each other and for viability of their hybrid meiotic progeny.

Burrascano et al. (1984) have studied crossability of five homothallic species of *Chlamydomonas, C.*

noctigama Korschikoff, *C. monoica, C. hindakii* Ettl, *C. pinicola* Ettl and *C. geitleri* Ettl, all of which are known to be classified into one group by functional similarity of sporangial wall autolysine (Schlösser, 1976, 1984). To select against intraspecific mating, they establish zygote maturation-defective mutants *(zym)* by UV irradiation of each species. Self-mating of a *zym* mutant produces zygotes that fail to mature properly and do not germinate, whereas crosses between complementary *zym* mutants produce viable heterozygous zygotes. They report the recovery of viable zygotes from all pairwise crosses between *zym* mutants of the five species. These five homothallic species are, therefore, sexually compatible with each other. Nevertheless, extensive post-meiotic lethality has been detected by dissecting tetrads obtained from interspecific hybrid zygotes. Four of the five species have been recorded hitherto only once from Czechoslovakia and the other species, *C. noctigama*, has several records of findings in Europe, mostly in Czechoslovakia (Ettl, 1976). From the evolutionary genetic view-point, we eagerly want ecological data concerning whether or not members of these closely related species do coexist in the same habitats. If any two coexisting members might reproduce sexually at the same time in nature, would their mating systems be diversified by natural selection, because of the inviable hybrid zygote formation, to develop a sexual isolation mechanism between them (cf. Butlin, 1987)? Alternatively, would recurrent hybridizations, if they might have occurred, amalgamate the two genomes of two morphologically distinguishable species into one intermediate form, by the appearance of rare hybrids that show lower postzygotic lethality, as shown by the study of Lee et al. (1990)? Although genetic and molecular studies in *Chlamydomonas* are in the most advanced state among algae, there has been very little information on geographic and ecological distributions of natural populations even of the most well studied species like *C. reinhardtii* (Harris, 1989).

Studying many different geographical isolates of the same and closely related mating groups (biological species), Ichimura and his colleague (Ichimura, 1982; Ichimura et al., 1984a, 1990a, unpublished data) have obtained F_1 viability data within and between mating groups. First, as shown in Table 1, crosses within mating groups result in high F_1 survival and the normal mating-type segregation, but crosses between mating groups yield very low F_1 survival with predominantly mt^- F_1 progeny. Second, rather low F_1 viabilities are recognized even in crosses within a mating group, when many strains from geographically wide areas are incorporated, as in Groups A and P. The reason for reduced F_1 viabilities within the same mating group may be due to genome diversification between geographically disjunct populations, as shown between Australian and Japanese populations of Group A. Unfortunately, as yet little analysis has been done on Australian populations and no information is available for populations in South East Asia and on the tropical Pacific islands stretching between Australia and Japan. Another reason for the reduced F_1 viability within a mating group is due to the presence of strains with unique genome rearrangements adapting to particular ecological conditions or particular strains with deleterious mutations maintained by populations in special habitats, as found in Group P (Ichimura et al., 1995).

Cells appearing from zygotes of haplontic microalgae are derived directly (in some cases, intercalated by one or more mitotic cycles) from meiotic products. Therefore, the lethality of intercross progeny of these algae is considered homologous to the gametic lethality, such as the pollen abortion in plants and the abnormality in spermatogenesis of animals that are, in general, highly correlated to degrees of chromosomal rearrangements between species and between populations within a species (King, 1993). The structural rearrangements such as translocations, inversions and deletions may be detected by cytological observations of meiotic chromosomal pairing, although this is very difficult for most freshwater algae. In cases of species in which complete tetrads can be recovered from a zygote, however, genetic analysis of the lethality patterns of tetrads will manifest such structural rearrangements of chromosomes (McBride et al., 1969; Perkins, 1974). In addition, for such hybrid plants that show apparently normal meiotic chromosome pairings but still some or considerable abortion of pollen grains, Stebbins (1950, 1971) has presented a hypothesis of cryptic structural hybridity. This hypothesis may well be adopted for most cases of intramating-group crosses of the *Closterium ehrenbergii* species complex in which some degree of reduced F_1 viability is present as shown in Table 1.

Modes and mechanisms of speciation

Coleman et al. (1977) have reported chromosomal polymorphism within the morphospecies *Pandorina morum*, with haploid numbers varying from 2 to 14. As they have suggested, however, it is likely that despite the chromosome number variation the genome size

Table 1. Geographic distribution and viability of F_1 progeny in crosses within and between mating groups of the *Closterium ehrenbergii* species complex.

Mating group	Distribution	Number of cross & F_1		% survival mean & range		Mating-type ratio
Within group						
A	Japan	20	3478	85	51–97	1:1
A	Japan × Australia	3	342	48	12–52	1:1
A	Haploid × diploid	18	3404	54	39–74	1:1 or 1:5
B	Japan & Taiwan	7	563	93	69–97	1:1
C	Japan	2	43	95	89, 100	1:1
P	Japan, Europe & USA	52	8214	64	8–95	1:1
Between group						
A × B	Japan × Japan	205	6651	9	0–74	+<<<-
A × H	Japan × Nepal	15	409	0.2	0–1.2	?
B × H	Japan × Nepal	23	2191	0.2	0–1.7	?

Compiled based on Ichimura (1982), Ichimura et al. (1984a, 1990a) and unpublished data. See Ichimura (1983) for the methods.

may differ little, with the karyotype of small chromosome number being composed of large size chromosomes and that of large number having small size chromosomes. The similarity in genome size may apparently correspond with the uniformity in size, colony form and life cycle among all syngens of *P. morum*. It is known, on the other hand, that the chromosome number is constant within collections of the same syngen, wherever their geographic origins (but a single exception found in one locality of Syngen II mentioned earlier) (Coleman, 1977; Coleman et al., 1977, 1994). Therefore, it is highly likely that the differences in chromosome numbers reflect differences in genetic systems between syngens, for instance different rates of genetic recombination or degrees of linkage disequilibrium between certain ecologically or sexually important gene loci.

Another finding of exceptional difference in chromosome numbers within a single syngen of *P. morum* has been obtained in a laboratory cross between strains from different continents, the United States and Thailand, of Syngen II, in which case F_1 progeny clones have twice the chromosome number, $n = 5$, of the parental strains (Coleman, 1977). Therefore, chromosomal rearrangements are suggested to exist between geographically disjunct populations of the same syngen (Coleman et al., 1977). Similarly, chromosomal rearrangements are suggested in *Volvox carteri* f. *nagariense* from Japan and India, which are sexually compatible but with partial postzygotic isolation (Adams et al., 1990). Although sexual isolation exists between two morphologically different formae, *nagariense* and *weismannia* (Powers) Iyengar of *V. carteri*, there seems to be no sexual isolation among three different geographic strains from the United States, Australia and India of the same forma, *V. carteri* f. *weismannia* (Starr, 1970). At present, we have no information on postzygotic isolation and chromosomal rearrangements among the strains from three different continents. Recently, Nozaki (1988) has described a new forma, *kawasakiensis* Nozaki of *V. carteri* from Japan, which is sexually isolated in one of the reciprocal crosses with *nagariense* from Japan. There is no further information whether or not the geographical distribution ranges of these two formae overlap with each other in Japan, except the two known localities are separated by more than 500 km; near Kobe for *nagariense* and Kawasaki for *kawasakiense*. Including three others, *carteri, manilina* (Shaw) Iyengar and *hazenii* Metzner, as yet to be studied in culture, all of the six described formae of *V. carterii* appear to be morphologically distinct from each other (Nozaki, 1988). Until recently, few workers have ever paid special attention and devotion to the speciation problems in *Volvox*., as in *Gonium, Pandorina* and *Closterium*, based on many natural populations (not merely on a few representative strains). The taxa, whether specific or infraspecific, of *Volvox* have been characterized by the morphology of developmental patterns of both asexual and sexual colonies (Smith, 1944). Following the Starr's pioneering work (Starr, 1970), genetic, biochemical and molecular studies on various aspects of asexual and sexual differentiation of *Volvox* have been reported by many workers, for instance Al-Hasani et al. (1992), Kirk et al. (1987, 1993), Kochert et al. (1974) and Starr et al. (1989). Therefore, further studies on *Volvox* will greatly contribute to speciation studies, particularly by uncovering both the aspects of reproductive isolation and morphological diversification.

Inhabiting similar freshwater habitats distributed worldwide, *Gonium pectorale* and *Pandorina morum* show a marked difference in degrees of reproductive isolation in that the former shows partial or no sexual isolation (possibility of sexual compatibility through indirect populations) but the latter is composed of many completely sexually isolated entities, and yet both of them are scarcely diversified in gross morphology. Coleman et al. (1994) have suggested "that *G. pectorale* is an evolutionary younger species that has rapidly become worldwide in distribution, whereas *P. morum*, with its many syngens, is older and more sedentary." Alternatively, however, we may also suppose different modes of sexual interactions between them. The sexual differentiation in *Closterium*, in which many reproductively isolated entities are known within a single taxonomic species, requires interactions of two complementary mating-type strains (Ichimura et al., 1984b, 1987). In *Closterium*, however, evolution of reproductive isolation has correlated in many cases with morphological diversification, although in so slight and intergrading degrees that entities are not accepted as formerly described taxa (Ichimura et al., 1990). Physiologic, biochemical and molecular aspects of the sexual reproduction have been investigated in *Closterium* (Ichimura, 1971, 1973; Kato et al., 1981, 1985; Sekimoto et al., 1992, 1990, 1993, 1994), although little is known about the details of the *Pandorina* sexual interactions.

The genome of freshwater algal species contains several sets of genes and gene systems, those for the characteristic of vegetative form, for specially structured zygospores and their unique mode of germination, and also those to control the cell size, asexual reproduction and interaction between the two sexual individuals. The genome, however, is not merely a mixture of genes and gene systems, but has a well-organized structure of chromosomes that properly transmit them without disrupting their coadapted relationships. The cohesiveness of a species is supported by the well-organized genome structures that are consistently shaped by natural selection through the vegetative and sexual cycles.

According to the allopatric speciation model (Mayr, 1963), speciation can occur only when such a genome structure is drastically transformed by being completely disrupted and then reformed into a new one in geographically isolated founding populations (Mayr, 1976, p. 188–217). In sexually reproducing animals and plants, one of the main difficulties for establishing a new isolating mechanism between two sympatric populations is how mutations that are candidates for such a mechanism can be fixed in the majority of one or the other population. Mayr has concluded that the genetic homeostasis forming the cohesive unity of a species would not be easily broken under normal conditions of a large geographically contiguous population and hence a small founding population that is protected from gene flow from the parental population is a necessary prerequisite to speciation. This theory of founder-effect speciation has profoundly influenced many evolutionary biologists and several theories (Coyne, 1994). Speciation in freshwater algae may, therefore, occur only when a special change occurs in the regulatory systems, especially mate-recognition system and meiotic events, that control and maintain the unique and coherent genome of a species. It seems, however, to be premature to discuss further these problems relating to the sexuality of algae until we gain more profound insights into the material basis that is currently being attacked by the recent powerful genetic and molecular techniques.

Acknowledgements

I wish to thank Dr J. Kristiansen for the invitation to the Workship of the International Phycological Congress, Qingdao, China and for the opportunity to prepare this essay and Dr A.W. Coleman for reading the manuscript and correcting my English. This study was supported by the grants 04640636 from the Scientific Research Fund of the Ministry of Education, Science and Culture, Japan.

References

Adams, C. R., K. A. Stamer, J. K. Miller, J. G. McNally, M. M. Kirk & D. L. Kirk, 1990. Patterns of organellar and nuclear inheritance among progeny of two geographically isolated strains of *Volvox carteri*. Curr. Genet. 18: 141–153.

Al-Hasani, H. & L. Jaenike, 1992. Characterization of the sex-inducer glycoprotein of *Volvox carteri* f. *weismannia*. Sex. Plant Reprod. 5: 8–12.

Allen, M. A., 1985. The biology of a species complex in *Spirogyra*. Ph. D. Thesis, Indiana University.
Bell, G., 1992. The emergence of gender and the nature of species in eukaryotic microbes. Verh. dt. zool. Ges. 85: 161–175.
Bell, R. A. & J. R. Cain, 1983. Sexual reproduction and hybridization in *Chlamydomonas smithii* and *C. reinhardtii* (Chlorophyceae, Volvocales), Phycologia 22: 243–247.
Berliner, M. D. & K.A. Wenc, 1976. Osmotic pressure effects and protoplast formation in *Cosmarium turpinii*. Microbios Letters 2: 39–45.
Brooks, A. E., 1966. The sexual cycle and intercrossing in the genus *Astrephomene*. J. Protozool. 13: 367–375.
Brown, R. M., Jr., 1971. Studies of Hawaiian fresh-water and soil algae. I. The atmospheric dispersal of algae and fern spores across the island of Oahu, Hawaii. In B. C. Parker & R. M. J. Brown (eds), Contributions in Phycology. Allen Press, Lawrence, Kansas, 175–188.
Burrascano, C. G. & K. P. Van Winkle-Swift, 1984. Interspecific matings of five *Chlamydomonas* species. Genetics 107: s15.
Butlin, R., 1987. Speciation by reinforcement. TREE 2: 8–13.
Cain, J. R., 1979. Survival and mating behaviour of progeny and germination of zygospores from intra- and interspecific crosses of *Chlamydomonas eugametos* and *C. moewusii* (Chlorophyceae, Volvocales). Phycologia 18: 24–29.
Carefoot, J. R., 1966. Sexual reproduction and intercrossing in *Volvulina steinii*. J. Phycol. 2: 150–156.
Carson, H. L., 1975. The genetics of speciation at the diploid level. Am. Nat. 109: 83–92.
Coesel, P. F. M., 1988. Biosystematic studies on the *Closterium moniliferum/ehrenbergii* complex (Chlorophyta, Conjugatophyceae) in western Europe. II. Sexual compatibility. Phycologia 27: 431–424.
Coesel, P. F. M., 1989. Biosystematic studies on the *Closterium moniliferum/ehrenbergii* complex (Chlorophyta, Conjugatophyceae) in western Europe. IV. Distributional aspects. Cryptogamie, Algol. 19: 133–141.
Coesel, P. F. M., S. R. Duque & G. Arango, 1988. Distributional patterns in some neotropical desmid species (Algae, Chlorophyta) in relation to migratory bird routes. Revue. Hydrobiol. Trop. 21: 197–205.
Coleman, A. W., 1959. Sexual isolation in *Pandorina morum*. J. Protozool. 6: 249–264.
Coleman, A. W., 1962. Sexuality. In R. A. Lewin (ed.), Physiology and Biochemistry of Algae. Academic Press, New York, 711–729.
Coleman, A. W., 1977. Sexual and genetic isolation in the cosmopolitan algal species *Pandorina morum*. Am. J. Bot. 64: 361–368.
Coleman, A. W., A. Suarez & L. J. Goff, 1994. Molecular delineation of species and syngens in volvocalean green algae (Chlorophyta). J. Phycol. 30: 80–90.
Coleman, A. W. & J. Zollner, 1977. Cytogenetic polymorphism within the species *Pandorina morum* Bory de St. Vincent (Volvocaceae). Arch. Protistenk. 119: 224–232.
Coyne, J. A., 1992. Genetics and speciation. Nature 355: 511–515.
Coyne, J. A., 1994. Ernst Mayr and the origin of species. Evolution 48: 19–30.
Da Silva, J. & G. Bell, 1992. The ecology and genetics of fitness in *Chlamydomonas* VI. Antagonism between natural selection and sexual selection. Proc. R. Soc. London B. 249: 227–233.
Dell'Uomo, A. & M. A. Masi, 1985. Sexual reproduction in nature of some filamentous conjugating algae. Nova Hedwigia 41: 177–189.
Dobzhansky, T., 1951. Genetics and the Origin of Species. Columbia University Press, New York.
Dutcher, S. K. & W. Gibbons, 1988. Isolation and characterization of dominant tunicamycin resistance mutations in *Chlamydomonas reinhardtii* (Chlorophyta). J. Phycol. 24: 230–236.
Ebersold, W. J., 1967. Heterozygous diploid strains of *Chlamydomonas reinhardi*. Science 157: 447–449.
Ehrlich, P. R. & P. H. Raven, 1969. Differentiation of populations. Science 165: 1228–1231.
Endler, J. A., 1977. Geographic Variation, Speciation, and Clines. Princeton University Press, Princeton, ix–246.
Ettl, H., 1976. Die Gattung *Chlamydomonas* Ehrenberg. Beih. Nova Hedwigia 49: 1–1122.
Eves, E. M. & K.-S. Chiang, 1982. Genetics of *Chlamydomonas reinhardtii* diploids. I. Isolation and Characterization and meiotic segregation pattern of a homozygous diploid. Genetics 100: 35–60.
Eves, E. M. & K.-S. Chiang, 1984. Genetics of *Chlamydomonas reinhardtii* diploids. II. The effects of diploidy and aneuploidy on the transmission of non-Mendelian markers. Genetics 107: 563–576.
Felsenstein, J., 1981. Skepticism towards Santa Rosalia, or why are there so few kinds of animals? Evolution 35: 124–138.
Francke, J. A. & P. F. M. Coesel, 1985. Isozyme variation within and between Dutch populations of *Chlosterium ehrenbergii* and *C. moniliferum* (Chlorophyta, Conjugatophyceae). Br. Phycol. J. 20: 201–209.
Fryxell, G. A., 1983. Survival Strategies of the Algae. Cambridge University Press, Cambridge, x–144.
Geitler, L., 1975. Formwechsel, sippenspezifischer Paarungsmodus und Systematik bei einigen pennaten Diatomeen. Pl. Syst. Evol. 124: 7–30.
Gerassimoff, J. J., 1905. Ueber die Grösse des Zellkerns. Beih. Bot. Zentr. 18: 45–118.
Gerrath, J. F., 1993. The biology of desmids: a decade of progress. In F. E. Round & D. J. Chapman (eds), Progress in Phycological Research. Biopress, Bristol 9: 79–192.
Gillham, N. W., 1969. Uniparental inheritance in *Chlamydomonas reinhardi*. Am. Nat. 103: 355–388.
Goldstein, M. E., 1964. Speciation and mating behaviour in *Eudorina*. J. Protozool. 11: 317–344.
Goodenough, U. W., 1985. An essay on the origins and evolution of eukaryotic sex. In H. O. Halvorson & A. Monroy (eds), The Origin and Evolution of Sex. Alan R. Liss, New York: 123–140.
Gowans, C. S., 1963. The conspecificity of *Chlamydomonas eugametos* and *Chlamydomonas moewusii*: An experimental approach. Phycologia 3: 37–44.
Grant, V., 1981. Plant Speciation. Columbia University Press, New York: 563 pp.
Gross, C. H., L. P. W. Ranum & P. A. Lefebvre, 1988. Extensive restriction fragment length polymorphisms in a new isolate of *Chlamydomonas reinhardtii*. Curr. Genet. 13: 503–508.
Harris, D. O. & R. C. Starr, 1969. Life history and physiology of reproduction of *Platydorina caudata* Kofoid. Arch. Protistenk. 111: 138–155.
Harris, E. H., 1989. The *Chlamydomonas* Sourcebook. Academic Press, San Diego, xiv–780.
Hoham, R. W., 1975. The life history and ecology of the snow alga *Chloromonas pichinchae* (Chlorophyta, Volvocales). Phycologia 14: 213–226.
Hoham, R. W. & J. M. Mullet, 1977. The life history and ecology of the snow alga *Chloromonas cryophila* sp. nov. (Chlorophyta, Volvocales). Phycologia 16: 53–68.
Homfeld, H., 1929. Beitrag zur Kenntnis der Desmidiaceen Nordwestdeutschlands, besonders ihrer Zygoten. Pflanzenforschung 12: 1–96.

Hoshaw, R. W., 1965. Mating types of *Chlamydomonas* from the collection of Gilbert M. Smith. J. Phycol. 1: 194–196.

Hoshaw, R. W. & H. Ettl, 1966. *Chlamydomonas smithii* sp. nov. – A Chlamydomonad interfertile with *Chlamydomonas reinhardtii*. J. Phycol. 2: 93–96.

Hoshaw, R. W. & R. M. McCourt, 1988. The Zygnemataceae (Chlorophyta); a twenty-year update of research. Phycologia 27: 511–548.

Hoshaw, R. W., J.-C. Wang & H. M. Hull, 1985. Ploidal changes in clonal cultures of *Spirogyra communis* and implications for species definition. Am. J. Bot. 72: 1005–1011.

Hoshaw, R. W., C. V. Wells & R. M. McCourt, 1987. A polyploid species complex in *Spirogyra maxima* (Chlorophyta, Zygnemataceae), a species with large chromosomes. J. Phycol. 23: 267–273.

Hutchinson, G. E., 1957. Concluding remarks. Cold Spring Harbor Symp. Quant. Biol. 22: 415–427.

Hutchinson, G. E., 1959. Homage to Santa Rosalia or why so many kinds of animals? Am. Nat. 93: 145–159.

Hutchinson, G. E., 1961. The paradox of the plankton. Am. Nat. 95: 137–145.

Hutchinson, G. E., 1968, When are species necessary: In R. C. Lewontin (ed.), Population Biology and Evolution. Syracuse University Press, Syracuse: 177–186.

Ichimura, T., 1971. Sexual cell division and conjugation-papilla formation in sexual reproduction of *Closterium strigosum*. Internat. Seaweed Symp. 7: 208–214.

Ichimura, T., 1973. The life cycle and its control in some species of *Closterium*, with special reference to the biological species problems. D. Sc. Thesis, University of Tokyo: 117 pp.

Ichimura, T., 1981. Mating types and reproductive isolation in *Closterium ehrenbergii* Meneghini. Bot. Mag. Tokyo 94: 325–334.

Ichimura, T., 1982. Isolating mechanisms in speciation of *Closterium*. Jpn. J. Phycol. 30: 332–343.

Ichimura, T., 1983. Hybrid inviability and predominant survival of mating type minus progeny in laboratory crosses between two closely related mating groups of *Closterium ehrenbergii*. Evolution 37: 252–260.

Ichimura, T., 1985. Geographic distribution and isolating mechanisms in the *Closterium ehrenbergii* species complex (Chlorophyceae, Closteriaceae). In H. Hara (ed.), Origin and Evolution of Diversity in Plants and Plant Communities. Academia Scientific Book INC., Tokyo: 295–303.

Ichimura, T. & F. Kasai, 1984a. Post-zygotic isolation between allopatric mating groups of *Closterium ehrenbergii* Meneghini (Chlorophyta). Phycologia 23: 77–85.

Ichimura, T. & F. Kasai, 1984b. Time-lapse analyses of sexual reproduction in *Closterium ehrenbergii* (Conjugatophyceae). J. Phycol. 20: 258–265.

Ichimura, T. & F. Kasai, 1987. Time-lapse analyses of sexual isolation between two closely related mating groups of the *Closterium ehrenbergii* species complex (Chlorphyta). J. Phycol. 23: 523–534.

Ichimura, T. & F. Kasai, 1990a. Genome differentiation in speciation of desmids (Chlorophyta). Jpn J. Phycol. 37: 305–319.

Ichimura, T. & F. Kasai, 1990b. Mating systems and speciation in haplontic unicellular algae, desmids. In S. Kawano (ed.), Biological Approaches and Evolutionary Trends in Plants. Academic Press, London: 309–332.

Ichimura, T. & F. Kasai, 1995. Morphological variation and zygotic defective mating-type plus in a population of *Closterium ehrenbergii* (Zygnematales, Chlorophyta) from Boest Brook in Denmark. Phycologia 34: 45–50.

Khan, M. & Y. S. R. K. Sarma, 1984. Cytogeography and Cytosystematics of Charophyta. In E. G. Ivrine & D. M. John (eds), Systematics of the Green Algae. Academic Press, London: 303–330.

Kallio, P., 1959. The relationship between nuclear quantity and cytoplasmic units in *Micrasterias*. Ann. Acad. Sci. Fenn. A IV Biol. 44: 5–44.

Kasai, F., 1986. Biosystematic Studies in the *Closterium ehrenbergii* species complex. D. Sc. Thesis, University of Tsukuba: 165 pp.

Kasai, F. & T. Ichimura, 1987. Stable diploids from intragroup zygospores of *Closterium ehrenbergii* Menegh. (Conjugatophyceae). J. Phycol. 23: 344–351.

Kasai, F. & T. Ichimura, 1990. A sex determining mechanism in the *Closterium ehrenbergii* (Chlorophyta) species complex. J. Phycol. 26: 195–201.

Kato, A., J. Obokata & K. Sasaki, 1981. Mating type interaction in *Closterium peracerosum-strigosum-littorale*: mating induced protoplast release. Pl. Cell Physiol. 22: 1215–1222.

Kato, A. & K. Sasaki, 1985. Sexual interaction in heterothallic strains of *Closterium peracerosum-strigosum-littorale*. Pl. Physiol. 77: 556–559.

King, M., 1993. Species Evolution – The Role of Chromosome Change. Cambridge University Press, Cambridge, xxi–336.

Kirk, D., G. J. Baran, J. F. Harper, R. J. Huskey, K. S. Huson & N. Zagris, 1987. Stage-specific hypermutability of the *regA* locus of *Volvox*, a gene regulating the germ-soma dichotomy. Cell. 48: 11–24.

Kirk, M. M., A. Ransick, S. E. McRae & D. L. Krik, 1993. The relationship between cell size and cell fate in *Volvox carteri*. J. Cell Biol. 123: 191–208.

Knoll, A. H., 1994. Proterozoic and Early Cambrian protists: Evidence for accelerating evolutionary tempo. Proc. natn. Acad. Sci. U.S.A. 91: 6743–6750.

Kochert, G. & I. Yates, 1974. Purification and partial characterization of a glycoprotein sexual inducer from *Volvox carteri*. Proc. natn. Acad. Sci. U.S.A. 71: 1211–1214.

Kowallik, K. V., 1989. Molecular aspects and phylogenetic implications of plastid genomes of certain chromophytes. In J. C. Green, B. S. C. Leadbeater & W. L. Diver (eds), The Chromophyte Algae, Problems and Perspectives. Claredon Press, Oxford: 101–124.

Kristiansen, J., 1981. Distribution problems in the Synuraceae. Verh. int. Ver. Limnol. 21: 1444–1448.

Lee, R. W. & C. Lemieux, 1990. Loss of hybrid lethality during backcross programs involving *Chlamydomonas eugametos* and *Chlamydomonas moweusii* (Chlorophyceae). J. Phycol. 26: 376–380.

Lemieux, C., M. Turmel & R. W. Lee, 1980. Characterization of chloroplast DNA in *Chlamydomonas eugametos* and *C. moewusii* and its inheritance in hybrid progeny. Curr. Genet. 2: 139–147.

Lemieux, C., M. Turmel & R. W. Lee, 1981. Physical evidence for recombination of chloroplast DNA in hybrid progeny of *Chlamydomonas eugametos* and *C. moewusii*. Curr. Genet. 3: 97–103.

Lemieux, C., M. Turmel, V. L. Seligy & R. W. Lee, 1984. Chloroplast DNA recombination in interspecific hybrids of *Chlamydomonas*: Linkage between a nonmendelian locus for streptomycin resistance restriction fragments coding for 16s rRNA. Proc. natn. Acad. Sci. U.S.A. 81: 1164–1168.

Ling, H. U. & P. A. Tyler, 1974. Interspecific hybridity in the desmid genus *Pleurotaenium*. J. Phycol. 10: 225–230.

Ling, H. U. & P. A. Tyler, 1976. Meiosis, polyploidy and taxonomy of the *Pleurotaenium mamillatum* complex (Desmidiaceae). Br. Phycol. J. 11: 315–330.

Lippert, B. E., 1967. Sexual reproduction in *Closterium moniliferum* and *Cl. ehrenbergii*. J. Phycol. 3: 182–198.

Maguire, B., Jr., 1963. The passive dispersal of small aquatic organism and their colonization of isolated bodies of water. Ecol. Monogr. 33: 161–185.
Manhart, J. R. & R. M. McCourt, 1992. Molecular data and species concepts in the algae. J. Phycol. 28: 730–737.
Mann, D. G., 1988. Why didn't Lund see sex in *Asterionella?* A discussion of the diatom life cycle in nature. In F. E. Round (ed.), Algae and the Aquatic Environment. Biopress, Bristol.
Mann, D. G., 1989. The species concept in diatoms: Evidence for morphologically distinct, sympatric gamodemes in four epipelic species. Pl. Syst. Evol. 164: 215–237.
Matagne, R. F. & C. Vincenzott, 1979. Triallelic complementation and hybrid enzyme formation in *Chlamydomonas reinhardi.* Molec. Gen. GEnet. 177: 121–127.
Maynard Smith, J., 1966. Sympatric speciation. Am. Nat. 100: 637–650.
Mayr, E., 1942. Systematics and the Origin of Species. Columbia University Press, New York.
Mayr, E., 1959. Isolation as an evolutionary factor. Proc. Am. Phil. Soc. 103: 221–230.
Mayr, E., 1963. Animal Species and Evolution. Belknap Press of Harvard University Press, Cambridge, Massachusetts, xiv–797.
Mayr, E., 1975. The unity of the genotype. Biol. Zentralbl. 94: 377–388.
Mayr, E., 1976. Evolution and the Diversity of Life: Selected Essays. Belknap Press of Harvard University Press, Cambridge, Massachusetts, ix–721.
Mayr, E., 1982. The Growth of Biological Thought: Diversity, Evolution and Inheritance. Harvard University Press, Cambridge.
McBride, A. C. & C. S. Gowans, 1969. The induction of gene mutation and chromosome aberration in *Chlamydomonas eugametos* by a phenylalanine analog. Genet. Res., Camb. 14: 121–126.
McCourt, R. M. & R. W. Hoshaw, 1990. Noncorrespondence of breeding groups, morphology, and monophyletic groups in Spirogyra (Zygnemataceae: Chlorophyta) and the application of species concepts. Syst. Bot. 15: 69–78.
Mets, L., 1980. Uniparental inheritance of Chloroplast DNA sequences in interspecific hybrids of *Chlamydomonas.* Curr. Genet. 2: 131–138.
Miller, R. D. & R. W. Hoshaw, 1974. Cell width as a taxonomic character with special reference to *Zygnema circumcarinum* Czurda. Br. Phycol. J. 9: 145–148.
Mishler, B. D. & M. J. Donogheu, 1982. Species concepts: a case for pluralism. Syst. Zool. 31: 491–503.
Nozaki, H., 1988. Morphology, sexual reproduction and taxonomy of *Volvox carteri* f. *kawasakiensis* f. nov. (Chlorophyta) from Japan. Phycologia 27: 209–220.
Nozaki, H., 1989. Morphological variation and reproduction in *Gonium viridistellatum* (Volvocales, Chlorophyta). Phycologia 28: 77–88.
Nozaki, H., 1991. *Pandorina colemaniae* sp. nov. (Volvocaceae, Chlorophyta) from Japan. Phycologia 30: 449–457.
Nozaki, H., 1993. Asexual and sexual reproduction in *Gonium quadratum* (Chlorophyta) with a discussion of phylogenetic relationships within the Goniaceae. J. Phycol. 29: 369–376.
Nozaki, H. & M. Itoh, 1994. Phylogenetic relationships within the colonial Volvocales (Chlorophyta) inferred from cladistic analysis based on morphological data. J. Phycol. 30: 353–365.
Ohno, S., 1970. Evolution by Gene Duplication. Springer-Verlag, Berlin, 160 pp.
Ohtani, S., 1985. Seasonal variation of desmids at a small marsh in Hiroshima, Japan. Jpn J. Phycol. 33: 190–198.
Otte, D. & J. A. Endler, 1989. Speciation and its Consequences. Sinauer Associates, Massachusetts, xiii–679.
Perkins, D. D., 1974. The manifestation of chromosome rearrangements in unordered asci of *Neurospora.* Genetics 77: 459–489.
Pocock, M. A., 1955. Studies in North American Volvocales. I. The genus *Gonium.* Madronon 13: 49–80.
Pringsheim, E. G., 1959. Kleine Mitteilungen über Flagellaten und Algen. V. *Gonium quadratum* nov. sp. Arch. Mikrobiol. 32: 429–436.
Proctor, V. W., 1959. Dispersal of fresh-water algae by migratory water birds. Science 130: 623–624.
Proctor, V. W., 1962. Viability of *Chara* oospores taken from migratory water birds. Ecology 43: 528–529.
Proctor, V. W., 1965. Dispersal of desmids by water birds. Phycologia 5: 227–232.
Proctor, V. W. & C. R. Malone, 1965. Further evidence of passive dispersal of small aquatic organisms via the intestinal tract of birds. Ecology 46: 728–729.
Provsasoli, L. & I. J. Pintner, 1959. Artificial media for freshwater algae: problems and suggestions. In C. A. Tryon & R. T. Hartman (eds), The Ecology of Algae. University of Pittsburgh, Pittsburgh: 84–96.
Rice, W. R. & E. E. Hostert, 1993. Laboratory experiments on speciation: What have we learned in 40 years? Evolution 47: 1637–1653.
Round, F. E., 1981. The Ecology of Algae. Cambridge Univ. Press, Cambridge, 653 pp.
Ruzicka, J., 1977. Die Desmidiaceen Mitteleuropas. E. Schweizerbart'sche Verlagsbuchhandlung, Stuttgart.
Sack, L., C. Zeyl, G. Bell, T. Sharbel, X. Rebound, T. Bernhardt & H. Koelewyn, 1994. Isolation of four new strains of *Chlamydomonas reinhardtii* (Chlorophyta) from soil samples. J. Phycol. 30: 770–773.
Sager, R. & S. Granick, 1954. Nutritional control of sexuality in *Chlamydomonas reinhardtii.* J. Gen. Physiol. 37: 729–742.
Sako, Y., K. Shrestha, A. Uchida, Y. Ishida & S. Saito, 1991. Isozyme analysis of mating populations of *Gonium pectorale* (Chlorophyta). J. Phycol. 27: 309–315.
Sarma, Y. S. R. K., 1982. Chromosome number in algae. Nucleus 25: 66–108.
Schlösser, U. G., 1976. Entwicklungsstadien- und sippenspezifische Zellwand-Autolysine bei der Freisetzung von Fortpflanzungszellen in der Gattung *Chlamydomonas.* Ber. dt. bot. Ges. 89: 1–56.
Schlösser, U. G., 1984. Species-specific sporangium autolysins (Cell-wall-dissolving enzymes) in the genus *Chlamydomonas.* In D. E. G. Irvine & D. M. John (eds), Systematics of the Green Algae. Academic Press, New York: 409–418.
Schopf, T. J. M., D. M. Raup. S. J. Gould & D. S. Simberloff, 1975. Genomic versus morphologic rates of evolution: influence of morphologic complexity. Paleobiol. 1: 63–70.
Sekimoto, H. & T. Fujii, 1992. Analysis of gametic protoplast release in the *Closterium peracerosum-strigosum-littorale* complex. J. Phycol. 28: 615–619.
Sekimoto, H., S. Satoh & T. Fujii, 1990. Biochemical and physiological properties of a protein inducing protoplast release during conjugation in the *Closterium peracerosum-strigosum-littorale* complex. Planta 182: 348–354.
Sekimoto, H., S. Satoh & T. Fujii, 1993. Analysis of binding of biotinylated protoplast-release-inducing protein that induces release of gametic protoplasts in the *Closterium peracerosum-strigosum-littorale* complex. Planta 189: 468–474.
Sekimoto, H. Y. Sone & T. Fujii, 1994. Regulation of expression of the genes for a sex pheromone by an inducer of the sex pheromone in the *Closterium peracerosum-strigosum-littorale* complex. Planta 193: 137–144.

Smith, G. M., 1944. A comparative study of the species of *Volvox*. Trans. Am. Microsc. Soc. 63: 265–310.

Soltis, P. S., E. E. Soltis & J. J. Doyle, 1992. Molecular Systematics of Plants. Chapman and Hall, New York.

Sonneborn, T. M., 1957. Breeding systems, reproductive methods and species problems in Protozoa. In E. Mayr (ed.), The Species Problem. Amer. Ass. for the Advancement of Science, Washington, D.C.: 125–324.

Spanier, J. G., J. E. Graham & J. W. Jarvik, 1992. Isolation and preliminary characterization of three *Chlamydomonas* strains interfiertile with *Chlamydomonas reinhardtii* (Chlorophyta). J. Phycol. 28: 822–828.

Starr, R. C., 1955. Isolation of sexual strains of placoderm desmids. Bull. Torrey Bot. Club. 82: 261–265.

Starr, R. C., 1959. Sexual reproduction in certain species of *Cosmarium*. Arch. Protistenk. 104: 155–164.

Starr, R. C., 1970. Control of differentiation of *Volvox*. Develop. Biol. 4 (suppl.): 204–222.

Starr, R. C. & L. Jaenicke, 1989. Cell differentiation in *Volvox carteri* (Chlorophyta): the use of mutants in understanding patterns and control. In A. W. Coleman, L. J. Goff & J. R. Stein-Taylor (eds), Algae as Experimental Systems. Alan R. Liss, New York: 135–147.

Stebbins, G. L., 1950. Variation and Evolution in Plants. Columbia University Press, New York.

Stebbins, G. L., 1971. Chromosome Evolution in Higher Plants. Addison-Wesley, Reading, Massachusetts, 216 pp.

Stein, J. R., 1958. A morphologic and genetic study of *Gonium pectorale*. Am. J. Bot. 45: 664–672.

Stein, J. R., 1965. Sexual populations of *Gonium pectorale* (Volvocales). Am. J. Bot. 52: 379–388.

Stein, J. R., 1966a. Effect of temperature on sexual populations of *Gonium pectorale* (Volvocales). Am. J. Bot. 53: 941–944.

Stein, J. R., 1966b. Growth and mating of *Gonium pectorale* (Volvocales) in defined media. J. Phycol. 2: 23–28.

Stein, J. R. & M. J. McCauley, 1976. Sexual compatibility in *Gonium pectorale* (Volvocales: Chlorophyceae) from soil of a single pond. Can. J. Bot. 54: 1126–1130.

Stewart, J. W. & H. E. Schlichiting Jr., 1966. Dispersal of algae and protozoa by selected aquatic insects. J. Ecol. 54: 551–562.

Strehlow, K., 1929. Über die Sexualität einiger Volvocales. Z. Bot. 21: 625–692.

Tappan, H., 1980. The Paleobiology of Plant Protists. Freeman, San Francisco, xxi–1028.

Van Valen, L. M., 1982. Integration of species: stasis and biogeography. Evol. Theory. 6: 99–112.

Van Winckle-Swift, K. P. & D. J. Thuerauf, 1991. The unusual sexual preferences of a *Chlamydomonas* mutant may provide insight into mating-type evolution. Genetics 127: 103–115.

Wang, J. C., R. W. Hoshaw & R. M. McCourt, 1986. A polyploid species complex of *Spirogyra communis* (Chlorophyta) occurring in nature. J. Phycol. 22: 102–107.

Waris, H., 1958. Splitting of the nucleus by centrifuging in *Micrasterias*. Ann. Acad. Sci. Fenn. A IV Biol. 40: 1–20.

Watanabe, M., 1977. A preliminary study of *Gonium viridistellatum* sp. nov. (Chlorophyta, Volvocaceae). Bull. Jap. Soc. Phycol. 25 (suppl. Mem. Iss. Yamada): 379–384.

White, J. D., 1978. Modes of Speciation. San Francisco, 455 pp.

Wiese, L., 1974. Nature of sex specific glycoprotein agglutinins in *Chlamydomonas*. Ann. New York Acad. Sci. 234: 383–395.

Wiese, L., 1984. Mating systems in unicellular algae. In H.-F. Linskens & J. Heslop-Harrison (eds), Encyclopedia of Plant Physiology. Springer-Verlag, Berlin, 238–260.

Wiese, L. & W. Wiese, 1977. On speciation by evolution of gametic incompatibility: a model case in *Chlamydomonas*. Am. Nat. 111: 733–742.

Wiese, L. & W. Wiese, 1978. Sex cell contact in *Chlamydomonas*, a model for cell recognition. In A. S. G. Curtis (ed.), Cell-cell Recognition. Cambridge University Press, Cambridge, 83–104.

Wood, A. M. & T. Leatham, 1992. The species concept in phytoplankton ecology. J. Phycol. 28: 723–729.

Wright, S., 1931. Evolution in Mendelian populations. Genetics 16: 97–159.

Yamagishi, T., M. Haga, Y. Nishihama, M. Watanabe & T. Ichimura, 1974. Taxonomic entities in the desmids – some experimental approaches. Gen. Ed. Rev., Col. of Agr. and Veter. Med., Nihon University 10: 54–83.

Hydrobiologia **336**: 19–32, 1996.
J. Kristiansen (ed.), Biogeography of Freshwater Algae.

3. Biodiversity, biogeography and conservation of diatoms

D. G. Mann & S. J. M. Droop
Royal Botanic Garden, Edinburgh EH3 5LR, United Kingdom

Key words: diatoms, biogeography, species concept, taxonomy, conservation

Abstract

Recent morphometric and breeding studies of diatoms show that the present species classification is too coarse and hides significant diversity. Many species are subdivided into phenodemes, which often differ in cell size, shape, stria density and pattern, but may also have different ultrastructural features. In raphid diatoms these can include the form of the raphe endings, details of the pore occlusions, and the structure of the girdle, while chloroplast structure can also vary. The phenodemes can be sympatric or allopatric. In *Sellaphora pupula* and other species, sympatric phenodemes are reproductively isolated. It is recommended that such demes are recognized as separate species; the total number of diatom species worldwide may thus be at least 2×10^5. Use of a fine-grained classification reveals that many diatom species may be endemics, some restricted to a single lake or catchment, others to wider areas. Environmental impact assessments and conservation strategies must begin to take account of endemism and rarity among microscopic algae and protists.

Introduction

The diatoms are a very widespread and ecologically important group of algae. Unlike the Rhodophyta and Phaeophyta, which are significant primary producers only along sea coasts, or the Haptophyta, significant only in the marine phytoplankton, diatoms are major constituents of benthic and planktonic algal communities worldwide, in terrestrial, freshwater and marine habitats. Werner (1977) suggested that diatoms may contribute 20 to 25% of world net primary production, although this estimate probably needs to be revised downwards in the light of recent data on the contribution made by marine picoplankton (Harris, 1986), which includes few if any diatoms. The numbers of diatom cells, even in a small lake, are enormous; any species occurring at a density of 1 cell. m^{-2} of lake surface would be essentially undetectable, yet there would still be 10^4 per hectare. Furthermore, many diatom species are claimed to be widely distributed within a climatic zone or even cosmopolitan (e.g. see the distributions recorded for freshwater species by Krammer & Lange-Bertalot, 1986, 1988, 1991a,b; for marine species by Guillard & Kilham, 1977). Hence it would be easy to conclude that local or national studies of diatom biodiversity are not a priority and that, for example, the same Floras and keys can be used in Europe and Asia as in the Americas; that diatom biogeography is worth studying only on a global scale or scarcely differs from diatom ecology (so that each species would occur everywhere its own ecological tolerances are met and where it is competitively superior); and that conservation is unlikely to be an important issue in diatoms. It may yet prove that these conclusions are justified, but we will present evidence that suggests otherwise.

The perception of diversity, the detection of biogeographical patterns and the assessment of rarity are inextricably linked with taxonomy. Using the recently completed *Süsswasserflora von Mitteleuropa* (Krammer & Lange-Bertalot, 1986, 1988, 1991a,b), one would have to conclude that no freshwater species is truly rare or endangered; that most if not all species are very widely distributed; and that the diatom biodiversity of an area is controlled primarily by the diversity of ecological niches within it.

However, the *Süsswasserflora* (at least in the first two volumes) uses a wide species concept, in which many taxa, formerly considered separate species, are placed in synonymy. For example, in the account of *Nitzschia* Hass. (Krammer & Lange-Bertalot, 1988, largely based on earlier revisions by Lange-Bertalot, 1977 and Lange-Bertalot & Simonsen, 1978), the new wider concept of *N. fonticola* Grun. in Cleve & Möll. includes *N. macedonica* Hust., *N. manca* Hust., *N. radicula* Hust., *N. romana* (Grun.) Grun. in Van Heurck, and *N. subromana* Hust. (Lange-Bertalot & Simonsen, 1978; Krammer & Lange-Bertalot, 1988). *N. fonticola sensu lato* is 'wahrscheinlich kosmopolitisch' (Krammer & Lange-Bertalot, 1988), whereas some of the taxa put in synonymy seem to have more restricted distributions: *N. macedonica*, for example, was considered by Hustedt (1945) to be endemic to the Balkans. *N. fonticola sensu stricto* and *N. macedonica* are morphologically very similar and the taxonomic changes made by Lange-Bertalot and his collaborators are therefore reasonable. However, Lange-Bertalot & Simonsen's plate of *N. fonticola* and its allies (1978, pl. 8, figures 126–138) show that there are subtle differences between them, which Hustedt, using a narrower species concept, evidently thought were enough to justify species status for *N. macedonica*.

Lange-Bertalot & Simonsen (1978) concluded 'that the traditional definition of species, because of the lack of a sufficient species concept, must lead to infinite separation and finally to individuals'. Nowhere in their work, however, is it shown why their wide concept of species is qualitatively different from, and intrinsically better than, the older, narrower concept used by Hustedt and others. In a recent work by Krammer (1992), there seems to be a change back to a more finely grained taxonomy (relative to that of Krammer & Lange-Bertalot, 1986); he recognizes 96 species of *Pinnularia* in Europe, and divides many of these into several 'morphotypes', instead of the 55 species given by Krammer & Lange-Bertalot (1986).

Thus, before we can discuss biogeography and conservation, we must examine the nature of diatom species further. To do this, we will refer to recent morphometric and breeding studies, paying particular attention to the common freshwater diatom *Sellaphora pupula* (Kütz.) Mereschk. (= *Navicula pupula*).

Materials and methods

Samples of epipelon were collected from various lakes in the United Kingdom. Material illustrated in this paper came from Blackford Pond (UK National Grid reference NT 253709, 70 m alt.) and Figgate Loch (NT 298737, 20 m alt.), both small eutrophic lakes in parkland within Edinburgh, Scotland; Threipmuir Reservoir, a mesotrophic lake draining peaty uplands near Edinburgh (NT 163634, 250 m alt.); the mesotrophic Loch of Craiglush, near Dunkeld, Scotland (NO 043446, 100 m alt.); and Fenemere, Shropshire, England (SJ 444231, 80 m alt.), a eutrophic lake surrounded by farmland (Reynolds, 1979). Other sites mentioned in the text include Loch Lubnaig, Central Region, Scotland (NN 587112, 125 m alt.); Loch Tulla, Highland Region, Scotland (NN 311437, 170 m alt.); Whinfell Tarn, Cumbria, England (SD 5598, 126 m alt.); Kates Cottage West Pond, near Sturston, Norfolk (grid reference not known). Sampling methods and processing have been described by Mann (1984, 1989b). Material from all of the sites mentioned is held in the herbarium of the Royal Botanic Garden, Edinburgh.

Microscopy was carried out using a Reichert Polyvar photomicroscope, using Kodak Technical Pan film.

Results and discussion

Variation and mating patterns in Sellaphora pupula

Sellaphora pupula is an almost ubiquitous freshwater species, occurring in the epipelon of lotic and lentic habitats (Hustedt, 1937a, 1961; see also Gasse, 1986), particularly in high conductivity waters (Krammer & Lange-Bertalot, 1986). Denys (1991) adds that it is somewhat euryhaline and prefers circumneutral to alkaline conditions. Gasse (1986), however, found contradictions within the literature, some authors claiming that *S. pupula* is halophilic, others that it is euryhalobous (*sensu* Simonsen, 1962); some that it is a good indicator of alkaline waters, others that it is pH-indifferent.

S. pupula is very variable in shape and striation pattern and many infraspecific taxa have been proposed to reflect this (VanLandingham, 1975 lists 36 names of varieties and forms). However, over the last 50 years these taxa have increasingly been merged or reduced in rank. For instance, Hustedt (1930) recognized 7 varieties (*elliptica, rectangularis, rostrata, capitata, pseudopupula, aquaeductae* and *mutata*: for full citations, see VanLandingham, 1975). Later, Hustedt decided that *rectangularis, capitata* and *rostrata* should be regarded as forms that 'gehen alle ineinander über und können deshalb nicht wie bisher als Varietäten im eigentlichen Sinne gewertet werden' (Hustedt, 1942). Hustedt (1961) recognized four forms (*elliptica, rectangularis, rostrata* and *capitata*) and three varieties (*pseudopupula, aquaeductae* and *mutata*). Continuing the trend towards amalgamation, Ross (1963) suggested that 'the various forms ... so ... intergrade that there seems little justification for the taxonomic recognition of them at any level'. Likewise, Schoeman & Archibald (1979), after studying many European and African samples, concluded that 'the distinction of different forms on the basis of valve outline alone cannot hold' and regarded *elliptica, rectangularis* and *capitata*, among others, as 'part of the range of variability in *N. pupula* in which no differentiation can be made'; the taxonomic status of *rostrata* and *mutata* was left undetermined. Finally, Krammer & Lange-Bertalot (1986) did not recognize *elliptica, rectangularis, capitata* and *rostrata* as distinct at any level, but maintained *pseudopupula, aquaeductae* and *mutata* as varieties; they also reduced *Sellaphora nyassensis* (O. Müll.) D. G. Mann (= *Navicula nyassensis*) to the status of a variety within *S. pupula*.

Krammer & Lange-Bertalot's (1986) comments are instructive. '*N. pupula* bietet ein Musterbeispiel für die subjektive Bewertung von Merkmalen bei der Artabgrenzung'. Of primary importance is the development of the terminal nodule into short transapical bars, a characteristic shared (in Krammer & Lange-Bertalot's classification) only by *Sellaphora* (= *Navicula*) *bacillum* (Ehrenb.) D. G. Mann. 'Demgegenüber werden andere Kriterien wie Größe, Form, Dichte und Neigung der Str[eifen], Begrenzung der Zentralarea hier als weitgehend bedeutungslos gewertet, obwohl sie bei anderen Taxa, denen es an einem übergeordneten Merkmal mangelt, sehr wohl als entscheidend betrachtet werden'. Given that Krammer & Lange-Bertalot (1986) do not recognize varieties or forms such as *rectangularis* and *elliptica*, it seems clear that they agree with this subjective down-weighting of size, shape and striation pattern and are not describing current practice. But, as with Lange-Bertalot & Simonsen's (1978) comments on species limits (in relation to *Nitzschia*; see Introduction), no justification is offered for the character weighting that has been applied.

The marginal sediments (at 1 m depth) of Blackford Pond, Edinburgh, usually bear a well-developed epipelic flora, except in very still conditions in summer or under ice during the winter, when even the top microlayer of mud may become anoxic; then the diatom populations are much depleted, leaving only a depauperate flora of euglenoid algae. Whenever significant diatom populations are present, *Sellaphora pupula* and *S. bacillum* can be detected. *S. bacillum* seems always to be present as a homogeneous population; most of the variation that exists can be attributed to the effects of size reduction during the life cycle (Mann, 1988a). *S. pupula*, on the other hand, is heterogeneous (Figures 1–6). There are six separate phenodemes (following Gilmour & Heslop-Harrison, 1954, a phenodeme is a group of individuals of a particular, specified taxon that differs phenotypically from other groups belonging to the same taxon). Four of the six exhibit a normal diatom life cycle, with an extended period of size reduction alternating with rapid size restitution via a sexually produced auxospore (Mann, 1984, 1988a, 1989a,b). These four have been given the informal names 'capitate', 'rectangular', 'small' and 'obese' (Mann, 1989b). The fifth phenodeme, 'neat', undergoes size reduction (and therefore must also auxosporulate), but it is usually the rarest of the demes and it is not yet known how it produces its auxospores; an isolated observation suggests it is autogamous or parthenogenetic (Mann, 1989b). The last phenodeme, 'lanceolate', always exhibits a narrow range of valve lengths and appears to avoid size reduction altogether. No diatom is known in which there is a complete absence of size reduction but which nevertheless exhibits sexual reproduction during the life cycle; hence 'lanceolate' is probably asexual. The alternative is that it reduces in size extremely slowly.

Representative valves of the Blackford phenodemes are shown in Figures 1–6. They differ in size (even after allowing for life cycle changes), shape, striation density and striation pattern (see also Mann, 1984, 1988a, 1989b). 'Rectangular' (Figure 1) and 'capitate' (Figure 2) have ± linear valves and 19–20 striae in 10 μm. They have similar striation patterns and central areas, but are clearly separated on length–width plots (Mann, 1988a, 1989b), 'rectangular' being consistently wider for any given length than 'capitate'. Other useful diagnostic features are the more obviously sinuous raphe in 'rectangular' and the slightly more capitate poles of 'capitate'. 'Small', 'lanceolate', 'neat' and 'obese' all have lanceolate valves. 'Small' (Figure 3) has ca. 22 strongly radiate striae in 10 μm, narrowly rostrate apices, and a bow-tie-shaped central area. 'Lanceolate' (Figure 4) has slightly fewer striae in 10 μm than 'small' and the poles are broadly rostrate. The striae are strongly radiate at the centre, where some of the striae continue in towards the central raphe endings as blind grooves ('shadow striae'); as a result, the central area is smaller and less well-defined

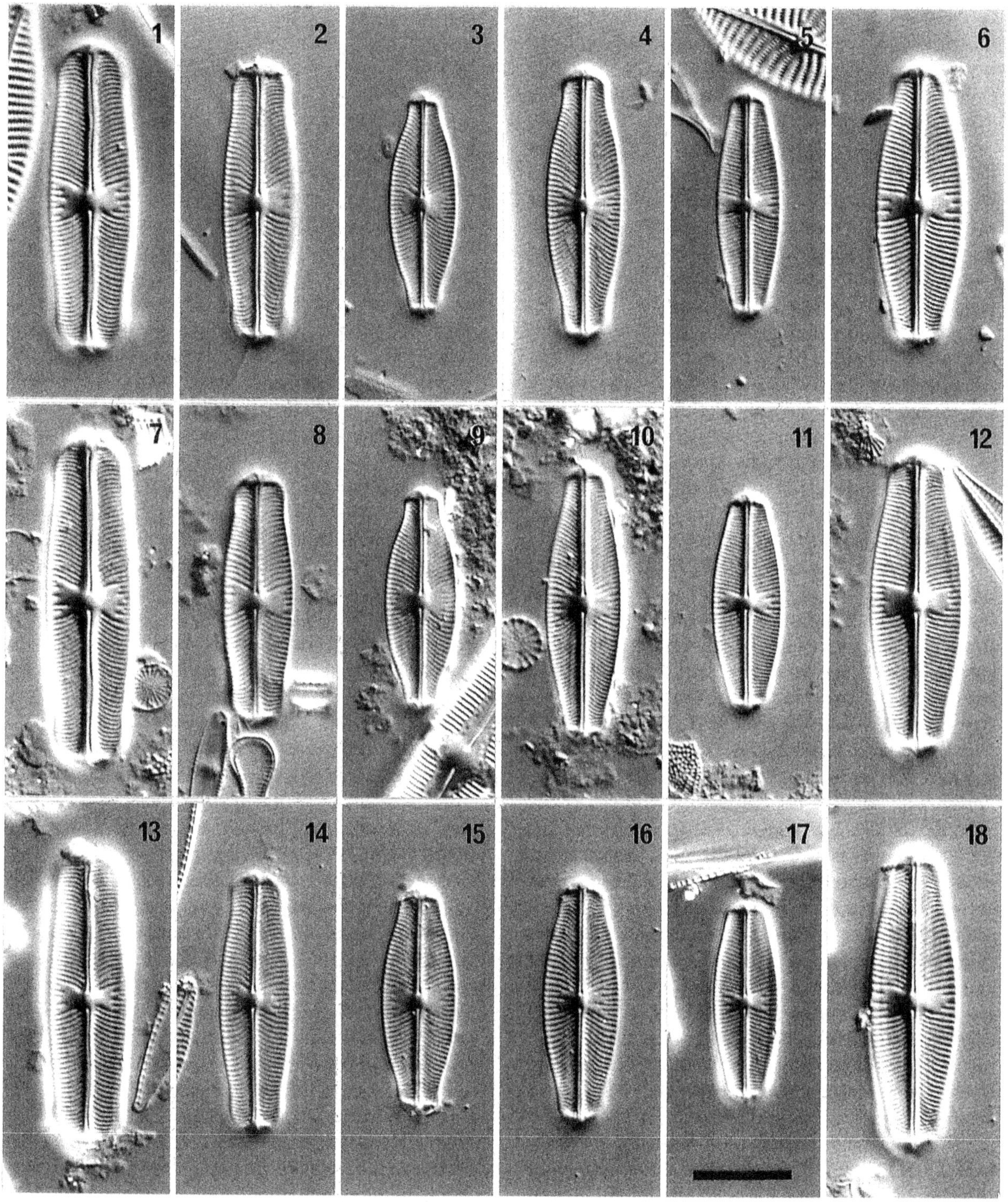

Figures 1–18. Sellaphora pupula demes from Blackford Pond (Figures 1–6), Fenemere (Figures 7–12) and Figgate Loch (Figures 13–18). Demes shown are 'rectangular' (Figures 1, 7, 13); 'capitate' (Figures 2, 8, 14); 'small' (Figures 3, 9) and cf. 'small' (Figure 15); 'lanceolate' (Figures 4, 10, 16); 'neat' (Figures 5, 11, 17); and 'obese' (Figures 6, 12, 18). All are shown to the same scale; scale bar (in Figure 17) = 10 μm.

than in the other demes. Towards the poles the striae become convergent and several are usually geniculate. 'Neat' (Figure 5) has only slightly radiate striae and these are more closely spaced than in any of the other demes (22–25 in 10 μm); the poles are scarcely if at all rostrate. 'Obese' (Figure 6) has 19–20 striae in 10 μm. As in 'neat', the striae are only slightly radiate at the centre, but towards the slightly rostrate poles they become convergent and geniculate striae are often present. The differences between the demes are maintained throughout size reduction (illustrated by Mann, 1989b).

The Blackford phenodemes can be identified even when alive, so that it is possible to examine mating patterns in mixed populations of sexualized cells. Such studies show that the four allogamous demes ('capitate', 'rectangular', 'obese' and 'small') are separated by substantial, perhaps inviolate, reproductive barriers (Mann, 1984, 1989b), which prevent them pairing. The same pattern of behaviour has been shown both in populations harvested from sediments brought into the laboratory and also *in situ* in Blackford Pond (Mann, 1989b). There is some indication that the demes have different ecological requirements and different susceptibilities to chytrid parasites (Mann, 1989b). Since the demes are not only phenotypically distinct but reproductively isolated, they are also gamodemes *sensu* Gilmour & Heslop-Harrison (1954). Culture studies show that the morphological differences between the demes have a genetic basis. This is confirmed also by the persistence of the demes and their characteristics in nature, despite seasonal changes in water chemistry, light and temperature.

One question at first left unanswered (Mann, 1984) was 'whether each pond has its own, different, reproductively isolated genodemes', though Mann (1989b) noted similarities between some of the Blackford demes and illustrations published by Hustedt (1961), Schoeman & Archibald (1979) and Krammer & Lange-Bertalot (1986). Subsequent observations have shown that each of the Blackford phenodemes does occur elsewhere. To illustrate this, Figures 7–12 show examples of 'rectangular', 'capitate', 'small', 'neat', 'lanceolate' and 'obese' valves from Fenemere, Shropshire, about 350 km from Blackford Pond. The obvious overall (*gestalt*) similarity between allopatric populations of the same deme can be analysed in terms of individual characters. The 'rectangular' valves, for example, have sinuous raphes (Figures 1, 7); the 'lanceolate' valves have radiating 'shadow striae' in the central area, extending in to near the raphe endings (Figures 4, 10); the 'neat' valves have closely spaced, almost parallel striae (Figures 5, 11); and so on.

'Rectangular', 'capitate', 'lanceolate', 'neat' and 'obese' are also present in Figgate Loch, Edinburgh, a few km from Blackford Pond (Figures 13, 14, 16–18). Again, 'neat' has the most finely striate valves, 'lanceolate' is the only one with 'shadow striae', 'obese' and 'rectangular' have the widest and most heavily silicified valves, and so on. The only questionable identification is of 'small'. Some valves in Figgate Loch have the same stria pattern, stria density, central area shape, and width as 'small' (Figure 15), but most (though not all) of them have broader apices than in the Blackford and Fenemere demes (compare Figures 3, 9). The congruity between allopatric populations is also illustrated by the contrast between any of the six demes shown in Figures 1–18 and those shown in Figures 19–30. The six clearly do not represent the whole range of variation within *Sellaphora pupula sensu lato*.

We are continuing to investigate the distribution of the demes first identified from Blackford Pond. Our preliminary conclusion is they are widespread (at least in the U.K.) in highly eutrophic lakes with fine, soft marginal sediments that are highly organic and strongly reduced below the top millimetre. We have detected some or all of the demes in Scotland (Blackford Pond, Figgate Loch), the English Lake District (Whinfell Tarn), the English Midlands (Fenemere) and East Anglia (Kates Pond). The accompanying epipelic flora generally includes such species as *Sellaphora bacillum, Craticula cuspidata* (Kütz.) D. G. Mann, *Navicula capitata* Ehrenb., *N. menisculus* Schumann, *N. protracta* (Grun.) Cleve, *Anomoeoneis sphaerophora* (Ehrenb.) Pfitzer, *Caloneis amphisbaena* (Bory) Cleve, *Stauroneis phoenicenteron* (Nitzsch) Ehrenb., *Tryblionella hungarica* (Grun.) D. G. Mann, (= *Nitzschia hungarica*), *Nitzschia recta* Hantzsch, *N. sigmoidea* (Nitzsch) W. Smith, and *Cymatopleura solea* (Bréb. & Godey) W. Smith.

It appears that all six Blackford demes do not always occur together. For example, the 'small' deme has not yet been found in Whinfell Tarn, although it is always difficult to be sure that a deme is truly absent, rather than very rare and hence undetected. What is certain, however, is that Figgate Loch and Fenemere contain other demes besides the six shared with Blackford. For example, Figure 23 shows a deme from Fenemere that has a similar outline to 'lanceolate' but has a quite different central area. And, although the accompanying epipelic flora is similar in Blackford Pond, Figgate Loch and Fenemere, there are also differences:

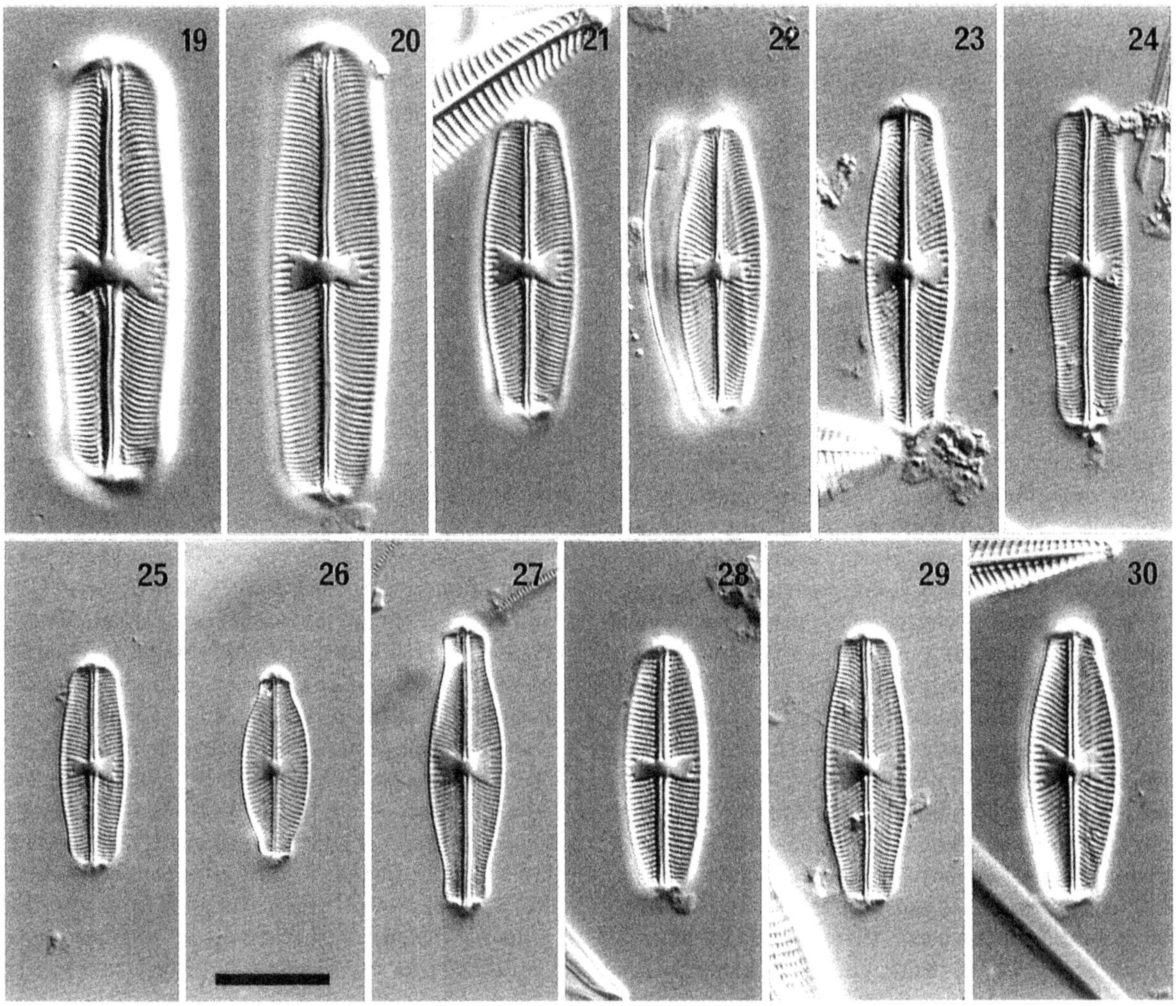

Figures 19–30. Sellaphora pupula demes. Figure 19. Threipmuir reservoir 'gross' deme; note the sinuous raphe bordered by prominent grooves, which flare outwards near the centre. Figure 20. Threipmuir reservoir 'large' deme: prominent grooves are absent (contrast Figure 19). Figures 21, 22. Demes from the Loch of Craiglush. Figure 23. Deme from Fenemere (contrast Figures 7–12). Figure 24. Deme from the Loch of Craiglush. Figures 25–29. Demes from Threipmuir reservoir (Figure 26 shows the 'tiny' deme). Figure 30. Deme from the Loch of Craiglush, possibly the same as that shown in Figure 22. All are shown to the same scale as in Figures 1–18; scale bar (in Figure 26) = 10 μm.

for example, *Cymatopleura elliptica* (Bréb.) W. Smith, *Cymbella ehrenbergii* Kütz. and *Tryblionella gracilis* W. Smith (= *Nitzschia tryblionella*) are quite common in Figgate Loch, but rare or absent in Blackford Pond.

When other types of lake are studied, the kinds of *Sellaphora pupula* present are often markedly different. In upland Scottish lochs, less eutrophic than Blackford Pond and draining peaty catchments, none of the Blackford demes may be present; this is true, for example, of Lochs Tulla and Lubnaig in the Grampian mountains of west-central Scotland. Threipmuir reservoir, near Edinburgh, has some demes similar (though perhaps not identical) to the Blackford demes, but it also possesses a suite of other demes (Figures 19, 20, 25–29). Some of these are very distinctive, such as the large-celled deme 'gross' shown in Figure 19, which has markedly sinuous external raphe fissures, bordered by deep grooves. During auxosporulation the gametangia of this deme remain unpaired and so, although meiosis takes place, reproduction is automictic (Mann, in preparation). There is also another large-celled deme in Threipmuir, 'large', which is always slightly narrower than 'gross', with less strongly radiate striae and much shallower grooves by the raphe (Figure 20); it also differs from most other demes in that it is possible to resolve the poroids using the light microscope (this is also true for the deme shown in Figure 28). 'Large', unlike 'gross', is allogamous. Threipmuir also contains two demes with very small valves (Figures 25, 26). 'Tiny' (Figure 26) resembles 'small' (Figures 3, 9) in its shape and striation pattern, but it is even smaller, over the whole life cycle (Mann, 1989b, illustrates size reduction in 'small' from Blackford Pond). The Threipmuir deme shown in Figure 27 has the highest striation density found so far (26 in 10 μm). Demes from the mesotrophic Loch of Craiglush are shown in Figures 21, 22, 24, 30. One (Figure 24) has a similar shape to 'rectangular' and 'capitate' (Figures 1, 2), but is narrower, with a much higher striation density (24–25 in 10 μm).

Altogether, from the 12 sites studied in some detail, over 20 phenodemes have been distinguished so far. In some cases (Figgate Loch, Threipmuir Reservoir) we have unpublished evidence of reproductive isolation between demes, just as in Blackford Pond. The number of demes per lake varies considerably, from one or two in Loch Tulla, through the six of Blackford, to ten or more in Threipmuir reservoir.

Variation and mating patterns in other diatom species

There is now abundant evidence (reviewed by Mann, 1989b) that many diatom species vary in the same way as *Sellaphora pupula*, being divided into two to many morphologically distinct entities (phenodemes or morphotypes), which may be sympatric or allopatric. In nine species at least, there is evidence of external isolation mechanisms (*sensu* Stebbins, 1950), operating at the diploid level, which prevent gene flow when the phenodemes are sympatric (see Mann, 1989b). In other species, where there is no information about breeding relationships, two or more phenodemes can often be demonstrated to coexist; each exhibits the kind of variation usually associated with size reduction, but remains quite distinct and separate from the other sympatric demes in one to several morphological characters. The fact that these phenodemes remain distinct suggests very strongly that they represent separate lineages, kept apart by external or internal barriers to interbreeding.

Thus, for example, Droop (1994) has carried out a morphometric analysis of 535 valves of the *Diploneis smithii* (Bréb. ex W. Smith) Cleve – *D. fusca* (Greg.) Cleve complex from the intertidal sands at Ganavan, near Oban, W. Scotland. This demonstrated the presence of 11 different phenodemes (morphotypes), most with a distinctive valve structure (the characters studied included the shape of the longitudinal canals, the structure of the striae and the shape of the central raphe endings), and all showing the kind of variation in length and width that is usually associated with size reduction during the life cycle. The two most similar of the demes, differing only in dimensions and shape (one has a higher rectangularity than the other), have also been studied at other sites (Droop, 1995). Both demes can be demonstrated to be present and separate in geographically remote areas of Britain. However, mean values of valve dimensions and rectangularity vary slightly between areas, so that an uncritical, non-quantitative analysis of variation might easily lead to the conclusion that only one, very variable deme exists. This could help explain why some authors claim that species like *Sellaphora pupula* are continuously variable, whereas results from Blackford and elsewhere clearly demonstrate the existence of discrete entities with different distributions and reproductive characteristics.

The studies of *Sellaphora pupula* and other freshwater epipelic species by Mann (1984, 1988a, 1989b) and those of *Diploneis* Ehrenb. by Droop (1994, 1995) have relied on analyses of characters observed or measured using the light microscope. As such, these investigations probably underestimate morphological differentiation between sympatric demes. Evidence that this is so comes from observations of *Hantzschia* Grun. species. At Aberdeen, NE Scotland, two demes of *H. marina* (Donkin) Grun. coexist in the intertidal sand (Mann, 1978). In the light microscope they can be separated by size and striation density, but SEM observations reveal extra differences. The smaller, more finely striate deme possesses elliptical or simply reniform poroids closed only by a hymen (Mann, 1978, 1981a); the external central raphe endings are straight. In the larger-celled deme the poroids contain cribra as well as hymenes and the external central raphe endings are displaced towards the ventral valve margin. At Tregantle Bay, Cornwall, intertidal sands again contain two sympatric demes of *H. marina*, though neither seems equivalent to the Aberdeen demes. Again, the demes differ in size. The larger one has reniform poroids, with branched struts extending in from the 'hilus' of the 'kidney', while the smaller one has elliptical poroids (Mann, 1978, 1981a). In addition, the most advalvar band of the cingulum is closed (i.e. it is a complete hoop) in the large-celled deme, but open in the small-celled deme; there are differences too in the polar raphe endings (Mann, 1978). Major differences in girdle structure also occur between members of the *H. amphioxys* (Ehrenb.) Grun. complex (recently revised in part by Lange-Bertalot, 1993), some having one closed band and at least five open bands (split rings) in the mature cingulum, while others have five or six closed bands and one or two open bands; in all cases, the closed bands are grouped at the advalvar end of the cingulum (Mann, 1978).

Mann (1981b) showed differences between varieties of *H. virgata* (Roper) Grun. in the central raphe endings (internal and external), the form and spacing of the fibulae, and the structure of the girdle. Subsequently, differences have also been discovered in chloroplast morphology, though all the varieties have two chloroplasts per cell, one on either side of the median transapical plane. Vars. *gracilis* Hust., *leptocephala* Østrup and *wittii* Grun. have relatively simple chloroplasts, each with a single central pyrenoid, though the pyrenoid of *gracilis* is elongate while in the others it is ± isodiametric. In var. *virgata*, on the other hand, at least some demes have complex chloroplasts. These are divided almost in two; each half has its own pyrenoid and is linked to the other half by a narrow lateral strand. Furthermore, even *within* each variety of *H. virgata* there is a considerable range of size and striation density (Mann, 1978, and unpublished), suggesting demic differentiation as in *Sellaphora pupula* and the other freshwater epipelic species studied by Mann (1989b). In *H. distinctepunctata* (Hust.) Hust. two different demes are present even on the type slide, but neither of them is the same as a deme found in Britain, differing in dimensions and striation density (Mann, 1980; see also Garcia-Baptista, 1993) (Lange-Bertalot, 1993, questions whether Hustedt ever validly published *distinctepunctata* as a species; this was not the opinion of Hartley, 1986, and clarification is necessary).

In the English Lake District, Knudson (1953a) showed that different lakes have different morphotypes of *Tabellaria flocculosa* (Roth) Kütz. In Blelham Tarn three morphotypes were found coexisting, two in the plankton, the other being epiphytic (1953b).

One of the most detailed studies of variation within and between closely related diatom species has been carried out by Theriot and Stoermer (e.g. Theriot & Stoermer, 1984; Theriot, 1992). A morphometric analysis of the *Stephanodiscus niagarae* Ehrenb. complex in N America revealed the existence of three subtly different entities, which were separated at species level, as *S. niagarae sensu stricto, S. superiorensis* Stoermer & Theriot and *S. yellowstonensis* Theriot & Stoermer (Theriot & Stoermer, 1984). Subsequently, a fourth species of the complex has been discovered: *S. reimerii* Theriot & Stoermer (Theriot, 1992). *S. reimerii* 'overlaps with *S. niagarae* in both time and space and yet remains morphologically distinct . . . Given that these two species each reproduce sexually, it seems simplest to infer that they are reproductively isolated and that *S. reimerii* is a biological species relative to *S. niagarae*' (Theriot, 1992). *S. superiorensis* and *S. niagarae* both occur in Lake Superior, but are spatially separated, *superiorensis* being found offshore and *niagarae* in nearshore waters. *S. yellowstonensis* is restricted to Yellowstone Lake, Wyoming, and is the only species of the *niagarae* complex found there. Studies of cores from Yellowstone Lake indicate that *yellowstonensis* has evolved from a *niagarae*-like ancestor in the last 12,000 years (Theriot, 1992). The origin of *S. superiorensis* is unknown but today it is restricted to Lake Superior, while *S. reimerii* occurs in a single drainage system in north Iowa. *S. niagarae*, on the other hand, is widespread, occurring in many lakes in the USA and Canada.

The need for a new species-level classification in diatoms

Recent studies demonstrate, therefore, that the existing species-level classification of diatoms is too coarse-grained, especially following the abandonment of most infraspecific taxa as worthless (compare, for instance, the number of varieties and forms recognized by Krammer & Lange-Bertalot, 1986, 1988, 1991a,b) with those given by Cleve-Euler (1951–55). The existing classification disguises variation that is neither continuous nor random, but which is organized discretely, in relation to ecology, geography and history. Theriot and Stoermer (Theriot & Stoermer, 1984; Theriot, 1992) have taken the radical step of recognizing these 'quanta' of variation as separate species (even though no single character separates *Stephanodiscus yellowstonensis* from *S. niagarae* – the pattern of variation in correlated characters has to be taken into account: see Theriot & Stoermer, 1984). This is the recommendation also of Mann (1989b) and Droop (1994), though they stopped short of putting it into effect by creating new binomials. Krammer (1992) also recognized that the present species-level taxonomy of diatoms lacks discrimination, but preferred to refer to many of the 'quanta' as morphotypes (e.g. six within *Pinnularia nodosa* (Ehrenb.) W. Smith; three within his new species *P. anglica*).

It seems to us that Theriot and Stoermer's solution is the best (see also discussion in Mann, 1989b). In higher plants, species are almost always distinguished on the basis of discontinuities in the pattern of morphological variation; most or all members of one species differ from most or all members of closely related species in several correlated characters. The two species of oak native in the British Isles, *Quercus robur* L. and *Q. petraea* (Mattuschka) Liebl., differ in the shape of the leaf at its base (with auricles in *robur*), the length of the peduncle (minimal in *petraea*), and the hairiness of the underside of the leaf (almost glabrous in *robur*, with stellate hairs in *petraea*; Clapham et al., 1962). The species also differ ecologically. *Q. petraea* is more characteristic of established woodland on nutrient-poor, acid soils, while *Q. robur* is a more widespread, opportunistic species (Rackham, 1980); in many places, however, the two species occur together. The morphological differences between them are such that most people do not notice them (all are 'oaks'). Furthermore, the two species hybridize and produce viable, fertile offspring (Stace, 1975), which are intermediate in phenotype and grow alongside the parents in nature. The consensus among taxonomists and ecologists, however, is that these species should be kept separate taxonomically, despite hybridization in nature, the occurrence of intermediate morphologies, and gene flow between them. Many similar examples are known from the British flora (Stace, 1975) and elsewhere. By analogy with higher plants, therefore, and assuming that some kind of uniformity in species concepts is desirable (otherwise, what possible meaning can there be in estimates of the numbers of species on earth, like those given by Groombridge, 1992?), demes like those we have described within *Sellaphora pupula* and other diatoms need to be treated and named as separate species; they seem to be at least as distinct as the two British oaks.

We would emphasize that we are not claiming that the ability or inability to interbreed (the biological species concept) is always or uniquely decisive in establishing species boundaries. Where demes do not encounter each other in nature (either because they grow at different times or in different places), no gene flow can occur. If the demes are asexual, again no gene flow can occur. In neither case is it inevitable that the genetic isolation will lead to divergence in morphological, ecological or physiological characteristics. However, wherever gene flow does occur it will tend to integrate populations and break down pre-existing gene combinations. Thus, if we are dealing with sexual organisms and there appears to be a discontinuity in the variation pattern between populations that live together (sympatric and synchronous), the biological species concept can be viewed as a test of significance. If the phenodemes cannot interbreed, we can have confidence that we are dealing with entities that have a biologically significant history, exist now, are worth searching for elsewhere and at other times, and will probably have a biologically interesting future. This is the position with the demes of *Sellaphora pupula* and the other species listed by Mann (1989b). If there are no barriers to interbreeding, then the nature of the discontinuity is called into question. Perhaps the discontinuity reflects different episodes of sexual reproduction, producing year classes like those discussed by Mann (1988b); perhaps it reflects disruptive selection (Thoday, 1972); perhaps it is entirely fortuitous.

A complication must be mentioned. There is no reason why diatom species should be *morphologically* distinct. The sibling species (syngens) of the ciliate protozoan *Paramecium aurelia* O.F. Müll. exhibit physiological and other differences between themselves and cannot interbreed, but they are so similar morphologically that it is either difficult or impossible to identify them by eye (Sonneborn, 1957). The significance of minor variations in shape and form in unicellular algae and protists is generally unknown. In some plants variation in shape and form is directly related to and necessary for speciation: the best examples are insect-pollinated orchids, where differences in floral morphology can lead directly to reproductive isolation, through selection of different pollinators. It is very unlikely indeed that this is so for diatoms – there is no lock-and-key fit of morphologies during mating – and we must anticipate that speciation sometimes is, and sometimes is not, accompanied by morphological change (see also Knowlton, 1993). Indeed, a number of studies have now revealed considerable cryptic variation within diatom species, for example in allozyme banding patterns (e.g. Gallagher, 1982; Soudek & Robinson, 1983), and differential sensitivity to parasites (Canter & Jaworski, 1978, 1979). Medlin et al. (1991) have used molecular sequence data to support the separation of a species (*Skeletonema pseudocostatum* Medlin) that is only minimally different morphologically from its nearest relative. Zechman et al. (1994), investigating internal transcribed spacers within ribosomal DNA in species of the *Stephanodiscus niagarae* complex, found no autapomorphic nucleotide sites in *S. yellowstonensis*, but did discover such autapomorphies in particular, morphologically indistinguishable populations of its presumed ancestor, *S. niagarae* (see above), which has a much longer fossil record. We can expect such examples to be the first of many.

Sonneborn delayed recognizing the *Paramecium aurelia* sibs formally (as *P. primaurelia, biaurelia, triaurelia,* etc.) until there was a practical means of identifying them, other than through comprehensive mating tests (Sonneborn, 1975). This may be a good policy. However, the demes of *Sellaphora pupula* or the *Diploneis smithii/fusca* complex, and the species of the *Stephanodiscus niagarae* complex are already identifiable, even if with some difficulty, and so Sonneborn's caution is irrelevant.

Consequences for biodiversity estimates

If a more finely grained species classification is adopted for diatoms, many new species will have to be described. Existing species, fossil or recent, will have to be examined for heterogeneity and split if necessary. The quantitative effects of this are not easily estimated. Theriot and Stoermer have (so far) distinguished four species in the *Stephanodiscus niagarae* complex, instead of the one recognized originally (Theriot, 1992). In one sample from W Scotland, Droop (1994) has found 11 morphotypes (presumed species) in the *Diploneis smithii–fusca* complex, instead of the two species recognized by Hustedt (1937b), and other morphotypes exist elsewhere in Britain (perhaps 30 in all). The *Sellaphora pupula* complex probably contains over 20 species in Britain alone. Some existing species are probably more narrowly and correctly circumscribed, although close inspection sometimes reveals heterogeneity where none was originally suspected. We initially thought that *S. laevissima* (Kütz.) D. G. Mann was unitary – an interesting counterpart to its more variable relative *S. pupula*. However, as soon as we began to study *S. laevissima* populations methodically, we discovered sympatric demes as in *S. pupula*, though not as many; Whinfell Tarn, for example, contains two demes that differ markedly in striation density and size. Overall, we estimate that existing species will need to be multiplied at least tenfold, in order to produce a classification that properly reflects diatom species diversity. The number of known diatom species is often given as ca. 10^4 (Hendey, 1964; Guillard & Kilham, 1977); with a narrower species concept, this would rise to 10^5. But this takes into account only those species that have already been described. To these must be added entirely new discoveries, which are likely to be many, especially from understudied areas and habitats. Tropical regions are still poorly explored and the marine benthos needs attention worldwide, especially in the subtidal zone; for some genera (e.g. *Nitzschia, Surirella* Turpin) the most recent overview of marine benthic species is still the account by Peragallo & Peragallo (1897–1908). Altogether, then, we suggest that the total number of diatom species worldwide is probably not less than 2×10^5. Diatoms would thus be confirmed as the most species-rich group of algae.

There is no need for the introduction of a narrower species concept to be accompanied by further splitting of existing genera, except where these are shown to be unnatural. Even so, the total number of diatom genera will undoubtedly grow. In the last 15 years, entirely new genera of living diatoms have been discovered (as opposed to being created through taxonomic revision) at a rate of roughly one per year. Examples include the raphid pennate diatom *Protokeelia* Reimer & Lee (1984), the araphid *Nephroneis* Amspoker (1989) and the centric *Cyclotubicoalitus* Stoermer, Kociolek & Cody (1990).

One likely consequence of a narrower species concept is greater precision of ecological monitoring and palaeoecological reconstruction. As mentioned before, *Sellaphora pupula sensu lato* occurs in so many different kinds of environment that one can make very few statements about its ecology, beyond that it is a freshwater species with some tolerance of brackish conditions, which avoids low pHs. However, our preliminary data suggest that the individual demes (segregate species) have more exacting requirements than *S. pupula sensu lato*; separating them should enhance the resolution of ecological and palaeoecological studies, though it will be incumbent on taxonomists to provide tools to make routine identification possible.

Whether or not they are separated as species, subspecies or varieties, or only informally, the demes within *S. pupula* and other species must be recognized and enumerated in any floristic or biodiversity studies. *S. pupula* is probably present in the epipelon of almost all meso- and eutrophic lakes, ponds and rivers in Britain, but in some cases '*S. pupula*' would refer to two demes, elsewhere perhaps 20, and even where the numbers are the same in two habitats, the identities of the demes are often quite different.

Biogeography and conservation

The fine-grained taxonomy we advocate helps support the view that a significant minority of diatom species may be narrow endemics, i.e. species restricted to a small geographical area, such as a single lake, drainage basin, or island. In the early part of this century, many diatoms were described as endemics. Skvortzow & Meyer (1928) described many new taxa, supposedly endemic to Lake Baikal, Siberia. Likewise, O. Müller and Hustedt described new endemic taxa from the East African Great Lakes (reviewed by Ross, 1983). However, as the *Sellaphora nyassensis* example illustrates (see earlier), changes in species concepts led to many of the 'endemics' being merged into very variable, cosmopolitan species. This trend must now be reversed.

Sensitive taxonomies reveal endemism of various kinds. In Africa, some taxa appear to be restricted to one lake and its catchment, such as *Surirella sparsipunctata* Hust., a species restricted to the Lake Tanganyika and its surrounds (Ross, 1983; Cocquyt & Vyverman, 1993). Other *Surirella* species, such as *S. gradifera* Hust. and *S. heidenii* Hust., are similarly restricted (Ross, 1983). The Baikal endemics are now being reinvestigated (e.g. Genkal, 1990) and confirmed as independent species; in some cases, Skvortzow & Meyer's taxonomy is itself being shown to be conservative (some of their varieties of *Cyclotella* have been elevated to species status: see Flower, 1993), even though they described 160 new taxa out of 450 reported from the lake (Foged, 1993).

Theriot and Stoermer's segregate species of *Stephanodiscus* Ehrenb. are even more striking examples of narrow endemics, since all the lakes in which they occur have come into being in the last 15,000 y. This sets an upper limit for existence of the species themselves, unless they evolved elsewhere, migrated into their present sites after the last ice retreat, and subsequently became extinct everywhere else. For *S. yellowstonensis* this scenario is virtually ruled out by Theriot's (1992) core studies, which show the gradual evolution of *yellowstonensis* morphology from a *niagarae* stock between 12,000 and 3,800 y bp. The East African Great Lakes, such as Lake Tanganyika, and Lake Baikal, are much older (Brooks, 1950) and so it is perhaps not surprising that they should contain suites of endemics.

Very restricted geographical distributions are also known or suspected for fossil taxa, such as *Biddulphia areolata* Hajós, *Caloneis hungarica* Hajós and the remarkable endemic genus *Dimidiata* Hajós, which were limited to the Paratethys inland sea, after its separation from the Mediterreanean basin the middle Miocene (Hajós, 1974, 1990).

Besides these narrow endemics, there are also species that are characteristic of part or all of a continent, but occur nowhere outside it. *Surirella engleri* O. Müll. seems to fall into this category, having been reported from Nigeria, Sierra Leone and Mali, as well as many of the East African Great Lakes (Ross, 1983; Gasse, 1986). *S. nyassae* O. Müll., is somewhat less widespread, occurring in Lakes Malawi and Victoria, and also in Angola (Ross, 1963; Gasse, 1986). *Gomphocymbella* O. Müll. species also show clear biogeographical patterns at this kind of scale (Kociolek & Stoermer, 1993). *Gomphonema ventricosum* Greg. is present in Europe, Asia and Alaska, but possibly not elsewhere (Kociolek & Stoermer, 1987), while *Gomphonitzschia ungeri* Grun. and *Gomphotheca sinensis* (Grun.) Hendey & Sims seems to be restricted to tropical Africa and East Asia, respectively (Hendey & Sims, 1982).

Endemism is not restricted to particular types of diatom community. *Cyclotella tasmanica* Haworth & Tyler, restricted to Tasmania but widely distributed there (Haworth & Tyler, 1993), is planktonic, as are Theriot's N American *Stephanodiscus* endemics. *Gomphocymbella* species are attached forms (Kociolek & Stoermer, 1993). *Rhopalodia iriomotensis* Kobayasi, Nagumo & Tanaka (1993), apparently endemic to Iriomote Island, Japan, was found on the prop roots of mangroves. Although it might be expected that marine diatoms would show less endemism than freshwater forms, because of the possibility of long distance transport in water rather than air, it appears that benthic marine taxa, if not their planktonic counterparts, do sometimes show interesting, restricted distributions. We think it unlikely, for instance, that *Hantzschia doigiana* Stidolph could have been overlooked if it occurred in Europe or N America. This species has very distinctive, long, beak-like apices and seems to be endemic to New Zealand, where it is widespread (Stidolph, 1993). Likewise, many of the curious and very distinctive taxa described from the marine littoral of Campèche Bay, Mexico (e.g. by Hustedt, 1952), have yet to be found elsewhere and may be restricted to the Gulf of Mexico.

Of course, until sufficient studies have been made using a sensitive, fine-grained taxonomy, it will not be clear how many species are narrow endemics and how many are truly cosmopolitan. Previously published information is of little use in determining distributions unless the identifications are supported by good photographs or voucher specimens.

Now that it is clear that there are narrow endemic species of diatoms, conservation becomes a significant issue. If Yellowstone Lake or Lake Superior contained no endemic *Stephanodiscus* species, but only

S. niagarae and other widespread species then, from a biodiversity standpoint, the loss of all *Stephanodiscus* (e.g. through pollution) might be tolerable, since *S. niagarae* etc. would still be abundant elsewhere. Similarly, if *Sellaphora pupula* is left as an undifferentiated, cosmopolitan species, then any changes in lake chemistry and biotic composition that permitted at least one of the demes to survive (even if this was not one of those present before disturbance), might be regarded as having had no significant effect, at least for that species. However, one point highlighted by the new fine-grained taxonomy is that, however subtle the differences between diatom taxa may seem to the layman, many lakes and regions probably contain unique, irreplaceable diatom species and germplasm. This can apply even when the lakes are geologically young, in regions deglaciated only ca. 10,000 years ago. There is already evidence too that some diatoms have become locally extinct, probably as a result of human activity: the nominate variety of *G. eriense* (Grun.) Skvortzow in Skvortzow & Meyer has disappeared from the Laurentian Great Lakes in the last 100 years and now occurs only in Western N America (Kociolek & Stoermer, 1988; see also Stoermer & Yang, 1971). We suggest, therefore, that it is time to consider whether environmental impact assessments of the effects of industrial or other developments on aquatic ecosystems should include an assessment of diatom biodiversity by a competent taxonomist. In addition, conservation agencies should begin to develop strategies for diatoms and other microscopic algae and protists.

Acknowledgements

This work was supported in part by SERC GR/C/68484, NERC GR 3/7923, and a grant from the Royal Society towards the purchase of a photomicroscope. We are grateful to Dr Jørgen Kristiansen for the opportunity to complete this paper, which is based on a presentation at the 5th International Phycological Congress in Qingdao, China; and Dr K. Clarke for slides of material from Sturston.

References

Amspoker, M. C., 1989. *Nephroneis macintirei*, gen. et sp. nov., a marine araphid diatom from California, U.S.A. Diatom Res. 4: 171–177.

Brooks, J. L., 1950. Speciation in ancient lakes. Quart. Rev. Biol. 25: 30–60, 131–176.

Canter, H. M. & G. H. M. Jaworski, 1978. The isolation, maintenance, and host range studies of a chytrid (*Rhizophydium planktonicum*) parasitic of *Asterionella formosa*. Ann. Bot. 42: 967–980.

Canter, H. M. & G. H. M. Jaworski, 1979. The occurrence of a hypersensitive reaction in the planktonic diatom *Asterionella formosa* Hass. parasitized by the chytrid *Rhizophydium planktonicum* Canter in culture. New Phytol. 82: 187–206.

Clapham, A. R., T. G. Tutin & E. F. Warburg, 1962. Flora of the British Isles. Ed. 2. Cambridge University Press, Cambridge, 1269 pp.

Cleve-Euler, A., 1951–55. Die Diatomeen von Schweden und Finnland. I–V. Kungl. svenska VetenskapsAkad. Handl., Ser. 4, 2(1): 1–163 (1951); 3(3): 1–153 (1952); 4(2): 1–158 (1953); 4(5): 1–255 (1953); 5(4): 1–232 (1955).

Cocquyt, C. & W. Vyverman, 1993. *Surirella sparsipunctata* Hustedt and *S. sparsipunctata* var. *laevis* Hustedt (Bacillariophyceae), a light and electron microscopical study. Hydrobiologia 269/270: 97–101.

Denys, L., 1991. A check-list of the diatoms in the Holocene deposits of the western Belgian plain with a survey of their apparent ecological requirements. I. Introduction, ecological code and complete list. Professional Paper 1991/2, no. 246. Belgische Geologische Dienst, Brussels, 41 pp.

Droop, S. J. M., 1994. Morphological variation in *Diploneis smithii* and *D. fusca* (Bacillariophyceae). Arch. Protistenk. 144: 249–270.

Droop, S. J. M., 1995. A morphometric and geographical analysis of two races of *Diploneis smithii/D. fusca* (Bacillariophyceae) in Britain. In D. Marino & M. Montresor (eds), Proceedings of the 13th International Diatom Symposium. Biopress Ltd, Bristol: 347–369.

Flower, R. J., 1993. A taxonomic re-evaluation of endemic *Cyclotella* taxa in Lake Baikal, Siberia. Nova Hedwigia, Beih. 106: 203–220.

Foged, N., 1993. Some diatoms from Siberia especially from Lake Baikal. Diatom Res. 8: 231–279.

Gallagher, J. C., 1982. Physiological variation and electrophoretic banding patterns of genetically different seasonal populations of *Skeletonema costatum* (Bacillariophyceae). J. Phycol. 18: 148–162.

Garcia-Baptista, M., 1993. Observations on the genus *Hantzschia* Grunow at a sandy beach in Rio Grande do Sul, Brazil. Diatom Res. 8: 31–43.

Gasse, F., 1986. East African diatoms. Taxonomy, ecological distribution. Bibl. Diatomol. 11: 1–202.

Genkal, S. I., 1990. Morphology and systematics of two endemic species of *Cyclotella* (Bacillariophyceae) from Lake Baikal. In H. Simola (ed.), Proceedings of the 10th International Diatom Symposium. Koeltz Scientific Books, Koenigstein: 55–60.

Gilmour, J. S. L. & J. Heslop-Harrison, 1954. The deme terminology and the units of micro-evolutionary change. Genetica 27: 147–161.

Groombridge, B. (ed.), 1992. Global biodiversity. Status of the Earth's living resources. Chapman & Hall, London, 585 pp.

Guillard, R. R. L. & P. Kilham, 1977. The ecology of marine planktonic diatoms. In D. Werner (ed.), The biology of diatoms (Botanical monographs 13). Blackwell Scientific Publications, Oxford: 372–469.

Hajós, M., 1974. Faciological and stratigraphic importance of the Miocene diatoms in Hungary. Nova Hedwigia, Beih. 45: 365–376.

Hajós, M., 1990. Succession of the fossil diatom floras in the Carpathian Basin, Hungary. Nova Hedwigia, Beih. 100: 227–236.

Harris, G. P., 1986. Phytoplankton ecology. Structure, function and fluctuation. Chapman & Hall, London & New York, 384 pp.

Hartley, B., 1986. A check-list of the freshwater, brackish and marine diatoms of the British Isles and adjoining coastal waters. J. mar. biol. Ass. U.K. 66: 531–610.

Haworth, E. Y. & P. A. Tyler, 1993. Morphology and taxonomy of *Cyclotella tasmanica* spec. nov., a newly described diatom from Tasmanian lakes. Hydrobiologia 269/270: 49–56.

Hendey, N. I., 1964. An introductory account of the smaller algae of British coastal waters. Part V: Bacillariophyceae (Diatoms). MAFF Fishery Investigations, series 4. Her Majesty's Stationery Office, London, 317 pp.

Hendey, N. I. & P. A. Sims, 1982. A review of the genus *Gomphonitzschia* Grunow and the description of *Gomphotheca* gen. nov., an unusual diatom group from tropical waters. Bacillaria 5: 191–212.

Hustedt, F., 1930. Bacillariophyta. In A. Pascher (ed.), Die Süsswasser-Flora Mitteleuropas, vol. 10 (2nd edn). G. Fischer, Jena, 466 pp.

Hustedt, F., 1937a. Systematische und ökologische Untersuchungen über die Diatomeenflora von Java, Bali und Sumatra nach dem Material der Deutschen Limnologischen Sunda-Expedition. Teil I. Systematischer Teil, Erste Fortsetzung. Arch. Hydrobiol., Suppl. 15: 187–295.

Hustedt, F., 1937b. Die Kieselalgen Deutschlands, Österreichs und der Schweiz unter Berücksichtigung der übrigen Länder Europas sowie der angrenzenden Meeresgebiete. In Dr L. Rabenhorst's Kryptogamen-Flora von Deutschland, Österreich und der Schweiz 7 (2:5). Akademische Verlagsgesellschaft, Leipzig: 577–736.

Hustedt, F., 1942. Süsswasser-Diatomeen des indomalayischen Archipels und der Hawaii-Inseln. Int. Revue ges. Hydrobiol. Hydrogr. 42: 1–252.

Hustedt, F., 1945. Diatomeen aus Seen und Quellgebieten der Balkan-Halbinsel. Arch. Hydrobiol. 40: 867–973.

Hustedt, F., 1952. Neue und wenig bekannte Diatomeen. II. Ber. dtsch. bot. Ges. 64: 305–315.

Hustedt, F., 1961. Die Kieselalgen Deutschlands, Österreichs und der Schweiz unter Berücksichtigung der übrigen Länder Europas sowie der angrenzenden Meeresgebiete. In Dr L. Rabenhorst's Kryptogamen-Flora von Deutschland, Österreich und der Schweiz 7 (3:1). Akademische Verlagsgesellschaft, Leipzig: 1–160.

Knowlton, N., 1993. Sibling species in the sea. Ann. Rev. Ecol. Syst. 24: 189–216.

Knudson, B. M., 1953a. The diatom genus *Tabellaria* II. Taxonomy and morphology of the plankton varieties. Ann. Bot. 17: 131–155.

Knudson, B. M., 1953b. The diatom genus *Tabellaria* III. Problems of infra-specific taxonomy and evolution in *T. flocculosa*. Ann. Bot. 17: 597–609.

Kobayasi, H., T. Nagumo & S. Tanaka, 1993. *Rhopalodia iriomotensis* sp. nov., a brackish diatom with shoehorn-shaped projections on the canal raphe (Bacillariophyceae). Nova Hedwigia, Beih. 106: 133–141.

Kociolek, J. P. & E. F. Stoermer, 1987. Geographic distribution and variability of the diatom (Bacillariophyceae) *Gomphonema ventricosum* Gregory. Nova Hedwigia 45: 223–236.

Kociolek, J. P. & E. F. Stoermer, 1988. Taxonomy, ultrastructure and distribution of *Gomphoneis herculeana, G. eriense* and closely related species (Naviculales: Gomphonemataceae). Proc. Acad. Nat. Sci. Philad. 140: 24–97.

Kociolek, J. P. & E. F. Stoermer, 1993. The diatom genus *Gomphocymbella* O. Müller: taxonomy, ultrastructure and phylogenetic relationships. Nova Hedwigia, Beih. 106: 71–91.

Krammer, K., 1992. *Pinnularia*, eine Monographie der europäischen Taxa. Bibl. Diatomol. 26: 1–353.

Krammer, K. & H. Lange-Bertalot, 1986. Bacillariophyceae 1. Teil: Naviculaceae. In H. Ettl, J. Gerloff, H. Heynig & D. Mollenhauer (eds), Süsswasserflora von Mitteleuropa 2(1). G. Fischer, Stuttgart & New York, 876 pp.

Krammer, K. & H. Lange-Bertalot, 1988. Bacillariophyceae 2. Teil: Bacillariaceae, Epithemiaceae, Surirellaceae. In H. Ettl, J. Gerloff, H. Heynig & D. Mollenhauer (eds), Süsswasserflora von Mitteleuropa 2(2). G. Fischer, Stuttgart & New York, 596 pp.

Krammer, K. & H. Lange-Bertalot, 1991a. Bacillariophyceae 3. Teil: Centrales, Fragilariaceae, Eunotiaceae. In H. Ettl, J. Gerloff, H. Heynig & D. Mollenhauer (eds), Süsswasserflora von Mitteleuropa 2(3). G. Fischer, Stuttgart & New York, 576 pp.

Krammer, K. & H. Lange-Bertalot, 1991b. Bacillariophyceae 4. Teil: Achnanthaceae. Kritische Ergänzungen zu *Navicula (Lineolatae)* und *Gomphonema*. In H. Ettl, G. Gärtner, J. Gerloff, H. Heynig & D. Mollenhauer (eds), Süsswasserflora von Mitteleuropa 2(4). G. Fischer, Stuttgart & New York, 437 pp.

Lange-Bertalot, H., 1977. Eine Revision zur Taxonomie der Nitzschiae Lanceolatae Grunow. Nova Hedwigia 28: 253–307.

Lange-Bertalot, H., 1993. 85 neue taxa und über 100 weitere neu definierte Taxa ergänzend zur Süßwasserflora von Mitteleuropa Vol. 2/1–4. Bibl. Diatomol. 27: 1–454.

Lange-Bertalot, H. & R. Simonsen, 1978. A taxonomic revision of the Nitzschiae lanceolatae Grunow 2. European and related extra-European fresh water and brackish water taxa. Bacillaria 1: 11–111.

Mann, D. G., 1978. Studies in the family Nitzschiaceae (Bacillariophyta). Ph.D. dissertation, University of Bristol.

Mann, D. G., 1980. Studies in the diatom genus *Hantzschia*. II. *H. distinctepunctata*. Nova Hedwigia 33: 341–352.

Mann, D. G., 1981a. Sieves and flaps: siliceous minutiae in the pores of raphid diatoms. In R. Ross (ed.), Proceedings of the 6th Symposium on Recent and Fossil Diatoms. O. Koeltz, Koenigstein: 279–300.

Mann, D. G., 1981b. Studies in the diatom genus *Hantzschia* 3. Infraspecific variation in *H. virgata*. Ann. Bot. 47: 377–395.

Mann, D. G., 1984. Observations on copulation in *Navicula pupula* and *Amphora ovalis* in relation to the nature of diatom species. Ann. Bot. 54: 429–438.

Mann, D. G., 1988a. The nature of diatom species: analyses of sympatric populations. In F. E. Round (ed.), Proceedings of the Ninth International Diatom Symposium. Biopress Ltd & Koeltz Scientific Books, Bristol & Koenigstein: 317–327.

Mann, D. G., 1988b. Why didn't Lund see sex in *Asterionella*? A discussion of the diatom life cycle in nature. In F. E. Round (ed.), Algae and the Aquatic Environment. Biopress, Bristol: 383–412.

Mann, D. G., 1989a. The diatom genus *Sellaphora*: separation from *Navicula*. Br. phycol. J. 24: 1–20.

Mann, D. G., 1989b. The species concept in diatoms: evidence for morphologically distinct, sympatric gamodemes in four epipelic species. Pl. Syst. Evol. 164: 215–237.

Medlin, L. K., H. J. Elwood, S. Stickel & M. L. Sogin, 1991. Morphological and genetic variation within the diatom *Skeletonema costatum* (Bacillariophyta): evidence for a new species, *Skeletonema pseudocostatum*. J. Phycol. 27: 514–524.

Peragallo, H. & Peragallo, M. (1897–1908). Diatomées Marines de France et des Districts Maritimes Voisins. Paris & Grez-sur-Loing.

Rackham, O., 1980. Ancient woodland, its history, vegetation and uses in England. Edward Arnold, London, 402 pp.

Reimer, C. W. & J. J. Lee, 1984. A new pennate diatom: *Protokeelia hottingeri* gen. et sp. nov. Proc. Acad. Nat. Sci. Philad. 136: 194–199.

Reynolds, C. S., 1979. The limnology of the eutrophic meres of the Shropshire-Cheshire plain: a review. Field Studies 5: 93–173.

Ross, R., 1963. The diatom genus *Capartogramma* and the identity of *Schizostauron*. Bull. Br. Mus. nat. Hist., Botany 3: 49–92.

Ross, R., 1983. Endemism and cosmopolitanism in the diatom flora of the East African Great Lakes. In R. W. Sims, J. H. Price & P. E. S. Whalley (eds), Evolution, time and space: the emergence of the biosphere. Systematics Association Special Volume 23. Academic Press, London: 157–177.

Schoeman, F. R. & R. E. M. Archibald, 1979. The diatom flora of southern Africa 5. CSIR Special Report WAT 50, Pretoria.

Simonsen, R., 1962. Untersuchungen zur Systematik und Ökologie der Bodendiatomeen der westlichen Ostsee. Int. Revue ges. Hydrobiol., Syst. Beih. 1: 1–144.

Skvortzow, B. W. & C. I. Meyer, 1928. A contribution to the diatoms of Baikal Lake. Proc. Sungaree River Biol Stat. 1: 1–55.

Sonneborn, T. M., 1957. Breeding systems, reproductive methods, and species problems in protozoa. In E. Mayr (ed.), The species problem. American Association for the Advancement of Science, Washington. Publication 50: 155–324.

Sonneborn, T. M., 1975. The *Paramecium aurelia* complex of fourteen sibling species. Trans. Am. microsc. Soc. 94: 155–178.

Soudek, D., Jr. & G. C. G. Robinson, 1983. Electrophoretic analysis of the species and population structure of the diatom *Asterionella formosa*. Can. J. Bot. 61: 418–433.

Stace, C. (ed.), 1975. Hybridization and the flora of the British Isles. Academic Press, London, 626 pp.

Stebbins, G. L., 1950. Variation and evolution in plants. Columbia University Press, New York, 643 pp.

Stidolph, S. R., 1993. *Hantzschia doigiana*, a new taxon of brackish-marine diatom from New Zealand coastal waters. Diatom Res. 8: 465–474.

Stoermer, E. F. & J. J. Yang, 1971. Contributions to the diatom flora of the Laurentian Great Lakes. I. New and little-known species of *Amphora* (Bacillariophyta Pennatibacillariophyceae). Phycologia 10: 397–409.

Stoermer, E. F., J. P. Kociolek & W. Cody, 1990. *Cyclotubicoalitus undatus*, genus et species nova. Diatom Res. 5: 171–177.

Theriot, E., 1992. Clusters, species concepts, and morphological evolution of diatoms. Syst. Biol. 41: 141–157.

Theriot, E. & E. Stoermer, 1984. Principal component analysis of *Stephanodiscus*: observations on two new species from the *Stephanodiscus niagarae* complex. Bacillaria 7: 37–58.

Thoday, J. M., 1972. Disruptive selection. Proc. R. Soc. Lond., B 182: 109–143.

VanLandingham, S. L., 1975. Catalogue of the fossil and recent genera and species of diatoms and their synonyms. Part V. *Navicula*. J. Cramer, Vaduz: 2386–2963.

Werner, D., 1977. Introduction with a note on taxonomy. In D. Werner (ed.), The biology of diatoms (Botanical monographs 13). Blackwell Scientific Publications, Oxford: 1–17.

Zechman, F. W., E. A. Zimmer & E. C. Theriot, 1994. Use of ribosomal DNA internal transcribed spacers for phylogenetic studies in diatoms. J. Phycol. 30: 507–512.

Hydrobiologia **336**: 33–40, 1996.
J. Kristiansen (ed.), Biogeography of Freshwater Algae.

4. Geographic distribution of freshwater blue-green algae

Lucien Hoffmann
University of Liège, Department of Botany (B.22), Sart Tilman, B-4000 Liège, Belgium

Key words: Cyanophyceae, Cyanobacteria, blue-green algae, biogeography, freshwater

Abstract

The analysis of the currently available data for morphologically unambiguously defined freshwater blue-green algae indicates that besides (sub-)cosmopolitan species, taxa with a more restricted distribution also exist. Many of these have a holarctic or pantropic distribution. It is hypothesized that, besides the distribution of ecological niches, temperature is one of the main controlling factors restricting species to particular latitudinal zones. Furthermore, the presence of species with a regional distribution (endemics) can not be ruled out, indicating that other factors must be considered. The possible role of dispersal capacities and of dispersal rates in relation to the earth history and to the speciation of blue-green algae is discussed.

Introduction

For many groups of microalgae, our knowledge about species distribution and about the factors which govern their distribution is still fragmentary. The main reasons for this situation are, on one hand, the lack of floristic lists for many regions of the world, and, on the other hand, the low level of taxonomic knowledge in many groups. This note analyses the present knowledge about the biogeography of freshwater blue-green algae (cyanobacteria).

Methodological considerations

The study of the distribution of freshwater blue-green algae is hampered by difficulties in describing and recognizing the different taxa due to their often simple morphology. For the identification of freshwater blue-green algae, Geitler's monograph (1932) still remains the main reference, for temperate as well as for tropical regions; however, it mainly deals with species from temperate regions (especially Europe). Specimens collected in the tropics which could not be identified according to the traditional keys were thus often referred to the closest described species from temperate regions. However, the detailed morphological and ecological study of tropical specimens often leads to the recognition of these apparently cosmopolitan species as separate taxa (e.g. Frémy, 1930; Geitler & Ruttner, 1935; Komárek, 1985). Many records in lists from tropical regions thus require careful checking.

Problems in the delimitation of the taxa are further complicated by the use of two widely different taxonomic systems. Drouet's system (summarized in Drouet, 1981) lead to the recognition of only 62 taxa and it was followed by several authors for regional floristic studies (e.g. Brannon 1952; Nielsen & Madsen 1948; Zaneveld 1988). However, several experimental studies (e.g. Stam & Holleman, 1979; Stulp & Stam, 1984; Waterbury & Stanier, 1978) have shown that this approach does not reflect the biodiversity of blue-green algae. For the discussions in this paper, only publications following the classical approach were thus taken into consideration. In this system, species can be defined as a complex of natural populations with an overall morphological and cytological similarity and which occupy the same ecological niche (Anagnostidis & Komárek, 1985; Geitler, 1932).

Another problem comes from the fact that no floristic lists exist for many parts of the world (large areas of Africa, Asia, S America, Australia) or that only local collections are available, so that endemic taxa or cosmopolitan forms of low abundance may be overlooked. The floras of other regions, i.e. Europe and North America, are better documented. Gaps in the knowledge of the distribution of freshwater blue-green algae may also come from an underrepresentation of sampling of certain niches in which blue-green algae are often important. In tropical regions, few data are available for benthic habitats (epilithic and epiphytic blue-green algae), whereas relatively many data exist for the plankton. Benthic habitats seem, however, to harbour a high diversity of forms (especially within the Chroococcales and the Stigonematales).

To discuss the distribution of freshwater blue-green algae in the present paper, only reports of morphologically and ecologically well defined taxa (within the traditional system updated by Anagnostidis & Komárek, 1985) which can hardly be mistaken for others were taken into account. It is reasonable to admit that these data concern populations belonging to the same species. Experimental studies in the genus *Anabaena* s.l. showed indeed that, at least for morphologically complex groups, populations with an identical morphology are also genotypically similar (Stulp & Stam, 1984).

Distribution patterns

Because blue-green algae are an evolutionary very old group, they were often thought to have a cosmopolitan distribution (e.g. Fogg et al., 1973). Due to the lack of sampling in many regions, the question whether a non-uniform distribution of taxa exists can in general only be asked at the spatial scale of thousands of kilometers. The observed distribution patterns can thus only be discussed relative to the highest biogeographical divisions as are holarctic, neotropic, paleotropic... regions. Only for a few species, sufficient and critically evaluated data exist to give a more precise picture of their actual distribution.

(Sub-)cosmopolitan taxa

Morphologically well defined species with a wide sub-cosmopolitan distribution certainly exist in freshwater habitats. This is the case, for example, for *Mastigocladus laminosus* (Cohn ex Born. & Flah.) Kirchn., a branching blue-green alga occurring in thermal springs in a temperature range of 45–60 °C. This species thus lives within a narrow range of ecological factors, yet it is found in corresponding localities almost all over the world. Similar wide geographical distributions exist for other blue-green algal species from less extreme environments (e.g. *Chroococcus turgidus* (Küts.) Näg., *Microcystis aeruginosa* (Kütz.) Kütz., *M. wesenbergii* (Kom.) Kom, *Oscillatoria agardhii* Gem, *Oscillatoria splendida* Gom., *Phormidium autumnale* C. Ag. ex Gem., *Lyngbya major* Gem., *Nostoc commune* Vauch. ex Born. & Flah.). As is demonstrated in the case of *Mastigocladus* this does not mean that these species are ubiquists occurring as well on soil, in alkaline lakes, in brackish waters, for even these species have specific ecological demands. Their wide distribution is thus mosaic-like, reflecting the corresponding distribution of habitats. It could be demonstrated at least in one case, that taxa with a large distribution are also genotypically related. Thus DNA/DNA hybridization results between strains of *Trichormus variabilis* (Kütz. ex Born. & Flah.) Kom. & Anagn. [= *Anabaena variabilis* Kütz. ex Born. & Flah.] originating respectively from the United States and Europe show that they belong to the same species when bacteriological species criteria are applied (Stulp & Stam, 1984). It is clear, however, that a certain microheterogeneity exists within cosmopolitan freshwater taxa. This heterogeneity can sometimes even be shown at the level of the same population. Besides morphology (e.g. Golubic 1965; Fjerdingstad, 1969; Komárek, 1985), this could for example be shown for contents of pigments (e.g. Aakerman et al., 1992; Meffert & Krambeck, 1977), the ability of chromatic adaptation (e.g. Kohl & Nicklisch, 1981), toxin production (e.g. Kangatharalingam & Priscu, 1993; Llukkainen et al., 1993), odorous compounds production (e.g. Skulberg & Skulberg, 1985) or isozyme patterns (e.g. Kato et al., 1991; Komárek 1994; Stulp & Stam, 1984).

Pantrophic/holarctic taxa

The analysis of the currently available floristic data reveals that planktonic freshwater species with a more restricted distribution exist (Komárek, 1975, 1985; Meffert, 1989). Many of these are distributed in more or less large latitudinal bands around the world and have a pantropic or a (temperate) holarctic distribution. This is certainly the case for some tropical species (e.g. *Aphanizomenon volzii* (Lemm.) Kom., *Arthronema africanum* (Schwabe & Simons.) Kom. & Lukavský, *Cylindrospermopsis raciborskii* (Wolosz.) Seenayya & Subba Raju, *C. philippinensis* (Tayl.) Kom., *Gloeotrichia pilgeri* Schmidle (Figure 1), *Spirulina gigantea* Schmidle, *Anabaena fuellebornii* Schmidle, *A. iyengarii* Bharadw., *A. leonardii*

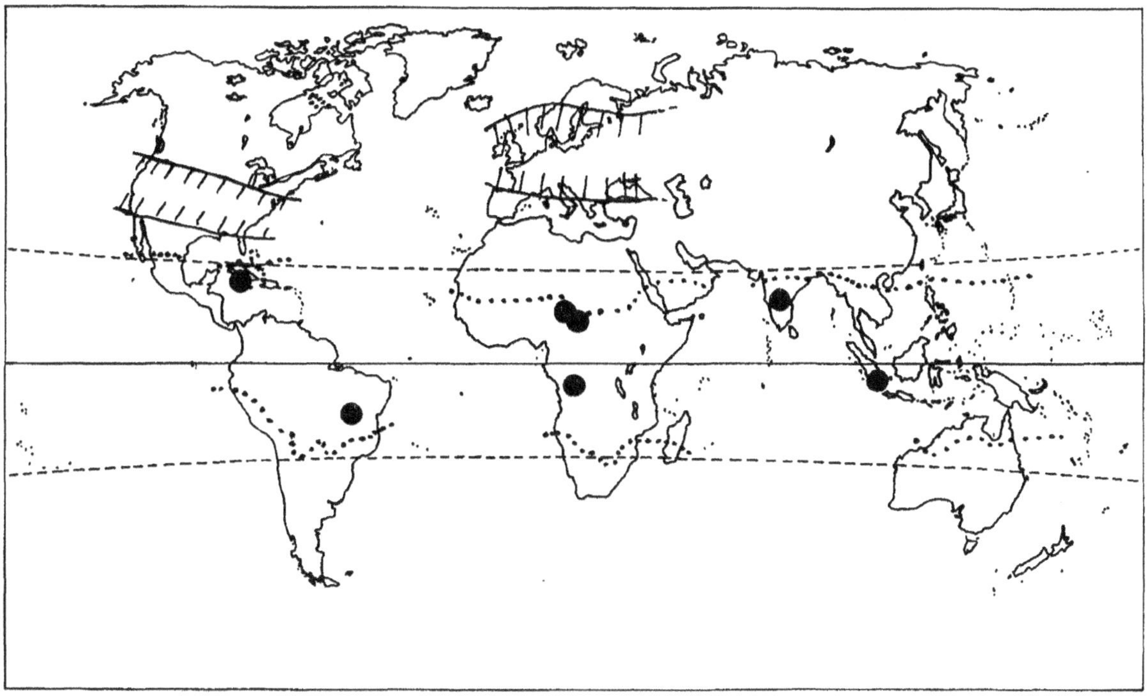

Figure 1. Examples of the distribution of pantropical/holarctic blue-green algal species. *Aphanizomenon flos-aquae* (after Komárek, 1975); ● *Gloeotrichia pilgeri* (after Komárek, 1985); ··· 20 °C winter isotherms in both hemispheres.

Comp., *A. oblonga* De Wildem., *A. recta* Geitl., *Anabaenopsis tanganyikae* (G.S. West) Wolosz. & Mill.) never found in algologically well studied temperate areas. *Oscillatoria limosa* Gem, *O. rubescens* Gem., *O. redekei* Van Goer, *Anabaena solitaria* Klebahn, *A. flos-aquae* Born. & Flah., *A. lemmermannii* Richt., *A. macrospora* Klebahn, *A. planctonica* Brunnth., *Anabaenopsis arnoldii* Aptekarj, *A. milleri* Voronikhin, *Aphanizomenon flos-aquae* Ralfs ex Born. & Flak (Figure 1), *A. issatschenkoi* Proschk.-Lavr. are examples of planktonic blue-green algal species probably restricted to the temperate zones. A discrete area restricted to the temperate zone could also be established for the benthic freshwater species *Nostoc pruniforme* C. Ag. ex Born. & Flah. (Mollenhauer, 1970).

This distribution pattern is also observed for a number of aerophytic (*Porphyrosiphon notarisii* (Menegh.) Kütz. ex Gem., *Scytonema millei* Born. & Thur. ex Born. & Flah., *S. guyanense* (Mont.) ex Born. & Flah., Hoffmann unpublished) and marine blue-green algae (Hoffmann, 1993, 1994).

Endemic taxa

Besides these widely distributed blue-green algae, species with a regional distribution exist. Even species endemic to a small area cannot be ruled out. Thus for example *Nostoc* known from the W. Baltic region (Mollenhauer, 1970; Komárek, 1985), *Pulvinularia suecica* Borzi is only recorded from Swedish lakes (Rott & Hernandez-Mariné, 1994). In tropical regions the existence of endemic species is less well established, especially due to the lack of floristic surveys in many regions. Thus many taxa instead of being endemics of a particular region probably have a pantropical (to warm temperate) distribution. Thus *Mastigocladopsis jogensis* Iyengar & Desik. (Figure 2) described in the forties in India (Iyengar & Desikachary, 1946) was not found again until a few years ago in Corsica (Hoffmann, 1990a). The same is true for other genera (e.g. *Doliocatella* (Figure 2), *Stauromatonema*) for a long time only known from the type locality. Other species, like *Aphanizomenon manguinii* Bourr. (Figure 2) and *Trichormus subtropicus* (Gardn.) Kom & Anagn. [= *Anabaena subtropica* Gardn.], recorded on several islands in the Caribbean region (Komárek,

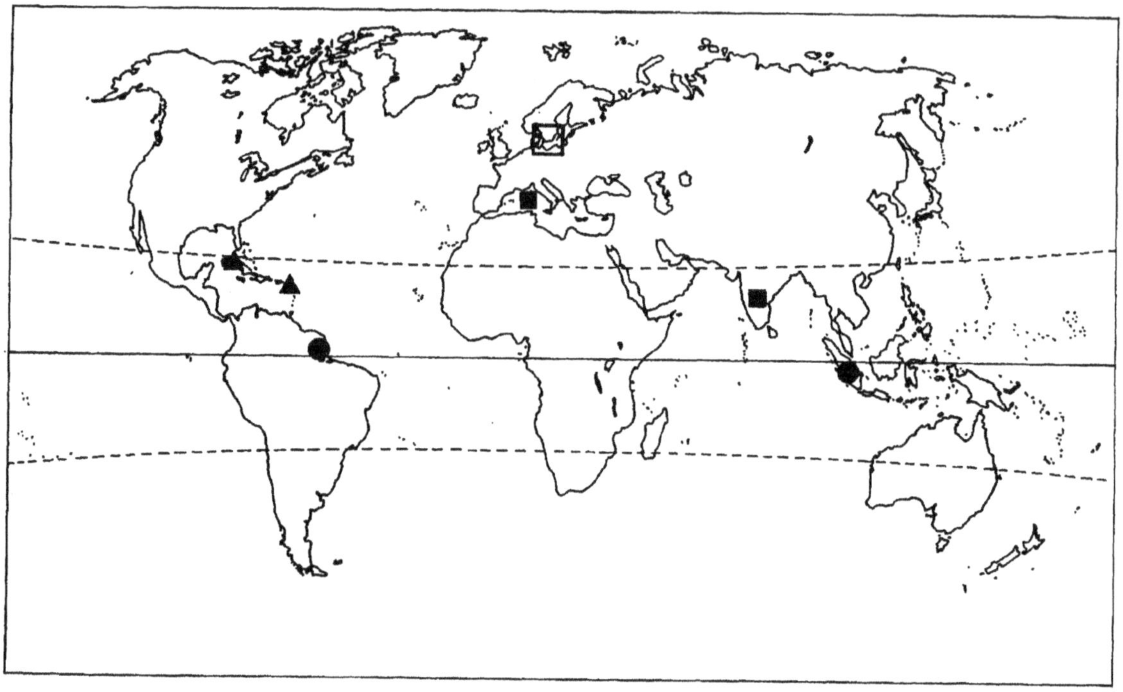

Figure 2. Examples of the distribution of endemic and pantropical, rare blue-green algal species. ▲ *Aphanizomenan manguinii*; ● *Doliocatella formosa*; ■ *Mastigocladopsis jogensis*; □ *Nostoc zetterstedtii.*

1985), may prove in the future to be true endemics. Similar restricted distribution areas also seem to exist for marine (Hoffmann, 1994) and terrestrial (Hoffmann, 1990b) blue-green algae.

In conclusion, the analysis of the currently available data of morphologically unambiguously defined freshwater blue-green algae indicates that beside cosmopolitan taxa, taxa with a more restricted holarctic or pantropical distribution exist. Even the presence of endemic species cannot be ruled out.

Possible explanations of distributions

Due to the absence of precise distribution limits for almost all blue-green algal species, it is difficult to formulate hypotheses about the factors which control their distribution. Several possible explanations addressed below are related to the ecophysiology and dispersal capacities of the different species.

Temperature

At the present level of precision, differences in the distribution of blue-green algae can almost only be considered in relation to the topographic features of the earth such as the position of the continents, or to latitude which overall determine the for algae important environmental factors that are light and temperature.

It is tempting to suggest that temperature is one of the controlling factors restricting certain species to a particular latitudinal zone. When the summer and winter isotherms are compared with the distributional boundaries of freshwater blue-green algae, a good agreement is observed for a number of species between the northern and southern boundaries of pantropical species and the 20 °C winter isotherm in both hemispheres (Figure 1). The same situation is observed for marine (Hoffmann, 1994) and terrestrial (Hoffmann, unpublished) species. Beside the absolute minimum and maximum temperatures, the annual fluctuations may also be of importance: these are smallest in polar and equatorial waters, greatest in the temperate zone. Species of the latter zone are thus expected to be the most tolerant of temperature variations, whereas strictly tropical or polar species should be stenothermal.

Experimental evidence for the inferred hypothetical limiting effect of temperature on survival and growth of freshwater blue-green algal species is, however, scanty (except for hot spring species) and partly contradictory. Results from studies of the effect of temperature on the growth of planktonic blue-green algae from temperate zones showed that these species (*Aphanizomenon flos-aquae*, *Oscillatoria redekei*) were unable to grow at 35 °C with *Anabaena flos-aquae* even showing poor growth at 25 °C (Foy et al., 1976), but this was also true for the cosmopolitan species *Oscillatoria agardhii*. In their study on freshwater *Anabaena* s.l. species, Stulp et al. (1984) showed that the strains attributed to the cosmopolitan species *A. cylindrica* Lemm. and *Trichormus variabilis* had a sharp high temperature boundary near 32 °C and a low temperature limit of 17 °C. All the strains originated from temperate regions (England, The Netherlands, The United States). On the other hand, a strain of the widespread *A. sphaerica* Born. & Flah. isolated from India did not grow below 25 °C, but grew up to 35 °C; a similar response (however with a high temperature limit of 40 °C) was observed for the tropical species *Trichormus randhawae* (Venkat.) Kom & Anagn. [= *A. randhawae* Venkat.]. The high temperature range for growth of these strains thus agrees with the temperature of the original habitats, but not necessarily with their global distribution pattern. On the contrary, in an investigation of ten *Phormidium autumnale* strains from different parts of the world (Komárek, 1972), a certain variability in the response of the strains was observed and the temperature requirements of individual strains did not depend on the temperature conditions of the original locality.

Distribution of habitats

We already mentioned the relation between the distribution of the freshwater blue-green algal species and the distribution of ecological niches. This relationship is especially demonstrated for extreme habitats, like hot springs. Thus the cosmopolitan distribution of *Mastigocladus laminosus* can be explained by the worldwide occurrence of thermal springs. But, this idea of limited ecological types is almost always supported by detailed morphological investigations (Komárek, 1994), also for species from less extreme habitats. The tropics are known for their high diversity of habitats, many of which are not present in temperate regions, e.g. rainforests, lateritic soils. It is possible that this has led to the evolution of morphological types adapted to these special biotopes (Komárek, 1985) which could explain the latitudinal distribution patterns of many species.

Dispersal

Although it is well established that different biotopes are inhabited by a specific blue-green algal flora, the opposite statement is not necessarily true, i.e. the distribution of blue-green algal species does not have to be the mirror of the distribution of corresponding niches or environmental conditions. Indeed, the absence of species on whole continents can generally not be explained on the basis of unsuitable habitats. Thus, the microbial dogma that everything is everywhere and that the environment only selects seems not to be verified for all blue-green algal species; other factors may be involved. One of these factors is dispersal.

Dispersal capacities. Whereas many aerophytic blue-green algal species are able to tolerate desiccation for long periods (e.g. Lipman, 1941) and are resistant to low as well as to high temperatures (Booth, 1946; Cameron & Blank, 1966), this is not the case for many aquatic species which live in an environment where temperature variations are generally small. It was shown for several freshwater species that desiccation and subsequent rewetting of the trichomes leads to a rapid lysis of the cells (e.g. Geitler, 1982; Stroh, 1938). The aquatic species, except for those which produce resting stages (akinetes), seem thus to be less adapted to long distance dispersal by air. On the basis of the dispersal capacities (tolerance of desiccation), it is possible to explain the extent of the distribution areas of the different *Nostoc* species (Mollenhauer, 1986).

Earth history and dispersal. Since the Precambrian, the earth surface has been in constant change. There were for example volcanic eruptions and glaciations creating new habitats and modifying the environmental conditions. If these creations of new habitats are recent enough or/and if the dispersal rate of freshwater blue-green algae is low, this could explain the absence of species in certain regions. An example of this phenomenon may be found within thermal water species.

In some regions, such as Iceland and the Azores, the blue-green algal flora found in thermal springs is species poor compared to hot springs of similar temperature and chemistry in other parts of the world (Castenholz, 1969, 1978). Thus, narrow-celled *Synechococcus* species are entirely missing in both areas.

Iceland was for a long time covered by an almost complete glacial cap and the thermal waters of this island have been exposed to colonization for no longer than 8000 to 10000 years. The Azores are of volcanic formations and the probable origin of hot springs dates from the late Tertiary ($20 \cdot 10^6$ year). The nearest hot springs are often several thousands of kilometers away. However, the inoculation with obligate thermophilic species generally requires, however, direct transport from another hot spring. Whereas numerous agents are available for short distance transports via insects (Brock et al., 1969), this is, except for transport by wind and birds, not true in the case of long distance dispersal. Airborne transport of certain of these species, like *Mastigocladus laminosus*, a species surviving room dryness at 25 °C for several months (Castenholz, 1970) and freezing for one year (Castenholz, 1983), over long distances seems possible; thus *M. laminosus* was already isolated after about 7 years following the formation of the volcanic island of Surtsey, probably brought in from hot springs located on Iceland at a distance of about 80 km (Castenholz, 1983). For other species (*Synechococcus* spp., (*Iscillatoria* spp.) this type of dispersal seems unlikely due to their degree of sensitivity to the environment outside of their actual habitat and to the improbability of viable inoculum being located anywhere else than in hot springs (Castenholz, 1983). The time available for the colonization of the hot springs on Iceland and on the Azores by the latter species may thus have been short enough to account for the poorer flora. Thus time seems a possible factor to explain the distribution of thermophilic blue-green algae and dispersal a major barrier for at least some thermophilic blue-green algae (Castenholz, 1978).

Speciation and dispersal. The establishment of a cosmopolitan, pantropic or holarctic flora depends on an efficient distribution in relation to the rate of evolution. The evolution of blue-green algae, asexually reproducing prokaryotes, depends on the rate of new genotype formation subject to natural selection. Thus if the dispersal rate of new genotypes is greater than the rate of evolution, a pantropic or holarctic flora will result. On the opposite, if the rate of dispersal of new genotypes precedes more slowly than the rate of evolution, floristic regionality will result and would explain regional or endemic species. The evolution of new taxa may thus have occurred during a time span insufficient to provide much likelihood yet of successful long distance colonization. The question when speciation took place in the blue-green algae and whether speciation is still going on today is unsettled (Castenholz, 1992). Blue-green algae are a very old group and already 3.5 milliard years ago forms occur which are morphologically similar to what we know now (Knell & Golubic, 1992). The general shapes of blue-green algal phylogenetic trees inferred from rRNA sequences (e.g. Giovannoni et al., 1988) indicate an early radiation of the group and a long separate evolution within each branch. The DNA homology studies within the genus *Anabaena* (Stulp & Stam, 1984) show a high degree of genetic relationship between the species which contradicts their apparently high evolutionary age.

The distribution of the 'thermal red' *Oscillatoria terebriformis* Gem. and of high temperature strains of *Synechococcus lividus* Copel. may reflect the limited spread of two relatively new taxa which may have evolved in specific springs in the Western part of the United States. Four temperature strains are distinguished within *Synechococcus lividus*. The high temperature strain is characterized by successfull growth at 70 °C and is only known from North America, whereas the other strains are also known from Guatemala, Japan, New Zealand, Central Africa. The red form of *Oscillatoria terebriformis* is also only known from Western United States (Oregon, North California, Nevada, South Idaho, Montana) and could not be localized in other parts of the world (Castenholz, 1983). The dispersal of these taxa is difficult because of their inability to survive desiccation or low temperatures near zero. If we admit that the dispersal of these organisms is similar to that of other thermal water species, a relatively recent origin of these taxa is probable. The alternative is that they represent relicts from a previously more global distribution (Castenholz, 1983).

Acknowledgments

The author is research associate of the Belgian National Fund for Scientific Research (F.N.R.S.). Drs V. Demoulin and P. Compère are thanked for commenting on the manuscript.

References

Aakerman, T., Skulberg, O. M. & S. Liaaen-Jensen, 1992. A comparison of the carotenoids of strains of *Oscillatoria* and *Spirulina* (Cyanobacteria). Biochem. Syst. Ecol. 20: 761–769.

Anagnostidis, K. & J. Komárek, 1985. Modern approach to the classification system of cyanophytes. 1. Introduction. Arch. Hydrobiol. Suppl. 71 (Algological Studies 38/39): 291–302.

Booth, W. E., 1946. The thermal death point of certain soil inhabiting algae. Proc. Montana Acad. Sci. 5/6: 21–23.

Brannon, M. A., 1952. Some Myxophyceae in Florida. Quart. J. Florida Acad. Sci. 15: 70–78.

Brock, M. L., R. G. Wiegert & T. D. Brock, 1969. Feeding by *Paracoenia* and *Ephydra* (Diptera: Ephydridae) on the microorganisms of hot springs. Ecology 50: 192–200.

Cameron, R. E. & G. B. Blank, 1966. Soil studies. Desert microflora. XI. Desert soil algae survival at extremely low temperatures. J. P. M. Space Programs Summary 37-37: 174–181.

Castenholz, R. W., 1969. The thermophilic cyanophytes of Iceland and the upper temperature limit. J. Phycol. 5: 360–368.

Castenholz, R. W., 1970. Laboratory culture of thermophilic cyanophytes. Schweiz. Z. Hydrol. 32: 538–551.

Castenholz, R. W., 1978. The biogeography of hot spring algae through enrichment cultures. Mitt. int. Ver. Limnol. 21: 296–315.

Castenholz, R. W., 1983. Ecology of blue-green algae in hot springs. In N. G. Carr & B. A. Whitton (eds), The Biology of Blue-Green Algae. Botanical Monographs 9: 379–414.

Castenholz, R. W., 1992. Species usage, concept, and evolution of the cyanobacteria (blue-green algae). J. Phycol. 28: 737–745.

Drouet, F. 1981. Revision of the Stigonemataceae with a summary of the classification of the blue-green algae. Beih. Nova Hedwigia 66: 1–221.

Fjerdingstad, E. 1969. Cell dimensions and taxonomy of *Anabaena variabilis* Kütz. emend. (Cyanophyceae). Schweiz. Z. Hydrol. 31: 59–80.

Fogg, G. E., W. D. P. Stewart, P. Fay & A. E. Walsby, 1973. The blue-green algae. Academic Press, London, 459 pp.

Foy, R. H., C. E. Gibson & R. V. Smith, 1976. The influence of daylength, light intensity and temperature on the growth rates of planktonic blue-green algae. Br. phycol. J. 11: 151–163.

Frémy, P. 1990. Les Myxophycées de l'Afrique équatoriale française Arch Bot. 3, mémoire 2: 1–508.

Geitler, L., 1932. Cyanophyceae. In Rabenhorst's Kryptogamen-Flora von Deutschland, Österreich und der Schweiz. Akad. Verlagsges., Leipzig: 14: 1–1196.

Geitler, L., 1982. Zur Kenntnis der Oscillatoriacee *Pseudanabaena galeata* (Cyanophyta). Pl. Syst. Evol. 141: 169–175.

Geitler, L. & F. Ruttner, 1935. Die Cyanophyceen der deutschen limnologischen Sunda-Expedition, ihre Morphologie, Systematik und Ökologie. Arch. Hydrobiol., Suppl. 14 (Tropische Binnengewässer 6): 308-369: 371–483.

Giovannoni, S. J., S. Turner, G. J. Olsen, S. Barns, D. J. Lane & N. R. Pace, 1988. Evolutionary relationships among cyanobacteria and green chloroplasts. J. Bacteriol. 170: 3584–3592.

Golubic, S., 1965. Über die Variabilität zweier mariner Oscillatorien aus der Sektion '*Margaritiferae*'. Schweiz. Z. Hydrol. 27: 233–237.

Hoffmann, L., 1990a. Presence of *Mastigocladopsis jogensis* (Cyanophyceae, Mastigocladopsidaceae) in Corsica (France). Cryptogamie, Algologie 11: 219–224.

Hoffmann, L., 1990b. Rediscovery of *Loriella osteophila* (Cyanophyceae). Br. Phycol. J. 25: 391–395.

Hoffmann, L., 1993. Marine Cyanophyceae of Papua New Guinea. V. Nostocales (except Rivulariaceae) and Stigonematales. Nova Hedwigia 57: 393–408.

Hoffmann, L., 1994. Biogeography of marine blue-green algae. Arch. Hydrobiol. Suppl. 105 (Algological Studies 75): 137–148.

Iyengar, M. O. P. & T. V. Desikachary, 1946. *Mastigocladopsis jogensis* gen. et sp. nov., a new member of the Stigonemataceae. Proc. Indian Acad. Sci. B 24: 55–59.

Kangatharalingam, N. & J. C. Priscu, 1993. Isolation and verification of anatoxin-a producing clones of *Anabaena flos-aquae* (Lyngb.) de Breb. from a eutrophic lake. FEMS Microbiol. Ecol. 12: 127–130.

Kato, T., M. F. Watanabe & M. Watanabe, 1991. Allozyme divergence in *Microcystis* (Cyanophyceae) and its taxonomic inference. Arch. Hydrobiol. Suppl. 92 (Algological Studies 64): 129–140.

Knell, A. H. & S. Golubic, 1992. Proterozoic and living cyanobacteria. In Schidlowski, M., Golubic, S., Kimberley, M. M., McKirdy, D. M. & P. A. Trudinger (eds), Early organic evolution: implications for mineral and energy resources, 450–462. Springer Verlag, Heidelberg.

Kohl, J.-G. & A. Nicklisch, 1981. Chromatic adaptation of the planktonic blue-green alga *Oscillatoria redekei* Van Goer and its ecological significance. Int. Rev. ges. Hydrobiol. 66: 83–94.

Komárek, J., 1972. Temperaturbedingte morphologische Variabilität bei drei *Phormidium*-Arten (Cyanophyceae) in Kulturen. Preslia 44: 293–307.

Komárek, J., 1975. Die Bedeutung der geographischen Areale der Süsswasseralgen (besonders planktischen Cyanophyceen) für das Studium der Balkanischen Mikroflora. In Problems of Balkan Flora & Vegetation, Sofia, 95–103.

Komárek, J., 1985. Do all cyanophytes have a cosmopolitan distribution? Survey of the freshwater cyanophyte flora of Cuba. Arch. Hydrobiol. Suppl. 71. (Algological Studies 56): 247–345.

Komárek, J., 1994. Current trends and species delimitation in the cyanoprokaryote taxonomy. Arch. Hydrobiol. Suppl. 105 (Algological Studies 75): 11–29.

Komárek, J. & J. Lukavský, 1988. *Arthronema*, a new cyanophyte genus from Afro-Asian deserts. Arch. Hydrobiol. Suppl. 80 (Algological Studies 50-53): 249–267.

Lipman, C. B., 1941. The successful revival of *Nostoc commune* from a herbarium specimen eighty-seven years old. Bull. Torrey Bot. Club 68: 664–666.

Llukkainen, R., Sivonen, K., Namikoshi, M., Färdig, M., Rinehart, K. L. & S. I. Niemelä, 1993. Isolation and identification of eight microcystins from thirteen *Oscillatoria agardhii* strains and structure of a microcystin. Appl. Envir. Microbiol. 59: 2204–2209.

Meffert, M.-E., 1989. Planktic unsheathed filaments (Cyanophyceae) with polar and central gas vacuoles. II. Biology, population dynamics and biotopes of *Limnothrix redekei* (Van Goor) Meffert. Arch. Hydrobiol. 116: 257–282.

Meffert, M.-E. & H.-J. Krambeck, 1977. Planktonic blue-green algae of the *Oscillatoria redekei* group. Arch. Hydrobiol. 79: 149–171.

Mollenhauer, D., 1970. Beiträge zur Kenntnis der Gattung *Nostoc*. Abh. Senckenb. naturf. Ges. 254: 1–75.

Mollenhauer, D., 1986. Blaualgen der Gattung *Nostoc* - ihre Rolle in Forschung und Wissenschaftsgeschichte IV. Natur und Museum 116: 104–120.

Nielsen, C. S. & G. L. Madsen, 1948. Preliminary checklist of the algae of the Tallahassee area. Quart. J. Florida Acad. Sci. 11: 63–66.

Rott, E. & M. Hernández-Mariné, 1994. *Pulvinularia suecica*, a rare stigonematalean cyanophyte. Arch. Hydrobiol. Suppl. 105 (Algological studies 75): 313–322.

Skulberg, O. M. & R. Skulberg, 1985. Planktic species of *Oscillatoria* (Cyanophyceae) from Norway. Characterization and classification. Arch. Hydrobiol. Suppl. 71 (Algological Studies 38/39): 157–174.

Stam, W. T. & H. C. Holleman, 1979. Cultures of *Phormidium, Plectonema, Lyngbya* and *Synechococcus* (Cyanophyceae) under different conditions: their growth and morphological variability. Acta Bot. Neerl. 28: 45–66.

Stroh, L., 1938. Über prämortale Mazeration bei Oscillatoriaceen. Arch. Protistenk. 91: 187–201.

Stulp, B. K. & W. T. Stam, 1984. Genotypic relationships between strains of *Anabaena* (Cyanophyceae) and their correlation with morphological affinities. Br. phycol. J. 19: 287–301.

Stulp, B. K., S. A. Bos, H. Stenveld & W. T.Stam, 1984. Growth of *Anabaena* strains (Cyanophyceae) exposed to crossed gradients of light and temperature. Cryptogamie, Algologie 5: 63–71.

Waterbury, J. B. & R. Y. Stanier, 1978. Patterns of growth and development in pleurocapsalean cyanobacteria. Bacteriol. Rev. 42: 2–44.

Zaneveld, J. S., 1988. The Cyanophyta of the Ross Sea islands and coastal Victoria land, Antarctica. Koeltz Scientific Books, Koenigstein, Germany, 85 pp.

Hydrobiologia **336:** 41–53, 1996.
J. Kristiansen (ed.), Biogeography of Freshwater Algae.

5. Biogeography of desmids

Peter F. M. Coesel
Department of Aquatic Ecology, University of Amsterdam, Kruislaan 320, NL–1098 SM Amsterdam, The Netherlands

Key words: Green algae, Desmidiaceae, distribution, biogeography

Abstract

Compared with other groups of unicellular freshwater algae, desmids lend themselves well to biogeographical studies since, at species level, identification is often relatively easy, whereas high ecological demands use to curtain their geographical distribution. Considering some ten desmid floral regions as distinguished in the beginning of this century, Indo-Malaysia/Northern Australia, tropical America, and equatorial Africa come to the fore as most pronounced. Also well typified are Eastern Asia, New Zealand/Southern Australia, and North America. Less endemic species are met with in Southern Africa and extratropical South America, whereas temperate Eurasia, with respect to the other continents, is mainly negatively characterized. The so-called arctic-alpine desmid flora may be encountered on all continents, provided that adequate minimum temperatures occur. Its distribution seems to be determined microclimatologically rather than macroclimatologically. Arguments for a tropical origin of the desmids as an algal group are adduced.

Introduction

Consisting of unicellular organisms, members of the freshwater green algal family Desmidiaceae tend to show cosmopolitic distribution patterns, in this respect resembling bacteria rather than, e.g., macroscopic seaweeds. Yet, among freshwater algae desmids are known for a high number of exceptions to this general rule. Already G. S. West (1909) stated that no group of freshwater algae exhibits such marked geographical peculiarities as the Desmidiaceae. West *(loc. cit.)* even suggested that these peculiarities would enable to recognize the rough geographical origin of any desmid collection. Presumable explanations for the particular position of the desmids in this respect are, in the first place, the high number of taxa characterized by a conspicuous cell shape facilitating a ready and reliable identification. Furthermore, since in most desmid species the formation of resistent, wind-transportable spores is a rare phenomenon (Coesel, 1974a), distribution is supposed to be realized mainly as vegetative cells by insects and birds (Brook, 1981: 208). Distances bridged in that way in general will be rather limited because of the chance of desiccation, or wash out in salt water. Finally, it is well-known that the vast majority of desmid taxa are confined to an oligo(-meso)trophic habitat. Consequently, most desmids may be considered K-strategists (Coesel & Kooijman-Van Blokland, 1991). Compared to r-strategists (like most chlorococcalean green algae) K-strategists make high demands upon their environment, requirements which usually may be only met in limited areas (MacArthur, 1972).

Krieger (1933, 1937) tentatively distinguished some ten desmid floral regions: Temperate Eurasia, The circumpolar regions, Eastern Asia, Indo-Malaysia/Northern Australia, New Zealand/Southern Australia, South Africa, Equatorial Africa, North America, the tropical part of America, and the extratropical part of South America. Although Krieger *(loc. cit.)* advanced that this classification would be refined as inventories intensified, so far nobody made a serious attempt to it. The reasons for this are not hard to find. First of all, the inventory of the various geographical regions is still extremely ill-balanced, so that geographical distribution maps often reflect the intensity of phycological investigations rather than real distribution patterns. Actually, the following statement by Prescott (1948) is still standing: 'The picture is still hazy and lines are not well drawn, partly because the literature is so bulky that summarizing analyses are difficult, whereas many species seem to be characteristic of geographical areas and indeed may be classed as endemic, generalizations are continuously broken down as information increases and as supposed endemics are reported from far away stations'. Another cause of the stagnating development in the discipline of desmid biogeography is in the confusion associated with many taxonomic delimitations. No doubt, there is a lot of synonymy to be cleared up. Since even at genus level taxa have been mixed up – see, e.g., the discussion in Scott & Prescott (1960) concerning the closely allied *Micrasterias moebii* (Borge) W. & G. S. West, and *Euastrum turgidum* Wall. – utmost carefulness is demanded when drawing biogeographical conclusions. As Brook (1981: 211) rightly stated, no real progress in the delimitation of distribution areas may be expected until more knowledge is available about the morphological variability of the taxa in question. Therefore, before trying to refine Krieger's classification it seems recommendable to discuss the different regions mentioned with the help of distribution data of a number of clearcut taxa.

Concise characterization of the main desmid floral regions

Temperate Eurasia

Although better investigated than any other region, Eurasia turns out to have but few species on its own. Considering some possible examples, *Euastrum vigrense* Rypp., known from N. Russia, Poland, Finland, and Sweden (Engels & Handke, 1994) shows a sub-arctic distribution, and *Staurastrum verticillatum* Arch., known from Ireland, Scotland, Norway, S.W. France, and Portugal (Heimans, 1969) a distinct atlantic one. *Cosmarium insigne* Schmidle has been reported from Germany, Switzerland, France, Italy and The Netherlands (Coesel, 1974b). *Cosmarium dilatatum* Lütkem., reported from the Czech Republic, Finland, and The Netherlands (Coesel, 1989b) meanwhile is also known from Poland (A. Oleksowicz, in manuscript), and France (F. Kouwets, pers. comm.).

The poorness in endemic species of the temperate Eurasian region at least partly may be ascribed to the Pleistocene glaciations. Most flora elements characteristic of the tropical climate ruling in a large part of Europe during the Paleocene and Eocene were driven away from the continent during the glacial peaks in the Pleistocene (Frenzel, 1968). Whereas a fair number of land plant taxa could find a refugium in the Mediterranean region, warmth-demanding desmid species probably were doomed to extinction since the dry and often calcareous soils prevailing in the Southern parts of Europe and W. Asia are reputedly unhospitable to this algal group.

The periods of glaciation were also responsible for a considerable lowering of the sea level resulting in the development of temporary land bridges. Biogeographically most important was the Bering connection between N. America and N.E. Asia, explaining the many related flora and fauna elements on both continents (Cox & Moore, 1993). With respect to desmids, a fair number of species may be designated as holarctic, thus restricted in their distribution to Eurasia and N. America. Apart from a series of so-called arctic-alpine taxa (see next section) can be mentioned: *Euastrum insigne* Ralfs, *Eu. verrucosum* Ralfs, *Micrasterias oscitans* Ralfs, *M. verrucosa* Biss., *Cosmarium cyclicum* Lund., *C. perforatum* Lund., *C. protractum* (Näg) De Bary, *Xanthidium subhastiferum* W. West, *X. brebissonii* Ralfs, *Staurastrum arctiscon* (Ralfs) Lund., *S. elongatum* Bark., and *S. ophiura* Lund. (Prescott et al., 1977, 1981, 1982; see also Figure 1).

The circumpolar and high mountain regions

Quite a lot of desmid species are said to show an arctic-alpine distribution pattern. However, a critical perusal learns that but a few of them are strictly confined to these regions. Apart from the mesotaeniaceous snow alga *Ancylonema nordenskioeldii* Berggr., also *Euastrum spetzbergense* (Nordst.) Krieg., *Eu. tetralobum* Nordst., *Cosmarium holmii* Wille, *Staurastrum novae-semliae* Wille, and *S. petsamoense* Järnef. so far were only found in the arctic parts of Eurasia and North America (Krieger, 1937; Kossinskaja, 1960; Prescott et al., 1981, 1982). Species like *Euastrum aboense* Elfv., *Eu. boldtii* Schmidle, *Eu. dissimile* (Nordst.) Schmidle, *Eu. tuddalense* Ström, *Cosmarium anceps* Lund., *C. caelatum* Ralfs, *C. costatum* Nordst., *C. crenatum* Ralfs, *C. cymatopleurum* Nordst., *C. hexalobum* Nordst., *C. holmiense* Lund., *C. nasutum* Nordst., *C. notabile* Bréb., *C. speciosum* Lund., *C. tatricum* Racib., *C. tetragonum* (Näg.) Arch., *Staurastrum acarides* Nordst., *S. capitulum* Bréb., *S. rhabdophorum* Nordst.,

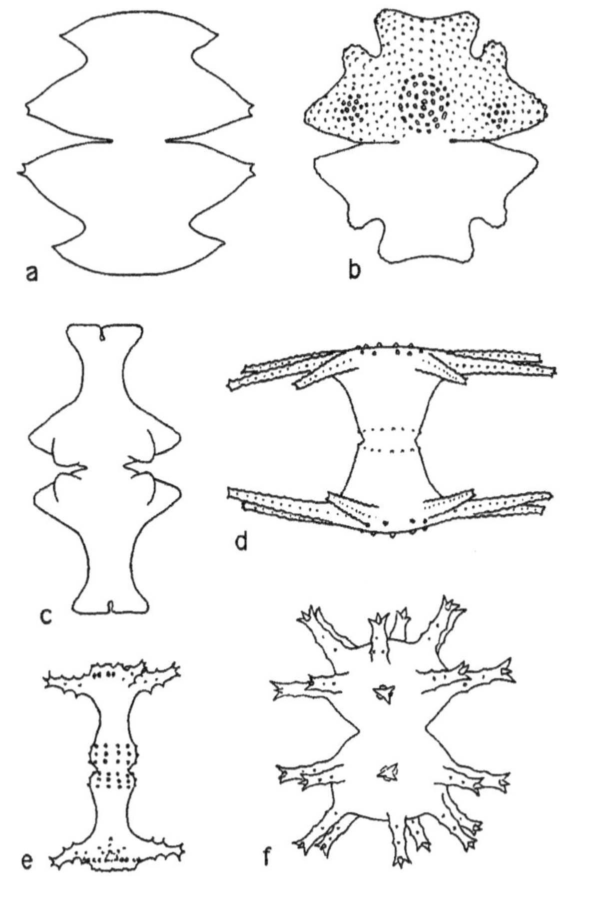

Figure 1. A selection of holarctic desmid species: a. *Micrasterias oscitans*, b. *Euastrum verrucosum*, c. *Euastrum insigne*, d. *Staurastrum ophiura*, e. *Staurastrum elongatum*, f. *Staurastrum arctiscon* (a, b. after Coesel, 1985; c. after West & West, 1905; d–f. after West et al., 1923).

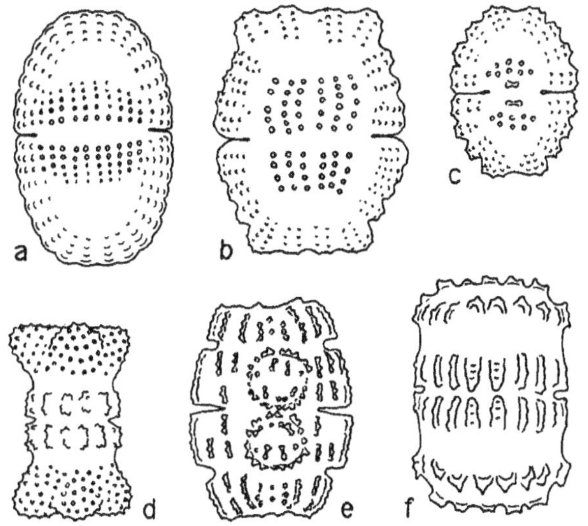

Figure 2. A selection of so-called arctic-alpine species: a. *Cosmarium speciosum*, b. *Cosmarium hexalobum*, c. *Cosmarium nasutum*, d. *Staurastrum capitulum*, e. *Staurastrum acarides*, f. *Staurastrum rhabdophorum* (a,c. after West & West, 1908; b. after Nordstedt, in West & West, 1908; d. after West & West, 1912; e. after West et al., 1923; f. after Nordstedt, in West & West, 1912).

and many others assigned to the group of arctic-alpine species (West & West, 1908, 1912; West et al., 1923; Ružička, 1981; see also Figure 2) are found predominantly under the rough climatic conditions of high latitude or altitude, but to a greater or less extent may be encountered in the temperate lowlands as well. Several authors (e.g. Grönblad, 1933; Krieger, 1937; Coesel, 1979) paid attention to the fact that so-called arctic-alpine species use to live hemi-atmophytically in water-saturated moss layers or on dripping rocks. Thomasson (1956), in a study on arctic and alpine lakes, concluded that among the euplanktic algae hardly any decidedly arctic or alpine forms could be distinguished, such in contradiction to the benthic algae. Obviously, the microclimate generated by the local environmental conditions to a large extent may compensate for the macroclimate.

Arctic-alpine species are particularly known from Eurasia and North America, but a number of them have been recorded from cold regions on other continents as well. For instance, from the Japanese Alps: *Cosmarium anceps, C. caelatum, C. nasutum, C. notabile, C. tetragonum* (Hirano, 1953), from Mount Wilhelm, in New Guinea: *Cosmarium caelatum, C. nasutum, Staurastrum capitulum* (Thomasson, 1967), from Mount Kenya, in equatorial Africa: *Cosmarium cymatopleurum, C. galeritum* Nordst., *C. garrolense* Roy & Biss. (Kusel-Fetzman, 1968), from Santa Marta Mountains, in Colombia: *Cosmarium caelatum, C. holmiense, C. nasutum, C. tetragonum* (Taylor, 1935), from the subantarctic Kerguelen Islands: *Cosmarium anceps, C. crenatum, C. holmiense, C. speciosum, C. tatricum, Staurastrum acarides* (Thérézien & Couté, 1977).

In contrast to the arctic regions, the desmid flora of Antarctica is extremely poor. Hirano (1965) in a review paper on Antarctic freshwater algae, mentions the mesotaeniaceous snow alga *Ancylonema nordenskioeldii* and, next to that, only five *Cosmarium* species: *C. crenatum, C. curtum* Bréb. (= *Actinotaenium curtum* (Bréb.) Teil.), *C. pseudoconnatum* Nordst., *C. undulatum* Ralfs, and the probably endemic *C.*

antarcticum Gain. In explanation of this poorness in species, Hirano *(loc. cit.)* refers to the strong isolation of Antarctica. Moreover, the absence of warm sea currents such as those operating near the northwesternmost parts of Eurasia and North America results in an almost complete and permanent covering of the Antarctic continent by pack-ice.

Eastern Asia

Although the temperate and colder parts of Eastern Asia have been relatively little searched for desmids a fair number of species seem to be characteristic of this geographical region. Illustrative are the many taxa newly described for Japan, e.g.: *Euastrum ozense* Hirano, *Cosmarium horomuiensis* Hirano, *C. margispinatum* Hirano, *C. pseudoquinarium* Hirano, *Xanthidium japonicum* Hirano, *Staurastrum asoensis* Hirano, *S. karasuensis* Hirano, *S. koidzumii* Hirano (Hirano, 1957, 1959a, 1959b). It is likely that a great deal of these taxa are present in China too. At the time of the Pleistocenic glaciations Japan was connected to China by a land bridge (De Lattin, 1967), and even nowadays the distance to China over sea may be thought to be bridged by migrating waterfowl. Unfortunately, so far only a few reports on desmids from China are known and, consequently, the number of species proper to both of these countries is low, e.g. *Euastrum horikawae* Hinode (Hinode, 1960; Wei, 1991) and *Staurastrum zahlbruckneri* Lütkem. (Lütkemüller, 1900; Hinode, 1977). The many new taxa recently described by Wei (1984, 1993) for some mountain areas in Eastern Asia make us believe that there is still much to be discovered in this geographical region.

Apart from a high number of presumable endemics, the biogeographical region under discussion is also characterized by quite a lot of species having the main point of their distribution in the tropics. Hirano's (1956, 1959a) Japanese desmid flora comprises species as *Ichthyocercus longispinus* (Borge) Krieg., *Pleurotaenium kayei* (Arch.) Rab., *P. ovatum* Nordst., *P. subcoronulatum* (Turn.) W. & G. S. West, *P. trochiscum* W. & G. S. West, *Euastrum gnatophorum* W. & G. S. West, *Eu. turgidum* Wall., all listed in Krieger's flora as tropical. No doubt, this has to do with the fact that, in contrast to the major part of Eurasia, the easternmost region of this continent remained free of pack-ice during the Pleistocenic glaciations (Pielou, 1979), and that there were open migration routes – unhampered by high mountain chains – to the tropical regions. This easy connection also explains the rather gradual transition between the Eastern Asian desmid region and the Indo-Malaysian/Northern Australian one.

North America

The North American continent is characterized by a rich desmid flora. As noticed already under the section Temperate Eurasia, North America shares a series of species with the European continent, among which quite a lot that seem to be confined to the northern hemisphere. However, in addition to that, the North American desmid flora counts many taxa which have the main point of their distribution in the tropics. Species like *Pleurotaenium ovatum, P. verrucosum* Lund., *Euastrum evolutum* (Nordst.) W. & G. S. West, *Micrasterias foliacea* Ralfs, *Staurastrum leptacanthum* Nordst., *S. leptocladum* Nordst., *S. rotula* Nordst. and *Phymatodocis nordstedtiana* Wolle, particularly known from various tropical regions, appear to penetrate into North America as far as Canada, or even Alaska (Prescott et al., 1975, 1977, 1982; Croasdale et al., 1983; see also Figure 3). The occurrence of these tropical elements in the colder parts of North America most probably has to do with the almost continuous series of freshwater bodies connecting tropical America with the northernmost regions (see, e.g., Figure 4 in Hoshaw & McCourt, 1988) and also with the N–S orientation of the main mountain chains, facilitating algal migration by waterfowl during and after the Pleistocenic glaciations.

Compared with Europe, the North American desmid flora not only is rich in tropical elements but also in endemic taxa. Of these latter, *Euastrum wollei* Lagerh., *Micrasterias muricata* Ralfs, *M. nordstedtiana* Wolle, *Cosmarium dentatum* Wolle, *C. eloiseanum* Wolle, and *Spinocosmarium quadridens* (Wood) Presc. & Scott are widely distributed throughout the United States and Canada (Prescott et al., 1977, 1981, 1982; see also Figure 4).

Tropical South and Central America

The geologically only recent (ca 3 million years ago) connection between Central and South America is well reflected in their respective desmid floras. Although treated as one geographical region by Krieger (1933), numerous taxa of common occurrence in tropical South America are unknown for Central America, e.g.: *Euastrum foersteri* Scott & Croasd., *Eu. grandiornatum* (Först.) Först., *Eu. pirassunungae* Borge, *Cosmarium cornigerum* (Nordst.) Först., *C. furcatum* Först.,

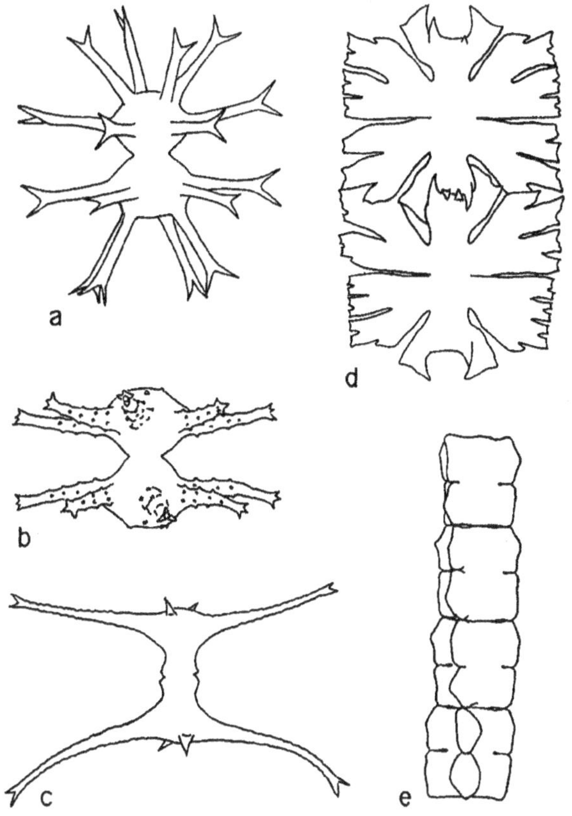

Figure 3. Desmid species with predominantly tropical distribution: a. *Staurastrum leptacanthum*, b. *Staurastrum rotula*, c. *Staurastrum leptocladum*, d. *Micrasterias foliacea*, e. *Phymatodocis nordstedtiana* (a. after Smith, in Prescott et al., 1982; b. after Förster, in Prescott et al., 1982; c. after Irénée-Marie, in Prescott et al., 1982; d. after Prescott et al., 1977; e. after Scott & Prescott, in Croasdale et al., 1983).

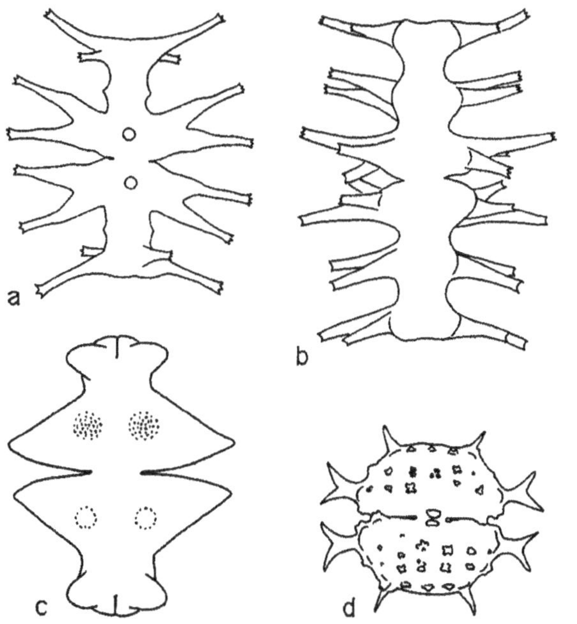

Figure 4. Some North American desmid species: a. *Micrasterias nordstedtiana*, b. *Micrasterias muricata*, c. *Euastrum wollei*, d. *Spinocosmarium quadridens* (a–c. after Prescott et al., 1977; d. after Prescott et al., 1982).

C. horridum Borge, *C. redimitum* Borge, *Xanthidium regulare* Nordst., *Staurastrum circulus* Grönbl., *S. foersteri* Coes., *S. spiculiferum* Borge (Förster, 1982; Coesel et al., 1988; see also Figure 5).

No doubt, the high number of endemic taxa in tropical South America has to do with the long period of isolation of this continent, from the Middle Cretaceous up to the Late Pliocene (Cox & Moore, 1993), whereas the absence of these taxa in the likewise tropical water bodies in nearby Central America could be explained by the upheaval of the Andean mountains in about the same period that the Panama land bridge was formed. Most probably, the Andes, by forming a barrier for migrating waterfowl, effectively restricts the exchange of freshwater algae between the above-mentioned tropical areas (Coesel et al., 1988).

Extratropical South America

The richness in endemic taxa in tropical South America is in distinct contrast to the rather poor and trivial desmid flora encountered in the cool-temperate parts of this continent. Thomasson (1963), who studied the plankton in a large number of Northern Patagonian lakes, did not find an obvious relationship between the desmid flora of these lakes and that of tropical South America. He ascribed this to the elimination of many biota during the quaternary glaciations when ice covered a large part of Northern Patagonia. Recolonization from tropical South American regions would be hampered by the Andean mountain chain and the deserts in between. Tell (1980) stated that, on the American continent, many tropical desmid species reach their southern distribution limits in Northeastern Argentina (province of Corrientes).

Some desmid species, e.g. *Cosmarium andinum* Couté & Iltis (1988), *C. araucarense* Thomasson (1963), *C. magdalenense* Taylor (1935), and *Staurastrum mayori* G. S. West (1914) are exclusively known from cool-temperate areas in South America notably from the Andean mountains, it is true, but as long as these taxa are only known from a single locality they

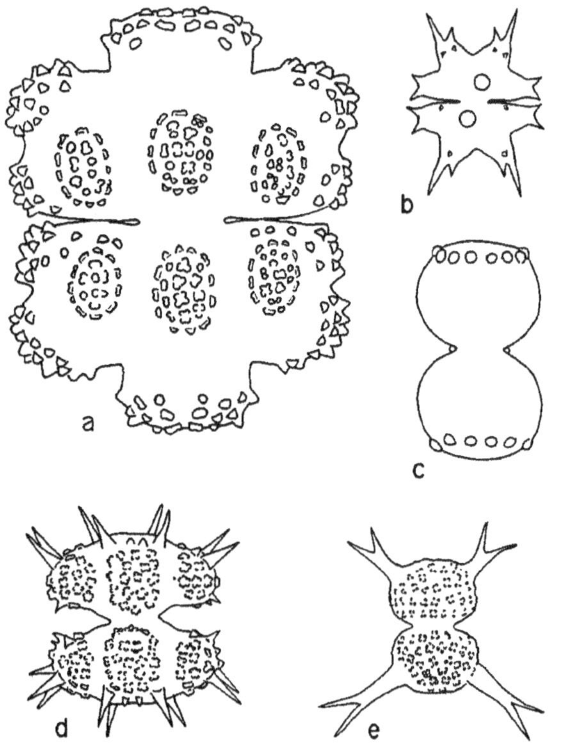

Figure 5. Some species characteristic of the Neotropics: a. *Euastrum grandiornatum*, b. *Euastrum foersteri*, c. *Cosmarium redimitum*, d. *Cosmarium cornigerum*, e. *Staurastrum circulus* (a–d. after Förster, 1982; e. after Coesel et al., 1988).

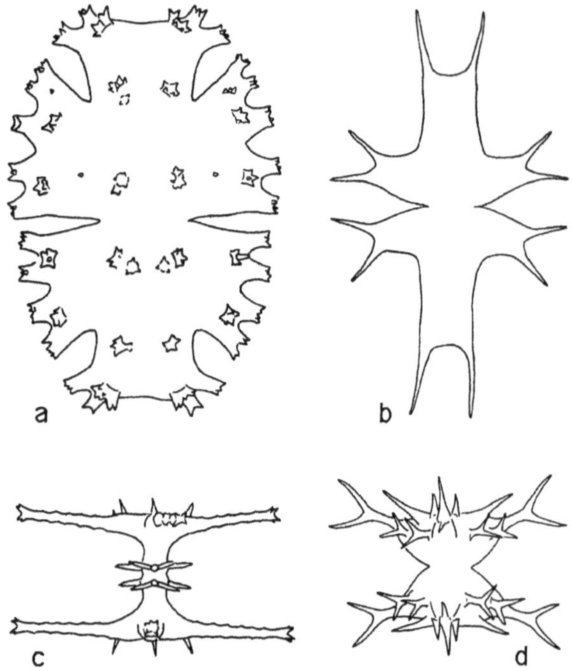

Figure 6. Some species characteristic of the Indo-Malaysian/North Australian region: a. *Micrasterias anomala*, b. *Micrasterias ceratofera*, c. *Staurastrum tauphorum*, d. *Staurastrum freemanii* (a–d. after Scott & Prescott, 1961).

are hardly to be considered characteristic of the region under discussion.

The discrepancy between the Central American and the South American desmid flora as signalized in the previous section of course does not imply an absolute partition. A few taxa might be considered American endemics, occurring on both the Southern and the Northern continent, but their number seems to be very limited, e.g.: *Staurastrum minnesotense* Wolle, and *S. novae-caesareae* Wolle (Förster, 1969; Couté & Tell, 1981; Prescott et al., 1982).

Indo-Malaysia/Northern Australia

The rich, tropical desmid floras of Indo-Malaysia and Northern Australia are closely related as has already pointed out by West & West (1902) and later on by e.g. Krieger (1933), Scott & Prescott (1958) and Thomasson (1986). Characteristic species of this regions are, e.g., *Triploceras splendens* Prowse, *Euastrum asperum* Borge, *Eu. moebii* (Borge) Scott & Presc., *Micrasterias anomala* Turn., *M. ceratofera* Josh., *Staurastrum freemanii* W. & G. S. West, *S. tauphorum* W. & G. S. West, and *Streptonema trilobatum* Wall. (Krieger, 1933; Scott & Prescott, 1961; Thomasson, 1986; Ling & Tyler, 1986; see also Figure 6).

The high degree of resemblance between the desmid flora of Indo-Malaysia and that of Northern Australia is not surprising in view of the fact that the Australian continental plate, as a result of continental drift, contacted the Asian one already some 5 million years ago (Cox & Moore, 1993). The resulting formation of the closely packed Indonesian island archipelago will have considerably facilitated the exchange of freshwater algae between these two continents. Nevertheless, a number of desmid species described from Northern Australia so far have not been encountered in S.E. Asia, so they might be endemic of Australia: *Pleurotaenium australianum* (Borge) Scott, *Cosmarium securiforme* Borge, *Xanthidium multicorne* Borge, and *Staurastrum elegans* Borge (Thomasson, 1986).

Southern Australia and New Zealand

Although West (1909) stated that the Australian desmid flora is largely tropical in character, 'even as far as Victoria', acknowledged tropical elements in the desmid flora so far described from Southern Australia are hardly known. Maybe, *Staurastrum sagittarium* Nordst. can be designated as such, being reported from tropical Africa (Bourrelly, 1957; Grönblad et al., 1958; Thomasson, 1965; Couté & Rousselin, 1975) and tropical Northern Australia (Scott & Prescott, 1958; Ling & Tyler, 1986), but also from Southern Australia (West, 1909; Prescott & Scott, 1952) as well as New Zealand from which the species originally was described (Nordstedt, 1888).

Rather than by pronounced tropical species the South Australian desmid flora is characterized by a number of (supposed) Australian endemics, such as *Micrasterias hardyi* G. S.West, *Cosmarium murrayi* Playf., *Staurastrum assurgens* Nordst., and *S. victoriense* G. S. West (West, 1909; Thomasson & Tyler, 1971; Thomasson, 1980; Ling & Tyler, 1986; Tyler & Wickham, 1988). However, these endemics are not at all common and in the southern part of Australia thrown into the shade by species widely distributed in temperate regions all over the world. The quite distinct character of the South Australian desmid flora as compared with the North Australian one probably has not only to do with a difference in climate but also with the absence of interconnecting series of suitable freshwater bodies, most of the land in between even being a desert.

The predominantly cosmopolitic character shown by the South Australian desmid flora is even more pronounced in the flora described from New Zealand, particularly where planktic habitats are concerned (e.g. Thomasson, 1973). Yet, a number of taxa may be designated as linking this island biogeographically to the Australian continent; e.g.: *Euastrum longicolle* Nordst., *Eu. sphyroides* Nordst., *Xanthidium octonarium* Nordst. (Croasdale & Flint, 1986, 1988), *Staurastrum rosei* Playf. (Thomasson, 1972), *S. sagittarium*, and *S. victoriense* (see above).

Although New Zealand, together with Southern Australia, originally made part of the Southern Gondwana province, it is unlikely that the distribution of the above-mentioned taxa argues a relict area. For New Zealand was split off from Gondwana already some 70 million years ago and since then has been isolated (Cox & Moore, 1993). Regarding the above-mentioned taxa, the close morphological resemblance between specimens from Australia and from New Zealand is indicative of a much more recent geographical contact. Moreover, if it does concern relict areas one would also expect one or more of the taxa under discussion in the southern part of South America which proceeded from Southern Gondwana too. However, in the phytoplankton of North Patagonian lakes no elements have been found indicating such a transantarctic connection (Thomasson, 1963).

In view of the long period of isolation it is not astonishing either that the New Zealand desmid flora is characterized by a fair number of taxa of its own. Some examples: *Euastrum euteles* Skuja, *Eu. haplos* Skuja, *Eu. lagynion* Skuja, *Cosmarium subcyclicum* Mask., *C. turnerianum* Mask., *Xanthidium multigibberum* (Nordst.) Skuja (Croasdale & Flint, 1986, 1988).

Equatorial Africa

Like those in Indo-Malaysia/Northern Australia and in Central-South America, the desmid flora in tropical Africa is highly varied and rich in peculiar forms. Among the species characteristic of the African continent, *Staurastrum fuellebornii* Schmidle probably is most widely distributed (e.g. Schmidle, 1902; Rich, 1932; Grönblad et al., 1958; Thomasson, 1960; Compère, 1967; Lind, 1971; Couté & Rousselin, 1975). Other prominent representatives of the African desmid flora are: *Allorgeia incredibilis* (Grönbl. & Scott) Thom., *Micrasterias ambadiensis* (Grönbl. & Scott) Thom., *M. cunningtonii* G. S. West, *M. sudanensis* Grönbl., Prowse & Scott, and *Staurastrum rzoskae* Grönbl. & Scott (Grönblad et al., 1958; Thomasson, 1960, 1966; Lind, 1971; see also Figure 7). The conspicuously shaped *Xanthidium calcarato-aculeatum* Hieron. and *X. sansibarense* Schmidle, originally described from Zanzibar (Schmidle, 1898), not only appear to occur in many a tropical African country (e.g. Bourrelly, 1957; Grönblad et al., 1958; Thomasson, 1960, 1966; Lind, 1971) but also in the Indo-Malaysian/North Australian region (Scott & Prescott, 1961; Ling & Tyler, 1986) suggesting them to be a typical element of the Old World tropics. On the other hand there seem to be species exclusively in common with the Neotropics, like *Xanthidium mucronulatum* (Nordst.) Först. (= *Arthrodesmus mucronulatus* Nordst.), see Nordstedt (1869), West & West (1895), Grönblad (1945), Bourrelly (1957), Grönblad et al. (1958), Lind (1971), Bicudo (1975), Coesel (1992).

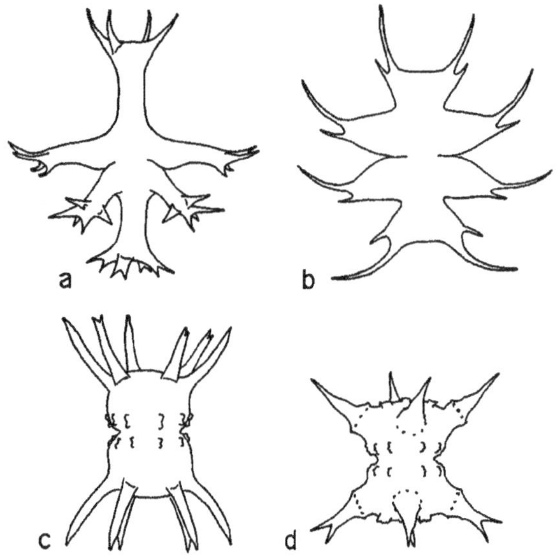

Figure 7. Some taxa characteristic of the equatorial African region: a. *Allorgeia incredibilis*, b. *Micrasterias sudanensis*, c. *Staurastrum rhoskae*, d. *Staurastrum fuellebornii* var. *evolutum* (a–d. after Grönblad et al., 1958).

In biogeographical considerations, the island of Madagascar uses to be dealt with as a separate unit because of the high degree of endemicity in its flora and fauna. Also with respect to desmids a fair number of species are exclusively known from this region, e.g.: *Cosmarium allorgei* Bourr. & Couté, *C. anax* W. & G. S. West, *C. baronii* W. & G. S. West, *C. beatum* W. & G. S. West, *C. elaboratum* W. & G. S. West, *C. notochondrum* W. & G. S. West, *Bambusina madagascariensis* Bourr. & Couté (West & West, 1895; Bourrelly & Couté, 1991). However, almost all reports refer to incidental observations at a single locality. So as yet it seems somewhat premature to class them as endemics. Of special interest is also the occurrence of *Haplozyga armata* (Nordst.) Racib. (see Bourrelly & Couté, 1991), a species admittedly known from tropical South America (Nordsted, 1889; Raciborski, 1895; Förster, 1966).

South Africa

The presentation by Krieger (1933) of South Africa as a separate biogeographical region presumably is co-inspired by the extremely high percentage of endemics in its macrophyte flora. However, with respect to desmids, the flora of Southern Africa rather has to be considered a depauperate variant of that in Equatorial Africa. Fritsch and Rich (e.g. Fritsch & Rich, 1937) described a number of new species from South Africa, it is true, but later on part of these appeared to be distributed in tropical Africa or even on other continents as well, e.g.: *Euastrum subhypochondrum* Fritsch & Rich (Bourrelly, 1957; Thomasson, 1960; Scott & Prescott, 1961), *Cosmarium salisburii* Fritsch & Rich (Grönblad et al., 1958; Lind, 1967; Gerrath & Denny, 1988), *Xanthidium decoratum* Fritsch & Rich (Thomasson, 1960, 1966; Lind, 1971; Bourrelly & Couté, 1991). Species not encountered elsewhere, like *Euastrum biceps* Fritsch & Rich, *Cosmarium multituberculatum* Fritsch & Rich, *C. subhumile* Rich (Rich, 1935), as far as known were not found again in southern Africa either, so they can hardly be presented as endemics for this region. Just as in Australia, the weak relationship of the southern African desmid flora with its equatorial counterpart could be explained by the position of a vast desert area in between, hampering an easy exchange of aquatic organisms.

Geographical distribution of some selected genera

Illustrative of a possible radiation from a tropical region of origin is the actual distribution of a number of less-known, more ore less related genera, i.e.: *Ichthyodontum* Scott & Presc., *Ichthyocercus* W. & G. S. West, *Triplastrum* Iyeng. & Raman., and *Triploceras* Bail. (Figure 8). Of these, *Ichthyodontum* so far has only been reported from the tropical Indo-Malaysian/Northern Australian region (Scott & Prescott, 1956, 1961; Ling & Tyler, 1986).

Ichthyocercus, obviously has a wider distribution, to be characterized as pantropical, i.e.: Indonesia (Krieger, 1933; Scott & Prescott, 1961; Lenzenweger, 1974), Northern Australia (Playfair, 1907; Croasdale & Scott, 1976), Equatorial Africa (West & West, 1897; Gauthier-Lièvre, 1960) and Equatorial South America (Borge, 1899; Förster, 1969; Couté & Tell, 1981; Thérézien, 1985). However, in addition to that, there are also reports from Southern Japan (Hirano, 1956), Madagascar (Bourrelly & Leboime, 1946; Bourrelly & Couté, 1991) and Northern Argentina (Tell, 1979).

The distribution of *Triplastrum* – as far as known limited to the Old World – shows a rather scattered pattern, but still with an accent on warmer climatic regions: Equatorial Africa (Gauthier-Lièvre, 1960; Couté & Rousselin, 1975), South Africa (Claassen, 1977), Southwestern France (Allorge, 1924; Capde-

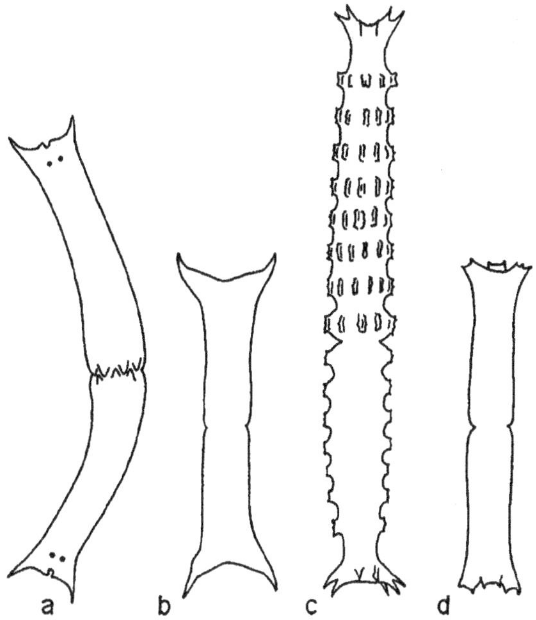

Figure 8. Four genera with a mainly tropical distribution, radiating to different extent into colder climatic regions. a. *Ichthyodontum (I. sachlanii)*, b. *Ichthyocercus (I. longispinus)*, c. *Triploceras (T. verticillatum* var. *minor)*, d. *Triplastrum (T. spinulosum* var. *africanum)* (a. after Scott & Prescott, 1956; b–d. after Gauthier-Lièvre, 1960).

vielle, 1978), Turkestan (Krieger, 1937), India (Iyengar & Ramanathan, 1942), Japan (Hinode, 1952).

Finally, of the four taxa under discussion, *Triploceras* is the most widely distributed and also the most succesful in penetrating into colder climatic regions. Being a common taxon in all tropical regions (Krieger, 1937), it is also known from, e.g., Labrador (Prescott et al., 1975), New Zealand (Croasdale & Flint, 1986) and Northern Eurasia (Kossinskaja, 1960). On last-mentioned continent it seems to reach its western limit in the Baltic States and Finnish Lapland (Kossinskaja, 1960; Wasylik, 1961). It has never been reported from Western Europe but probably its appearance is only a question of time, for its potential distribution area seems to be cosmopolitic.

Discussion

Because of their unique way of sexual reproduction, the conjugatophycean group of green algae, comprising the unicellular desmids, is hypothesized to have its evolutionary origin in an environment that may be characterized as terrestric rather than aquatic (Stebbins & Hill, 1980). That their occurrence, nowadays, in aquatic habitats is likely to be considered a secondary adaptation is supported by the fact that the vast majority of the desmid taxa still have a benthic or tychoplanktic way of life. Euplanktic species (which are hardly or never encountered in a sexual stage) are relatively scarce (Brook, 1981). As already stated in the Introduction section, in addition to that, most desmid taxa are confined to a low nutrient state of the habitat. It is worth mentioning that euplanktic desmid taxa, particularly those from eutrophic waters, biogeographically show much less differentiation than benthic, oligomesotrophic taxa (Thomasson, 1955; Coesel, 1992). This might be indicative of a more efficient dispersal and a more succesful distribution.

Presumably, the desmids as an algal group – recently estimated at some 3000 species (Gerrath, 1993) – have their evolutionary origin in the tropics. In the discipline of biogeography it is customary to consider the area richest in species or morphological diversity as the area of evolutionary origin of a given taxonomic group (e.g. Banarescu, 1990). Roughly speaking, from poles to equator there is a gradual increase in the number of desmid taxa (both species and genera), culminating into a morphologically highly diversified tropical flora. Bizarre forms as found in the asymmetrical genera *Allorgeia* Gauthier-Lièvre and *Amscottia* Grönbl. are confined to the equatorial zone, whereas in (sub)polar regions most taxa are characterized by simple, compact cell shapes (compare Figures 2 and 7). Also when regarding their contribution to the total algal biomass, desmids seem to play a more important role in tropical aquatic ecosystems than in cold-temperate ones. In temperate climatic regions, desmids are well-known for the high number of species, relative to other phytoplankton groups, in oligotrophic waters (Hutchinson, 1967). However, their quantitative contribution (in terms of cell numbers, or biomass) uses to be so small that they are not figuring in any plankton typology based on that criterion (Reynolds, 1984). In contrast to that, there are several authors reporting desmids as the quantitatively dominant group of primary producers in tropical aquatic ecosystems (e.g. Hegewald et al., 1976; Khan & Ejike, 1984; Biswas, 1984, 1992; Anton, 1994).

Finally, the hypothesis of a tropical origin of the desmidiaceous algae is also supported by the high optimum temperatures as experimentally determined in a series of planktic species isolated from Western Europe. Coesel & Wardenaar (1990) found optima ranging from 25 to 30 °C, whereas hardly any growth was measured at temperatures lower than 10 °C. These optimum temperatures are higher than in most other phytoplankton groups (Reynolds, 1984). In this light it is not amazing that Hirano (1965) reported only 8 desmid species from the Antarctic continent, as against 68 species of diatoms, which, in general, show distinctly lower temperature optima (Reynolds, *loc. cit.*). Accordingly, potential geographical distribution patterns of desmid species will be determined by minimum, rather than maximum temperatures in a given region. For even in arctic and high mountain regions, in summer period the water temperature in shallow pools may rise to over 25 °C at day time (Messikommer, 1942; Thomasson, 1956). In this context it is interesting to note that, as already discussed in a previous section, 'typical' arctic-alpine species are particularly encountered in thin water films covering bare rocks and moss vegetations: a substrate that readily will be frozen at night. Obviously, the species in question can well stand periodical (and also lengthened) freezing. Since these environmental conditions are also operating in temperate lowlands in winter time, it may explain the local occurrence of arctic-alpine species there, but only in such hemi-atmophytic habitats where they will be in competitive advantage over freezing-sensitive species. In bigger water bodies, desmid taxa not withstanding freezing may survive near the relatively warm mud bottom or in the thermically stable (4 °C) hypolimnion zone of deep lakes.

If the potential area of desmid taxa would be paramountly determined by minimum temperatures indeed, likely but three main geographical regions may be distinguished: tropical, cool-temperate and arctic-alpine. In that case, a further differentiation as shown by actual geographical distribution patterns has to be ascribed to migrational barriers (seas, arid areas, mountain chains) met with in the course of evolutionary development. The distribution of the genera *Ichthyodontum, Ichthyocercus, Triplastrum* and *Triploceras* as discussed in a previous section, could give an impression of various degrees of unfurling from a tropical region of origin. However, it is quite obscure whether the actual distribution patterns in question have to do with different times of origin at a phylogenetic tree, differences in ecophysiology determining their potential distribution area, or a different success in dispersal. A similar problem is raised by the question whether the occurrence of so-called alpine desmid taxa in high mountain areas in (sub)tropical regions is exogenic (long-distance dispersal from similar habitats elsewhere), or endogenic (evolutionary adaptation of related taxa in nearby lowlands) in origin, or the result of a drastic climatic change (relict area). Considering the above-discussed high upper limit of the temperature range in arctic-alpine species, and also in view of the remarkable morphological uniformity in many of these taxa, incompatible with a long period of geographical isolation, the first option seems to be the most likely.

However, it is not to be excluded that small differences in morphology as illustrated in certain desmid taxa along altitudinal gradients are caused by temperature. Such differences originally could be phenotypical in nature, but later on genotypically fixed. See, e.g., the length of cell processes in *Staurastrum leptacanthum* Nordst. from different altitudes in Colombia (Coesel, 1992), and in *Micrasterias crux-melitensis* Ralfs versus *M. radians* Turn. described from higher and lower altitudes in Papua New Guinea respectively (Vyverman, 1991).

In biogeography, one of the main indications of the splitting up of an originally connected distribution area is the occurrence of vicariant species. Because of the poor knowledge of the morphological variability of most desmid taxa it is quite a problem to designate such pairs of vicariant species. A possible example might be *Xanthidium trilobum* Nordst. and *X. subtrilobum* W. & G. S. West. So far, *S. trilobum* has only been reported from South and Central America (e.g., Nordstedt, 1869; Grönblad, 1945; Förster, 1964; Prescott, 1966; Thomasson, 1971; Couté & Tell, 1981; Thérézien, 1985; Coesel, 1992), whereas *X. subtrilobum* is known from Africa, S.E. Asia, and N. Australia (e.g., West & West, 1897; Rich, 1935; Jao, 1949; Skuja, 1949; Grönblad et al., 1958; Thomasson, 1960; Scott & Prescott, 1961; Couté & Rousselin, 1975; Croasdale & Scott, 1976; Islam & Haroon, 1980; Ling & Tyler, 1986). It has to be noted, however, that the differences in morphology between *X. trilobum* and *X. subtrilobum* seem far too small to render its plausible that the period of separate evolutionary development started already ca. 100 million years ago (i.e., the presumable extension of the Atlantic Ocean, see Pielou, 1979). So, more likely, the supposed allopatric speciation under discussion is to be described by the dispersal model (colonization) rather than by the vicariance model (fragmentation), see Craw (1988). In general, such desmid sister taxa will be detected more readily at infraspecific level since their mutual relationship is more obvious. A still higher degree of differentiation may be expected at submicroscopic level. Slight differences in genome structure whether or not bound up with a geographical isolation not necessarily need to be expressed in cell morphology but may result into sexual incompatibility, see the biological entities distinguished within *Closterium ehrenbergii* Ralfs (Ichimura, 1985; Coesel, 1989a). However, as yet, for getting a better insight into the main features of desmid biogeography a critical investigation of the morphological variability and geographical distribution of traditional taxa seems to have priority.

When trying to come to a very first, rough evaluation of Krieger's desmid floral regions as discussed in this paper we may conclude that Indo-Malaysia/Northern Australia, tropical America, and equatorial Africa are the most pronounded regions. Also well typified are Eastern Asia, New Zealand/Southern Australia, and North America. Less endemic species are met with in Southern Africa and extratropical South America, whereas temperate Eurasia, with respect to the other continents, is mainly negatively characterized. Finally, the so-called arctic-alpine desmid flora may be encountered on all continents, provided that adequate minimum water temperatures occur. Its distribution seems to be determined microclimatologically rather than macroclimatologically.

Acknowledgements

Many thanks are due to Dr W. N. Ellis (Zoological Museum Amsterdam) for valuable suggestions and comments on general biogeographical aspects.

References

Allorge, P., 1924. Desmidiées du Lac de Grand-Lieu. Rev. Algol. 1: 462–470.
Anton, A., 1994. Long-term phytoplankton changes in a tropical reservoir. Mitt. int. Ver. Limnol. 24: 243–249.
Banarescu, P., 1990. Zoogeography of fresh waters. 1. General distribution and dispersal of freshwater animals. Aula-Verlag, Wiesbaden, 511 pp.
Bicudo, C. E. M., 1975. Polymorphism in the desmid *Arthrodesmus mucronulatus* and its taxonomic implications. Phycologia 14: 145–148.
Biswas, S., 1984. Phytoplankton of Opi Lake, Anambra State, Nigeria. Verh. int. Ver. Limnol. 22: 1180–1184.
Biswas, S., 1992. Phytoplankton periodicity in Ogelube Lake, Anambra State, Nigeria during 1979–1980. Hydrobiologia 246: 169–172.
Borge, O., 1989. Über tropische und subtropische Süsswasser-Chlorophyceen. Bih. Kongl.Svenska Vetensk.-Akad. Handl. 24(12): 1–33.
Bourrelly, P., 1957. Algues d'eau douce du Soudan Français, région du Macina (A. O. F.). Bulletin de l'Institut Français d'Afrique Noire 19, Sér. A 4: 1047–1102.
Bourrelly, P. & A. Couté, 1991. Desmidiées de Madagascar (Chlorophyta, Zygophyceae). Biblioth. Phycol. 86: 1–349.
Bourrelly, P. & R. Leboime, 1946. Notes sur quelques algues d'eau douce de Madagascar (Mission H. Humbert 1937). Biol. Jaarb. 13: 75–111.
Brook, A. J., 1981. The biology of desmids. Blackwell Scientific Publications, Oxford, 276 pp.
Capdevielle, P., 1978. Recherches écologiques et systématiques sur le phytoplankton du Lac de Cazaux-Sanguinet-Biscarosse. Ann. Stat. Biol. Besse-en-Chandesse 12: 1–304.
Claassen, M. I., 1977. Freshwater algae of southern Africa, 2. *Triplastrum spinulosum* from the Transvaal. Bothalia 12: 231–237.
Coesel, P. F. M., 1974a. Notes on sexual reproduction in desmids, 1. Zygospore formation in nature (with special reference to some unusual records of zygotes). Acta bot. neerl. 23: 361–368.
Coesel, P. F. M., 1974b. Bijdragen tot de kennis der Nederlandse Desmidiaceeënflora, 1. Enige interessante soorten uit het plassengebied van N.W.-Overijssel. Gorteria 7: 20–26.
Coesel, P. F. M., 1979. Desmids of the broads area of N.W.-Overijssel (The Netherlands), 2. Acta bot. neerl. 28: 385–423.
Coesel, P. F. M., 1985. De Desmidiaceeën van Nederland, 3. Wetensch. Mededel. Koninkl. Nederl. Natuurhist. Ver. 170: 1–69.
Coesel, P. F. M., 1989a. Biosystematic studies on the *Closterium moniliferum/ehrenbergii* complex (Chlorophyta, Conjugatophyceae) in western Europe, 4. Distributional aspects. Cryptogamie, Algol. 10: 133–141.
Coesel, P. F. M., 1989b. Taxonomic notes on Dutch desmids. Cryptogamie, Algol. 10: 181–193.
Coesel, P. F. M., 1992. Desmid assemblies along altitude gradients in Colombia. Nova Hedwigia 55: 353–366.
Coesel, P. F. M. & H. Kooijman-Van Blokland, 1991. Seasonality of planktonic desmid species in Lake Maarsseveen (The Netherlands) related to experimentally determined growth rates in a temperature-light gradient. Verh. int. Ver. Limnol. 24: 763–767.
Coesel, P. F. M. & K. Wardenaar, 1990. Growth responses of planktonic desmid species in a temperature-light gradient. Freshwat. Biol. 23: 551–560.
Coesel, P. F. M., S. R. Duque & G. Arango, 1988. Distributional patterns in some neotropical desmid species (Algae, Chlorophyta) in relation to migratory bird routes. Revue Hydrobiol. trop. 21: 197–205.
Compère, P., 1967. Algues du Sahara et de la région du Lac Tchad. Bull. Jard. Bot. Nat. Belg. 37: 109–288.
Couté, A. & A. Iltis, 1988. Étude en microscopie électronique à balayage de quelques desmidiacées (Chlorophyta, Zygophyceae) des lacs andins Boliviens. Cryptogamie, Algol. 9: 13–26.
Couté, A. & G. Rousselin, 1975. Contribution à l'étude des algues d'eau douce du moyen Niger (Mali). Bull. Mus. Hist. Nat. (Paris), Sér. 2, 277: 73–175.
Couté, A. & G. Tell, 1981. Ultrastructure de la paroi cellulaire des desmidiacées au microscope électronique à balayage. Beih. Nova Hedwigia 68: 1–228.
Cox, C. B. & P. D. Moore, 1993. Biogeography: an ecological and evolutionary approach. Blackwell Scientific Publications, Oxford, 326 pp.
Craw, R., 1988. Panbiogeography: method and synthesis in biogeography. In A. A. Myers & P. S. Giller (eds), Analytical biogeography: an integrated approach to the study of animal and plant distributions. Chapman & Hall, London: 405–435.
Croasdale, H. T. & E. A. Flint, 1986. Flora of New Zealand. Freshwater algae, chlorophyta, desmids: with ecological comments on their habitats, 1. Government Printing Office, Wellington, 132 pp.

Croasdale, H. T. & E. A. Flint, 1988. Flora of New Zealand. Freshwater algae, chlorophyta, desmids: with ecological comments on their habitats, 2. The Caxton Press, Christchurch, 147 pp.
Croasdale, H. T. & A. M. Scott, 1976. New or otherwise interesting desmids from northern Australia. Nova Hedwigia 27: 501–595.
Croasdale, H. T., C. E. de M. Bicudo & G. W. Prescott, 1983. A synopsis of North American desmids, 2:5. University of Nebraska Press, Loncoln, 117 pp.
De Lattin, G., 1967. Grundriß der Zoogeographie. G. Fischer, Stuttgart, 602 pp.
Engels, M. & K. Handke, 1994. *Euastrum vigrense* Ryppowa: A light and electron microscopial approach. Biológia (Bratislava) 49: 511–518.
Förster, K., 1964. Desmidiaceen aus Brasilien, 2. Bahia, Goyaz, Piauhy and Nord-Brasilien. Hydrobiologia 23: 321–505.
Förster, K., 1966. Die Gattung *Haplozyga* (Nordst.) Racib. in Brasilien. Revue Algol., N. S. 8: 151–157.
Förster, K., 1969. Amazonische Desmidieen. 1. Teil: Areal Santarém. Amazoniana 2: 5–116.
Förster, K., 1982. Conjugatophyceae: Zygnematales und Desmidiales (excl. Zynemataceae). Binnengewässer, 16: 8, 543 pp.
Brenzel, B., 1968. Grundzüge der pleistozänen Vegetationsgeschichte Nord-Eurasiens. F. Steiner, Wiesbaden, 326 pp.
Fritsch, F. E. & F. Rich, 1937. Contributions to our knowledge of the freshwater algae of Africa, 13. Algae from the Belfast Pan, Transvaal. Trans. r. Soc. S. Afr. 25: 153–228.
Gauthier-Lièvre, L., 1960. Les genres *Ichthyocercus, Triploceras* et *Triplastrum* en Afrique. Rev. Algol. N. S. 5: 55–65.
Gerrath, J. F., 1993. The biology of desmids: a decade of progress. Progress in phycological research 9: 79–192.
Gerrath, J. F. & P. Denny, 1988. Freshwater algae of Sierra Leone, 5. Desmids from the lake Sonfon region. Nova Hedwigia 47: 39–58.
Grönblad, R., 1933. A contribution to the knowledge of sub-aërial desmids. Soc. Sci. Fenn. Comm. Biol. 4(4): 1–7.
Grönblad, R., 1945. De algis brasiliensibus,, praecipue Desmidiaceis, in regione inferiore fluminis Amazonas a professore August Ginzberger (Wien) anno MCMXXVII collectis. Acta Soc. Sci. Fenn., Nov. Ser. B, 2: 1–43.
Grönblad, R., G. A. Prowse & A. M. Scott, 1958. Sudanese desmids. Acta bot. fenn. 58: 1–82.
Hegewald, E., A. Aldave & T. Hakuli, 1976. Investigations on the lakes of Peru and their phytoplankton, 1. Review of literature, description of the investigated waters and chemical data. Arch. Hydrobiol. 78: 494–506.
Heimans, J., 1969. Ecological, phytogeographical and taxonomic problems with desmids. Vegetatio 17: 50–82.
Hinode, T., 1952. On some Japanese desmids, 1. Hikobia 1: 145–149.
Hinode, T., 1960. On some Japanese demids, 3. Hikobia 2: 111–118.
Hinode, T., 1977. Desmids collected from Miyajima Island, Hiroshima Prefecture. Hikobia 8: 71–90.
Hirano, M., 1953. The alpine desmids from the Japanese Alps, 1. Bot. Mag. Tokyo 66: 125–134.
Hirano, M., 1956. Flora Desmidiarum Japonicarum, 2. Contr. Biol. Lab. Kyoto University 2: 57–106.
Hirano, M., 1957. Flora Desmidiarum Japonicarum, 4. Contr. Biol. Lab. Kyoto University 5: 166–225.
Hirano, M., 1959a. Flora Desmidiarum Japonicarum, 5. Contr. Biol. Lab. Kyoto University 7: 226–301.
Hirano, M., 1959b. Flora Desmidiarum Japonicarum, 6. Contr. Biol. Lab. Kyoto University 9: 302–386.
Hirano, M., 1965. Freshwater algae in the Antarctic regions. Monographiae Biologicae 15: 127–193.
Hoshaw, R. W. & R. M. McCourt, 1988. The Zygnemataceae (Chlorophyta): a twenty-year update of research. Phycologia 27: 511–548.
Hutchinson, G. E., 1967. A treatise on limnology, 2. Introduction to lake biology and the limnoplankton. J. Wiley & Sons, New York, 1115 pp.
Ichimura, T., 1985. Geographical distribution and isolating mechanisms in the *Closterium ehrenbergii* species complex (Chlorophyceae, Closteriaceae). In H. Hara (ed.), Origin and evolution of diversity in plants and plant communities. Academia Scientific Book, Tokyo: 295–303.
Islam, A. K. M. N. & A. K. Y. Haroon, 1980. Desmids of Bangladesh. Int. Revue ges. Hydrobiol. 65: 551–604.
Iyengar, M. O. P. & K. R. Ramanathan, 1942. *Triplastrum*, a new member of the Desmidiaceae from South India. J. Ind. Bot. Soc. 21: 225–229.
Jao, C. C., 1949. Studies on the freshwater algae of China, 19. Desmidiaceae from Kwangsi. Bot. Bull. Acad. Sinica 3: 37–95.
Khan, M. & C. Ejike, 1984. Limnological aspects of plankton variation and water characteristics of a tropical West African reservoir. Verh. int. Ver. Limnol. 22: 1584–1589.
Kossinskaja, C. C., 1960. Desmidiales, 1. Flora Plantarum Cryptogamarum URSS, 5., 511 pp.
Krieger, W., 1933. Die Desmidiaceen der Deutschen limnologischen Sunda-Expedition. Arch. Hydrobiol., Suppl. 11: 129–230.
Krieger, W., 1937. Die Desmidiaceen Europas mit Berücksichtigung der aussereuropäischen Arten. Rabenhorst's Kryptogamenflora, 13: 1, 712 pp.
Kusel-Fetzman, E., 1968. Beiträge zur Kenntnis der Algenflora ostafrikanischer Hochgebirgsseen. Hochgebirgsforschung 1: 69–100.
Lenzenweger, R., 1974. Beitrag zur Desmidiaceen-Flora von West-Borneo. Mitt. Bot. Arbeitsgem. Oberösterr. Landesmuseum Linz 6: 103–128.
Lind, E. M., 1967. Some East African desmids. Nova Hedwigia 13: 361–387.
Lind, E. M., 1971. Some desmids from Uganda. Nova Hedwigia 22: 535–585.
Ling, H. U. & P. A. Tyler, 1986. Freshwater algae, exclusive of diatoms. A limnological survey of the Alligator Rivers Region, 2. Australian Government Publishing Service, Canberra, 172 pp.
Lütkemüller, J., 1900. Desmidiaceen aus den Ningpo-Mountains in Zentral China. Ann. K. K. Naturhist. Hofmus. 15: 115–126.
MacArthur, R. H., 1972. Geographical ecology: patterns in the distribution of species. Harper & Row, New York, 268 pp.
Messikommer, E., 1942. Beitrag zur Kenntnis der Algenflora und Algenvegetation des Hochgebirges um Davos. Beitr. geobot. Landesaufn. Schweiz, 24: 1–452.
Nordstedt, O., 1869. Desmidiaceae. Symbolae ad floram Brasiliae centralis cognoscendam, 5. Vidensk. Meddel. Dansk Naturhist. Foren. Kjøbenhavn 14–15: 233–234.
Nordstedt, O., 1888. Fresh-water algae collected by Dr S. Berggren in New Zealand and Australia. Kongl. Svenska Vetensk.-Akad. Handl. 22: 1–98.
Nordstedt, O., 1889. De Algis et Characeis, 3–6. Acta Univ. Lund. 25: 1–41.
Pielou, E. C., 1979. Biogeography. J. Wiley, New York, 351 pp.
Playfair, G. I., 1907. Some new or less known desmids found in New South Wales. Proc. linn. Soc. New South Wales 32: 160–201.
Prescott, G. W., 1948. Desmids. Bot. Rev. 14: 644–676.
Prescott, G. W., 1966. Algae of the Panama Canal and its tributaries, 2. Conjugales. Phykos 5: 1–49.
Prescott, G. W. & A. M. Scott, 1952. Some South Australian desmids. Trans. r. Soc. S. Aust. 75: 55–69.

Prescott, G. W., C. E. M. Bicudo & W. C. Vinyard, 1982. A synopsis of North American desmids, 2:4. University of Nebraska Press, Lincoln, 700 pp.
Prescott, G. W., H. T. Croasdale & W. C. Vinyard, 1975. A synopsis of North American desmids, 2:1. University of Nebraska Press, Lincoln, 275 pp.
Prescott, G. W., H. T. Croasdale & W. C. Vinyard, 1977. A synopsis of North American desmids, 2:2. University of Nebraska Press, Lincoln, 413 pp.
Prescott, G. W., H. T. Croasdale, W. C. Vinyard & C. E. M. Bicudo, 1981. A synopsis of North American desmids, 2:3. University of Nebraska Press, Lincoln, 720 pp.
Raciborski, M., 1895. Die Desmidieenflora des Tapakoomasees. Flora 81: 30–35.
Reynolds, C. S., 1984. The ecology of freshwater phytoplankton. Cambridge University Press, Cambridge, 384 pp.
Rich, F., 1932. Contributions to our knowledge of the freshwater algae of Africa, 10. Phytoplankton from South African Pans and Vleis. Trans. r. Soc. S. Afr. 20: 149–188.
Rich, F., 1935. Contributions to our knowledge of the freshwater algae of Africa, 11. Algae from a pan in Southern Rhodesia. Trans. r. Soc. S. Afr. 23: 107–160.
Ružička, J., 1981. Die Desmidiaceen Mitteleuropas, 1:2. E. Schweizerbart, Stuttgart, 736 pp.
Schmidle, W., 1898. Die von Professor Dr. Volkens und Dr. Stuhlman in Ost-Afrika gesammelten Desmidiaceen, bearbeitet unter Benützung der Vorarbeiten von Prof. G. Hieronymus. Bot. Jahrb. Syst. 26: 1–59.
Schmidle, W., 1902. Algen, insbesondere solche des Plankton, aus dem Nyassa-See und seiner Umgebung, gesammelt von Dr. Fülleborn. Bot. Jahrb. Syst. 32: 56–88.
Scott, A. M. & G. W. Prescott, 1956. Notes on Indonesian freshwater algae, 2. *Ichthyodontum*, a new desmid genus from Sumatra. Reinwardtia 4: 105–112.
Scott, A. M. & G. W. Prescott, 1958. Some freshwater algae from Arnhem Land in the Northern Territory of Australia. Rec. Amer.-Aust. sci. Exp. Arnhem Land 3: 9–136.
Scott, A. M. & G. W. Prescott, 1960. Notes on Indonesian freshwater algae, 4. Concerning *Euastrum moebii* (Borge) Scott & Prescott comb. nov. Reinwardtia 5: 323–340.
Scott, A. M. & G. W. Prescott, 1961. Indonesian desmids. Hydrobiologia 17: 1–132.
Skuja, H., 1949. Zur Süsswasseralgenflora Burmas. Nova Acta Regiae Soc. Sci. Upsal., Ser. 4, 14: 1–188.
Stebbins, G. L. & G. J. C. Hill, 1980. Did multicellular plants invade the land? Am. Nat. 115: 342–353.
Taylor, W. R., 1935. Alpine algae from the Santa Marta Mountains, Colombia. Am. J. Bot. 22: 763–781.
Tell, G., 1979. Chlorophyceae d'eau douce rares et nouvelles de la République Argentine. Revue Algol., N. S. 14: 39–48.
Tell, G., 1980. Le genre *Staurastrum* (Algues Chlorophycées, Desmidiées) dans le nord-est de l'Argentine. Bull. Mus. natn. Hist. nat., Paris, Sér. 4:2, Sect. B: 145–207.
Thérézien, Y., 1985. Contribution à l'étude des algues d'eau douce de la Guyane Française à l'exclusion des diatomées. Bibliotheca Phycologia 72: 1–275.
Thérézien, Y. & A. Couté, 1977. Algues d'eau douce des Iles Kerguelen et Crozet (à l'exclusion des diatomées). Com. nat. Franc. Rech. Antarct. 43: 1–91.
Thomasson, K., 1955. Studies on South American freshwater plankton, 3. Plankton from Tierra del Fuego and Valdivia. Acta Horti Gothob. 19: 193–225.
Thomasson, K., 1956. Reflections on arctic and alpine lakes. Oikos 7: 117–143.
Thomasson, K., 1960. Notes on the plankton of Lake Bangweulu, 2. Nova Acta Regiae Soc. Sci. Upsal., Ser. 4, 17: 1–43.
Thomasson, K., 1963. Araucanian lakes. Acta Phytogeographica Suecica 47: 1–139.
Thomasson, K., 1965. Notes on algal vegetation of Lake Kariba. Nova Acta Regiae Soc. Sci. Upsal., Ser. 4, 19: 1–34.
Thomasson, K., 1966. Phytoplankton of Lake Shiwa Ngandu. Hydrobiological survey of the Lake Bangweulu – Luapula River basin, 4(2). Cercle Hydrobiologique de Bruxelles, Brussels, 91 pp.
Thomasson, K., 1967. Phytoplankton from some lakes on Mt Wilhelm, East New Guinea. Blumea 15: 285–296.
Thomasson, K., 1971. Amazonian algae. Inst. r. Sci. Nat. Belg., Mém., Sér. 2, 86: 1–57.
Thomasson, K., 1973. *Actinotaenium*, *Cosmarium*, and *Staurodesmus* in the plankton of Rotorua Lakes. Svensk Bot. Tidskr. 67: 127–141.
Thomasson, K., 1980. Antipodal algal annotations. Nova Hedwigia 33: 919–931.
Thomasson, K., 1986. Algal vegetation in North Australian billabongs. Nova Hedwigia 42: 301–378.
Thomasson, K. & P. Tyler, 1971. Taxonomy of Australian freshwater algae, 2. Planktic Staurastra from Tasmania. Nova Hedwigia 21: 287–319.
Tyler, P. & R. Wickham, 1988. Yan Yean revisited – a bicentennial window on Australian freshwater algae. Br. phycol. J. 23: 105–114.
Vyverman, W., 1991. Desmids from Papua New Guinea. Bibliotheca Phycologica 87: 1–200.
Wasylik, K., 1961. Desmidiaceen der Moore in der Umgebung von Korvanen in Sodankylä, Finnisch-Lappland. Commentat. Biol. 23: 1–46.
Wei, Y., 1984. Some new green algae from Xizang (Tibet). Acta Phytotaxonomica Sinica 22: 321–336.
Wei, Y., 1991. SEM study of cell walls of 24 desmids (Desmidiaceae, Chlorophyta) from China. Chin. J. Oceanol. Limnol. 9: 263–272.
Wei, Y., 1993. Some new taxa of Desmidiaceae from the Hengduan Mountains region of China. Acta Phytotaxonomica Sinica 31: 477–486.
West, G. S., 1909. The algae of the Yan Yean Reservoir, Victoria: a biological and ecological study. J. linn. Soc. Bot. 39: 1–88.
West, G. S., 1914. A contribution to our knowledge of the freshwater algae of Columbia. Mém. Soc. Sci. Nat. Neuchâtel 5: 1013–1051.
West, W. & G. S. West, 1895. A contribution to our knowledge of the freshwater algae of Madagascar. Trans. linn. Soc. London, Bot. 5: 41–90.
West, W. & G. S. West, 1897. Welwitsch's African freshwater algae, 11. Desmidiaceae. J. Bot. 25: 77–89.
West, W. & G. S. West, 1902. A contribution to the freshwater algae of Ceylon. Trans. linn. Soc. London, Bot. 6: 123–215.
West, W. & G. S. West, 1905. A monograph of the British Desmidiaceae, 2. Ray Society, Monographs 84: 1–206.
West, W. & G. S. West, 1908. A monograph of the British Desmidiaceae, 3. Ray Society, Monographs 88: 1–274.
West, W. & G. S. West, 1912. A monograph of the British Desmidiaceae, 4. Ray Society, Monographs 92: 1–194.
West, W., G. S. West & N. Carter, 1923. A monograph of the British Desmidiaceae, 5. Ray Society, Monographs 108: 1–300.

Hydrobiologia **336:** 55–65, 1996.
J. Kristiansen (ed.), Biogeography of Freshwater Algae.

6. Quantitative analysis on euglenoid distribution in seven regions of China

Shi Zhi-Xin
Institute of Hydrobiology, Chinese Academy of Sciences, Wuhan 430072, China

Key words: euglenoids, quantitative relationships, frequency distributions, regional differences, relationship of 'species–abundance', progressive changes

Abstract

Investigations on the qualitative and quantitative distribution of euglenoids in Jianghan Plain, Western Hubei, Wuling Mountains, Hengduan Mountains, South-eastern Xizang (Tibet), Southern Xizang and Northern Xizang show that the species composition and flora characters of the euglenoids are not very different, but the quantitative distribution varies in different regions.

1. The similarity comparisons between every two regional floras have been analysed by similarity coefficient except Northern Xizang. Most of the coefficients except a few are higher, general range in 0.6–0.7, the highest 0.81. It is reflected that there are better similarities of floras in these regions except Northern Xizang.

2. Frequency distribution for euglenoid abundance (i.e. species numbers of the euglenoids per sample) is approximate to a normal distribution in Jianghan Plain, whereas in the other regions, the distribution curves are all conspicuously exponential.

3. Regional difference: on the basis of indices of qualitative and quantitative distribution as well as of ecological factors, a cluster analysis and the Duncan multiple comparison show three distribution types of the euglenoids: (1) Jianghan Plain has the best, it differs obviously from other regions by its favourable range of natural conditions which have a warm and humid climate, low altitude and abundant rainfall etc.; (2) Northern Xizang, the least favourable type, is distinct from others by its extremely harsh environment which has a extremely frigid and dry climate, very high altitude, little rainfall, a lot of water bodies which are brackish or saline etc.; (3) The others are intermediate, despite their different range of natural conditions, the variance does not cause significant differences except that Western Hubei differs from Southern Xizang.

4. The relationship of 'species–abundance' appears to be logarithmic except in Northern Xizang, as expressed by $S = \alpha b^r/r$. By analyses, it may be postulated that the species diversity (α-value) possibly relates to the complexity and diversity of environment, principally to microhabitats and microclimates, but not closely to natural macro-conditions.

5. The seven regions show a progressive geographical change from east to west. The qualitative and quantitative distributions of the euglenoids likewise change from east to west, in accordance with the change in location and natural conditions (mainly temperature, altitude and rainfall etc.) of the regions.

Introduction

Previous research concerning euglenoid distribution has been purely descriptive, without use of mathematical methods (e.g. Gerrath & Deey, 1979; Jao, 1964; Prescott, 1951; Shi, 1988, 1989 and 1992b; Skuja, 1939). Recently, mathematical statistics have been used in some preliminary studies on other freshwater algae. Yung & Stokes (1985) have made an introduction to distribution and floristic similarities of algae by cluster analysis. The fluctuation of the species number in water bodies was expressed by the frequency distribution, but this method was imperfect because the power function could not include the species number value '0' (Shi, 1992a, 1994a and 1994b). The relationships of 'species–abundance' have been tried by a

Figure 1. Sketch map showing the localities of the seven regions in China. 1. Jianghan Plain; 2. Western Hubei; 3. Wuling Mountains; 4. Hengduan Mountains; 5. South-eastern Xizang; 6. Southern Xizang; 7. Northern Xizang.

mathematical formula that is also a power function, but perhaps this was also not perfect enough (Shi, op. cit.). In the present study, the author then applied mathematical statistics and ecological principles in analysing the quantitative relationships between the euglenoids and the macro-environments, and tried to improve the methods.

The euglenoids to a great extent can adapt themselves to different environments, and they prefer habitats such as warm, nutrient-rich, small water bodies. The relationship between the euglenoids and their microhabitats is very close; i.e. despite the type of macro-environment, the euglenoids can grow and multiply so long as there are such water bodies suitable for them. Most of them have a wide ecological range except a few that only exist in ice, snow, acid water and other special habitats. As a result, the species composition and flora characters of the euglenoids in the seven regions examined in this paper (Figure 1) were not obviously different despite the different environmental conditions of the regions. It is difficult to appraise the characteristics of the euglenoids floras in these regions according to their composition. However, the changes in the distribution of species number and other quantitative indices per locality are obvious as the macro-environments still have an influence on distribution of the euglenoids. The quantitative changes are mainly reflected in species number, appearance frequency, appearance probability and diversity index, etc.

Materials and methods

General description of the data

Regions

The seven regions discussed in this paper are located in the southern part of China (Figure 1). Their natural conditions are summarized below.

1. Jianghan Plain (J. P.), situated in Hubei province, has a warm and humid subtropical climate, annual mean temperature 15–17°C, altitude about 50 m, annual rainfall about 1160 mm.

2. Western Hubei (W. H.), situated in Hubei province, has a warm and humid subtropical climate, annual mean temperature 11–13°C, altitude 50–1000 m (which is the height range of the samples collected, the same below), annual rainfall about 1140 mm.

3. Wuling Mountains (W. M.), mainly situated in north-eastern Guizhou province and western Hunan province, have a warm and humid subtropical climate, annual mean temperature about 16°C, altitude 20–2000 m, annual rainfall about 1250 mm.

4. Hengduan Mountains (H. M.), mainly situated in north-western Yunnan province, western Sichuan province and eastern Xizang (Tibet, the same below) autonomous region, are divided into two natural regions: tropical and subtropical mountain forests, and coniferous forests, annual mean temperature 6–12°C altitude 800–4000 m, annual rainfall about 747 mm.

5. South-eastern Xizang (S. E. X.), situated in Xizang autonomous region, according to its natural environment it may be divided into tropic, subtropical mountain forests and coniferous forests, annual mean temperature 8–18°C, altitude 800–4000 m, annual rainfall about 980 mm.

6. Southern Xizang (S. X.), is situated in Xizang autonomous region, has a temperate semi-arid plateau climate, annual mean temperature 4–8°C, altitude 2000–4000 m, annual rainfail about 450 mm, has a lot of water bodies which are brackish or saline.

7. Northern Xizang (N. X.), situated in Xizang autonomous region, the climate is extremely frigid and dry, belonging to the subfrigid semi-arid and frigid arid plateau climate, annual mean temperature −4 – −2°C, altitude 4000–5000 m, annual rainfall about 75 mm, has a lot of water bodies which are brackish or saline.

Euglenoid samples
The samples, altogether about 900, were collected in diverse ecological conditions from above regions, and preserved in the Herbarium of the Institute of Hydrobiology (HBI). They include:

1. HP8450–8523, collected from J. P.
2. HP7121–7282, from W. H.
3. HN88001–88043, KC88001–88142, from W. M.
4. HD81001–81105, HD82001–82033, from H. M.
5. TB73001–73029, TB76001–76125, from S. E. X.
6. TB75009–75115, TB101–139, TB1–24, from S. X.
7. TB762001–762081, from N. X.
8. TB74001–74169, from S. E. X., and S. X.

The plankton algae in the above samples have been taken by means of a nylon net no. 25, in surface water (depth about 0.5 m). Most of them have been collected in small water bodies (ponds, pools, ditches and marshes) and a few in lakes and rivers. The following data of the habitats were recorded: temperature, pH, area, depth, altitude and vegatation. The samples were fixed with formalin, c. 4%.

The euglenoids contained in the plankton samples (about 500) have been identified. Fifty plankton samples were selected randomly in every region, so that the quantitative relationships of the euglenoids could be analysed, except the calculations of 'frequency distribution' and 'relationship of species–abundance'.

Calculation methods

Similarity coefficient
Calculated by the formula below (Odum, 1971):

$$S = 2c/a + b \tag{1}$$

S = similarity coefficient (the range of the values are from 0 to 1), a = species number of 'A' sample or community (region in the paper), b = species number of 'B' sample or community (region in the paper), c = number of the same species in two samples or communities (regions in the paper).

Normal distribution analysis
A simplified formula is adopted:

$$F = (N/S') \cdot z \tag{2}$$

F = distribution frequency of euglenoid abundance (it is equal to the number of samples with the same species number of euglenoids. If the number of samples with two species of euglenoids is 10, the frequency is equal to 10. The euglenoid abundance is the species number of euglenoids per sample); N = total frequency (i.e. ΣF actually it is the total number of samples); S' = standard deviation/distance between classes; z may be found in the table on normal distribution.

Regression analysis
In the paper, some common formulas are adopted:

$$Y = ae^{bX} \tag{3}$$

$$Y = a + \ln X \tag{4}$$

$$Y = a + bX \tag{5}$$

in above formulas, X and Y respectively are two kinds of data analysed, a is one constant, b is regression coefficient. Regression analysis is calculated by the "Scientific Calculator" (model: Casio fx–180p). When correlation coefficient $(r) > r_{0.05}$ (i.e. probability: $P < 0.05$), the result of any regression analysis can be established.

Duncan multiple comparison
Calculation R_k value:

$$R_k = r_a(k, df) \cdot S_x \tag{6}$$

r_a can be found in the Duncan table according to k and df, $k = 1, 2, \ldots$, df = degree of freedom; $S_x = (MSe/n)^{1/2}$, MSe = mean square error, n = sample number.

Cluster analysis
The steps are as follows:

1. Standardization treatment

$$X' = (\bar{X}_1 - \bar{X}_i)/S \tag{7}$$

X' = standardized value, $\bar{X}_1$ = mean value of 1 term, $\bar{X}_i$ = mean value of i term, S = standard deviation.

2. Distance coefficient

$$d_{jk} = [\Sigma(X_{ij} - X_{ik})]^{1/2} \tag{8}$$

d_{jk} = distance coefficient between j and k, X_{ij} = one value of i index in j region, X_{ik} = one value of i index in k region.

3. Clustering by UPGMA method (Sneath & Sokal, 1973).

Relationship of 'species–abundance'
Analysing by logarithmic progressive (Plieou, 1969), it is expressed as:

$$S = \alpha b^r / r \tag{9}$$

S = number of species which are equal in appearance frequency, if there are 5 species of which appearance frequencies all are the same, the S value = 5; r = species abundance (expressing it in individual number or appearance frequency of every species, and it is expressed by appearance frequence in this paper. The species abundance is different from another term "euglenoid abundance" which is the species number of euglenoid per sample); α = diversity index; b = constant.

Calculation and illustration of indices of quantity and distribution

In order to analyse and illustrate the quantitative relationships between euglenoids and environment, following indices were adopted:

1. Mean species number = species number of every region / total number of samples.
2. Mean euglenoid abundance = the sum of euglenoid abundance of all samples / total number of samples. The euglenoid abundance = species number of euglenoids occurring in every sample, and is different from the above term 'species–abundance' in the formula 9.
3. Mean appearance frequency = the sum of appearance frequency of all species / total number of species. The appearance frequency of a species is equal to the number of samples containing this species. If a species is found in 8 samples, its appearance frequency is equal to 8.
4. Appearance probability = (number of samples containing euglenoids / total number of samples) ×100%.
5. The b value = constant b of the formula 10: $F = ae^{-ba}$.
6. Diversity index = constant α of the formula 9: $S = \alpha b^r/r$.
7. Largest abundance = the largest r value of formula 9: $S = \alpha b^r/r$.

In addition, calculation of some ecological factors:

1. Mean water temperature = the sum of temperature of all samples / total number of samples.
2. Mean altitude = the sum of altitude of all collected samples / total number of samples.
3. Annual mean temperature and annual rainfall refer to local meteorological data.

Results

Analysis of flora characteristics

The euglenoid floras of these regions except J. P. have been analysed (Shi, 1988, 1992b, 1994a and 1994b). Their characteristics are briefly compared below:

1. The similarity of the euglenoid flora: A comparison has been made by similarity coefficient (formula 1, Odum 1971) according to main species distributed more widely in these regions except N. X. where the euglenoids not only are scarce but never dominant as the environment is so harsh (the data analysed from these papers published, i.e. Shi, *op. cit.*). Most of the coefficients are higher, general range in 0.6–0.7, the highest 0.81 except a few that mainly are those between S. X. (or S. E. X.) and J. P. (or W. M.) (Table 1). It is reflected that there are better similarities of most floras in these regions.
2. The analysis of different species in these regions: Although there are better similarities of their floras, a part of species are still different among these regions. Most of them, except a few (i.e. rare and new species), possess the peculiarity of wide distribution, therefore they all may be present in any of these regions. They have not been found temporarily because of the limitation and random of samples collected. Most of the species found even in N. X. also have a wide ecological range and may be present in other regions.

To sum up, it is difficult to appraise the characteristics of the euglenoid floras in these regions according to their compositions.

Frequency distributions for euglenoid abundance

The frequency distribution for euglenoid abundance is a reflection of the fluctuation of the species number in water bodies (the euglenoid abundance = species number of euglenoids per sample). The frequency distribution is a statistical term, and here equal to the number of the samples with the same euglenoid abundance). It may be determined that the environmental conditions in different regions exert an influence on the quantity and distribution of the euglenoids. The frequency distribution in these regions are as follows:

It is approximate to a normal distribution in J. P. (the data used from the paper: Shi, 1992a), but in other regions, the distributive curves are all conspicuously exponential (the data from these papers: Shi, 1988, 1992b, 1994a and 1994b). The peak value of frequen-

Table 1. Species number of the euglenoids in water bodies in the seven regions (random selection of 50 samples)

Regions	Specific number of the euglenoids																							
J. P.	11	13	8	15	18	14	14	8	5	9	4	2	2	15	15	8	6	20	13	7	9	4	17	7	5
	1	6	4	26	19	16	10	6	15	33	10	10	8	14	8	6	9	30	3	11	20	24	6	21	5
W. H.	2	1	0	0	0	0	1	2	0	0	0	0	0	1	2	0	15	4	3	2	2	20	0	0	1
	2	3	18	20	3	36	2	3	19	3	6	7	14	7	23	8	7	15	2	12	22	7	25	0	18
W. M.	1	17	0	2	0	0	1	0	2	4	2	3	4	0	1	0	7	4	1	33	2	20	0	0	4
	7	5	5	0	15	1	0	5	0	0	0	10	0	0	15	0	0	5	2	8	1	3	5	0	4
H. M.	0	0	0	26	1	0	3	1	3	14	1	2	4	7	3	4	0	0	4	5	0	6	4	0	3
	4	24	8	1	1	1	16	9	4	4	2	2	18	2	2	2	25	1	8	1	2	0	0	0	0
S. E. X.	2	2	1	0	4	0	0	1	0	1	0	0	1	1	5	3	0	0	2	8	4	0	18	1	0
	1	7	1	0	17	0	2	0	1	3	16	10	4	0	6	0	2	1	1	28	6	2	0	5	3
S. X.	0	1	16	2	0	0	12	1	0	3	1	0	5	0	6	0	1	0	5	3	11	8	2	0	0
	1	0	11	1	0	5	0	16	3	0	3	0	0	2	3	0	9	2	0	3	5	1	0	0	9
N. X.	0	0	0	0	0	0	0	0	0	7	2	0	0	0	0	0	0	0	1	2	1	1	9	1	0
	6	1	0	0	0	5	2	0	3	1	0	0	0	0	0	0	0	2	1	6	0	2	0	0	0

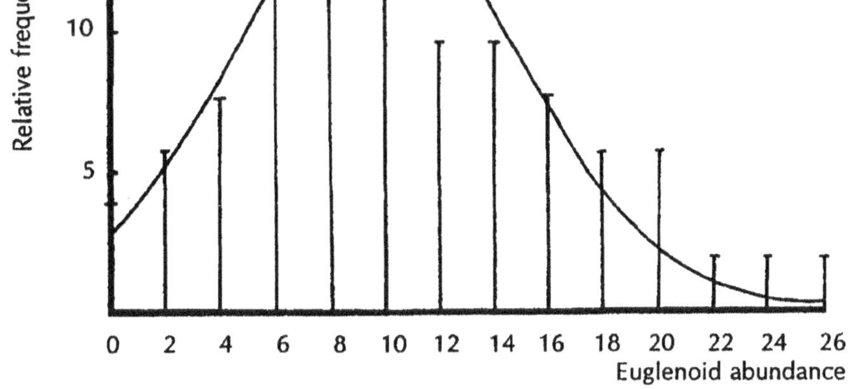

Figure 2. Normal distributions for the euglenoid abundance in Jianghan Plain.

cy distribution in J. P. is in the middle of the curve, reflecting that species numbers in most water bodies fall in the medial range (4–16 species, Figure 2). In other regions, most of the values in exponential distributions are '0', which reflects that species numbers are located to the left in the diagrams (0–5 species, Figure 3).

It can be seen from this that the quantity and distribution of the euglenoids in J. P. is approximate to a normal distribution. In other regions, there also exist evident differences, even if their curves all belong to the same type. The differences were analysed as follows:

The exponential equation is

$$F = ae^{-bS} \tag{10}$$

of which F is distribution frequency of euglenoid abundance, S = the euglenoid abundance, (if the samples with the same euglenoid abundance are 8, the F value = 8); a and b are the constants of the equation. When $S = 0$, $F = a$ and also is the peak value in frequency distribution. If we let $a = F_0$, the exponential can be expressed as

$$F = F_0 e^{-bS}. \tag{11}$$

After the formula 11 is calculated, it becomes

$$b = (\ln F_0 - \ln F)/S. \tag{12}$$

In the formula 12, the b value can reflect the change of curve to a certain degree, i.e. when the b value is high, the curve is very steep; and conversely, when the b value is low, it has a gentle slope. In fact, the b value of W. H. is the smallest, its curve is also the gentlest; conversely, N. X., the highest, the steepest. In other regions, the b values progressively increase from east to west, as do the corresponding curves (Table 4, Figure 3).

It was found that the patterns of frequency distributions (containing normal and exponential) from east to west, obviously bear relation to their environmental factors.

Regional difference

In order to examine the different degrees of the euglenoid distribution in these regions, the author made the following statistical analyses and comparisons regarding their random distributions, see Table 1 and 4.

Table 2. The similarity coefficients between the seven regions

Regions	N. X.	S. X.	S. E. X.	H. M.	W. M.	W. H.
J. P.	0	0.34	0.49	0.63	0.72	0.63
W. H.	0	0.60	0.60	0.81	0.68	
W. M.	0	0.38	0.44	0.65		
H. M.	0	0.50	0.67			
S. E. X.	0	0.57				
S. X.	0					

Table 3. Duncan multiple comparison for species number of the euglenoids

Regions	N. X.	S. X.	S. E. X.	H. M.	W. M.	W. H.
J. P.	**	**	**	**	**	**
W. H.	**	**				
W. M.	**					
H. M.	**					
S. E. X.	**					
S. X.	*					

'**' shows very significant difference between every two regions ($P < 0.01$); '*', marked difference ($P < 0.05$); no star sign, no difference ($P \geq 0.05$).

Analysing the difference between them by Duncan multiple comparison

After treating numbers in Table 1 with extraction of the square root (i.e. $X^{1/2}$, the X is any number in Table 1), a Duncan multiple comparison was analysed. The steps analysed are briefly as follows:

1. Calculating the values R_k by formula 6, the results are:

K:	2	3	4	5	6	7
$r_{0.05}$:	2.77	3.31	3.63	3.86	4.03	4.17
$R_{0.05}$:	0.510	0.609	0.668	0.710	0.724	0.767
$r_{0.01}$:	3.64	4.12	4.40	4.60	4.76	4.88
$R_{0.01}$:	0.670	0.758	0.810	0.846	0.876	0.898

2. The variance significances between every two regions were compared by the following principles:

If $\bar{X}_1 - \bar{X}_2 < R_{k(r0.05)}$, the variance is not marked.

If $\bar{X}_1 - \bar{X}_2 > R_{k(r0,01)}$, the variance is very significant, (showed with '**').

If $R_{k(r0,01)} > \bar{X}_1 - \bar{X}_2 > R_{k(r0.05)}$, the variance is marked (showed with '*'). Here, $\bar{X}_1$ and $\bar{X}_2$ are respectively the mean values of every two regions in Table 1. By comparison, J. P. and N. X. respectively exhibit different features from other regions, while there are no differences among the other regions, except that

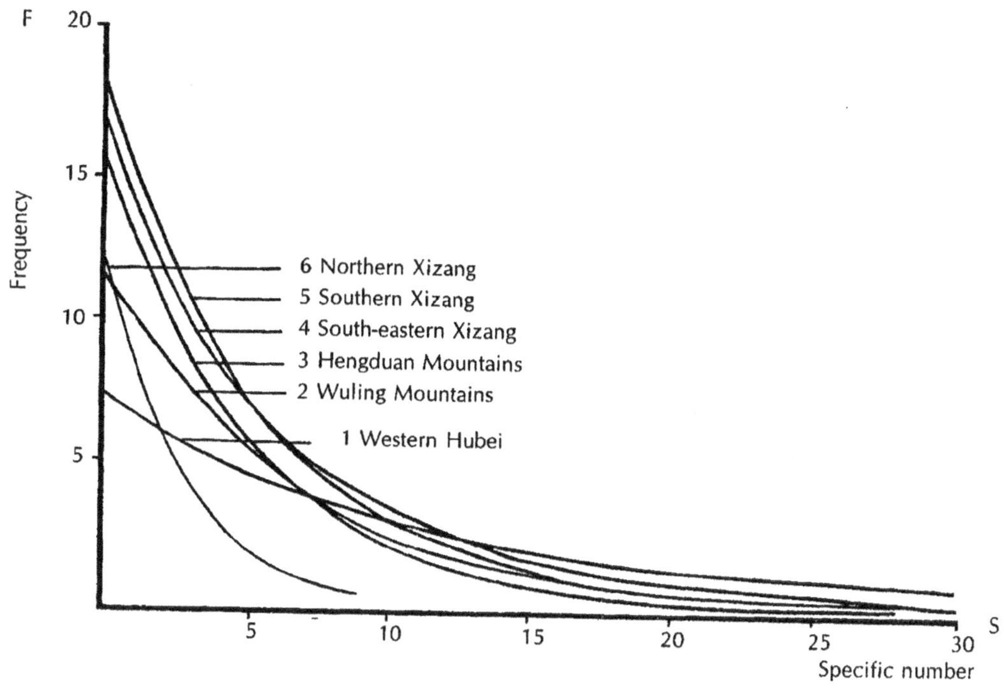

Figure 3. Frequency distributions for the euglenoid abundance in the other six regions. 1. $F = 7.030e^{-0.083S}$ ($r = -0.728, p > 0.01$); 2. $F = 11.342e^{-0.147S}$ ($r = -0.785, p > 0.01$); 3. $F = 13.236e^{-0.189S}$ ($r = -0.844, p > 0.001$); 4. $F = 16.705e^{-0.154S}$ ($r = -0.904, p > 0.001$); 5. $F = 17.132e^{-0.161S}$ ($r = -0.888, p > 0.01$); 6. $F = 12.089e^{-0.356S}$ ($r = -0.787, p > 0.05$).

Table 4. Indices on qualitative and quantitative distribution of the euglenoids as well as ecological factors

Regions	E	S	F	P	b	α	R	T	A	W
J. P.	11.4	3.10	3.00	96	0.002	84.61	22	26	50	1160
W. H.	6.8	2.75	2.46	74.5	0.083	80.03	14	23.1	288	1140
W. M.	4.0	1.25	2.72	58	0.147	60.03	17	21.3	524	1250
H. M.	4.5	1.70	2.00	70	0.189	106.92	19	17.5	2695	747
S. E. X.	3.4	1.65	2.42	65.3	0.152	111.80	11	17.6	3357	980
S. X.	3.0	1.01	1.96	49.3	0.161	81.99	6	16	3962	450
N. X.	1.1	0.58	1.17	36	0.356	28.28	5	12.4	760	75

E = mean euglenoid abundance, S = mean species number, F = mean appearance frequence, P = appearance probability (%), b = constant b of the exponential formula: $F = ae^{-bS}$, α = diversity index, R = the largest r value of the formula: $S = \alpha b^r/r$, T = mean water temperature (°C), A = mean altitude (m), W = annual rainfall (mm).

the variance between W. H. and S. X. is marked (see Table 3).

Because of the favourable natural conditions in J. P., it is evident that euglenoid quantity and their random distribution there is distinctly different from the other regions, supporting assemblages with high average species number. Conversely, N. X. differs also from the others regions by its extremely harsh environment with a low number of co-occurring species. In the other regions, despite their different natural conditions, the variance does not cause significant difference in qualitative and quantitative distribution except that W. H. differs from S. X. Because W. H. and S. X. are very far away from each other, their environmental variance further enhances the difference.

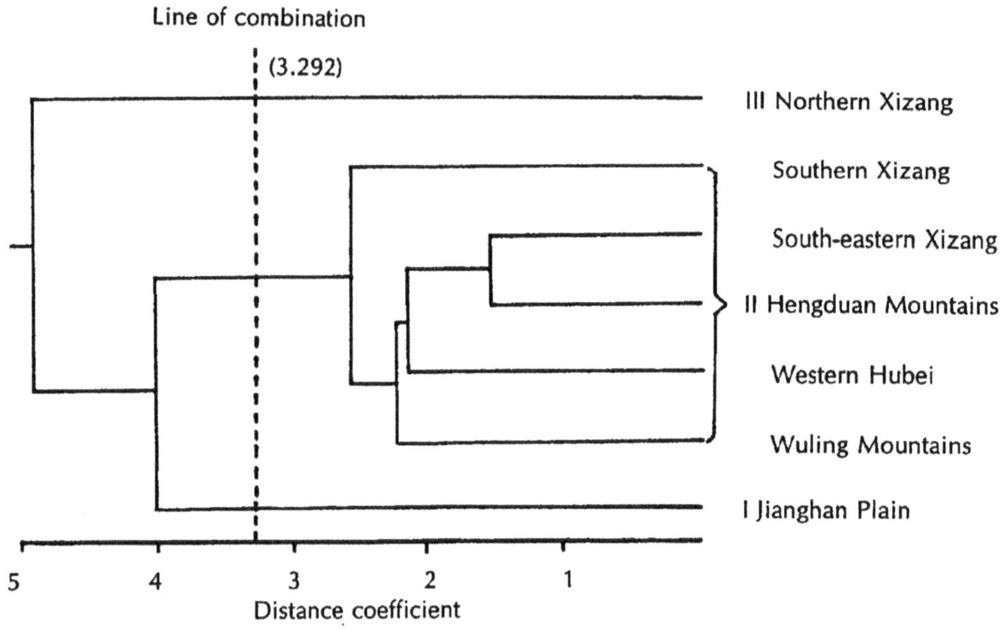

Figure 4. Cluster analysis for the euglenoid qualitative and quantitative distributions in the seven regions.

Making an induction by cluster analysis

On the basis of the indices in the Table 4, a cluster analysis was made by distance coefficients (formula 8) and UPGMA method (Sneath & Sokal, 1973). The result is that these regions can be clearly divided into three types (Figure 4):

1. J. P. is the most suitable region for supporting higher numbers of euglenoids.

2. All plateaus and mountainous regions except N. X. Plateau. Though their overall natural conditions are obviously less favourable than in J. P., there are still a lot of microclimates and microhabitats, suitable for the euglenoids. For this reason, the species numbers of these regions are relatively high and the diversity indices of the regions are basically similar to or even higher than that of J. P. Of course, other indices on qualitative and quantitative distribution are evidently lower than those of J. P. (Table 4).

3. N. X., all indices in Table 4 are the lowest, so that the qualitative and quantitative distribution of the euglenoids is the most limited in this region compared to the other regions.

The relationship of 'species–abundance'

The relationships between species number and species abundance were analysed by logarithmic progression (formula 9; Plieou, 1969). Here, the species number (S) and the species abundance (r) were used; see explanations of the formula 9, and the data based on these papers published (Shi, 1988, 1992b, 1994a and 1994b). The results indicate that the relationships of every region except N. X. all conform to the distributions of the logarithmic progression, and the curves all present obvious concavity (Figure 5).

On the basis of an ecological law, the dominant species with high abundance are few, but rare species with low abundance are present in most environments. This relationship of 'species–abundance' results in a high α value and the resultant curve is concave. If the environment is polluted or extremely harsh, the α value is low, and the curve tends to be flat and gentle (Odum, 1971). By analysing the 'species–abundance' of the seven regions, the diversity index of S. E. X. is the highest; H. M., the second; N. X., the lowest; other regions, intermediate. According to the α values and curve shapes, it may be suggested that all regions except N. X. belong to normal environments.

Having compared the α values and the highest abundance (i.e., the maximal lengths of the curves)

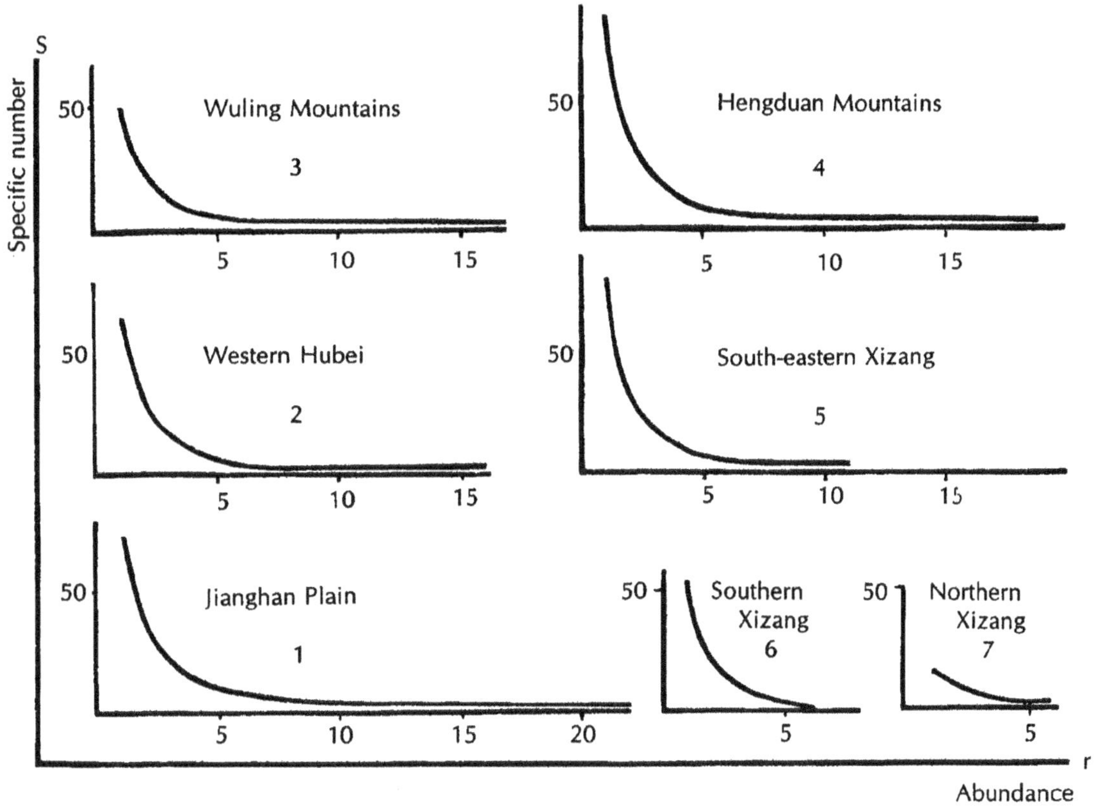

Figure 5. 'Species–abundance' relationships of the euglenoids in the seve regions. 1. $S = (84.61)\ (0.877)^r/r$; 2. $S = (80.03)\ (0.808)^r/r$; 3. $S = (60.03)\ (0.844)^r/r$; 4. $S = (106.92)\ (0.809)^r/r$; 5. $S = (111.80)\ (0.729)^r/r$; 6. $S = (81.99)\ (0.640)^r/r$; 7. $S = (28.28)\ (0.546)^r/r$.

in these regions, they may be divided into four main types.

1. H. M., W. H. and W. M., with high α values and long curve, especially H. M. in which the euglenoid distribution is the best in accordance with the relationship of 'species-abundance' in normal environment. It is possible that the natural environments of these regions are very complex and diversified; it provides favourable conditions not only for the euglenoid growth, which result in high α values, but also for the extensive distribution of some euglenoids to bring about obviously dominant species with high abundance (long curve).

2. S. E. X. and S. X., with the highest or higher α values but short curves, although the euglenoid compositions show high diversity (α value) by virtue of their complex environment (mainly microclimates and microhabitats); yet unfavourable macro-environments limit some species in their distribution and in becoming dominant (short curve).

3. J. P., with higher α values and particularly long curve, has a natural environment of very simple structure; it is still favourable to euglenoids so that the species diversity (α value) is lower than in H. M. and S. E. X., but higher than in other regions. On the other hand, it is possible that its macro-environment is so favourable that more euglenoids are extensively distributed in the region and become distinctly dominant species. As a result, the curve is very long.

4. N. X. with very low α value and very short curve. Probably, its environment is so harsh that the euglenoids not only are few in the region, but furthermore never dominant.

From the above-mentioned facts, it may be postulated that the species diversity (α value) may possibly relate to the complexity and diversity of environment, principally microhabitats and microclimates, but not closely relate to the natural macro-conditions though it has a certain connection with them (the macro-conditions express no significant relationship to α values by regression analysis). There are harsh macro-conditions in some regions, but the complexity and diversity of climate, topographies, altitude, vegetation, water bodies etc. cause these regions to contain a multitude of microclimates and microhabitats, many of which are in favourable to euglenoids. As a result, there are still a lot of euglenoid species in these regions, which is mainly reflected in the species diversity with high α values that are even higher than that in better macroscopic environment, i.e. the natural conditions in J. P. is better than those in H. M. and S. E. X., but the α value of the J. P. is definitely lower than in the latter.

Phenomenon of progressive change

The seven regions are approximately located from east to west in the southern part of China. The qualitative and quantitative distribution of the euglenoids change progressively from east to west, too (Table 4). The phenomenon of progressive change is obviously caused by environmental factors, and may be explained by the change in temperature, precipitation and altitude.

Temperature

The temperatures in the seven regions gradually decrease from east to west, see Table 4. Most of the euglenoids adapt themselves to warm environment, therefore other indices reflecting their quality and quantity of distribution progressively increase or decrease, too. The relationships between temperature and other indices are conspicuously exponential:

$$E = 0.233e^{0.148T} \quad (r = 0.942,\ p < 0.01)$$
$$S = 0.177e^{0.112T} \quad (r = 0.890,\ p < 0.01)$$
$$F = 0.688e^{0.599T} \quad (r = 0.888,\ p < 0.01)$$
$$P = 19.486e^{0.060T} \quad (r = 0.882,\ p < 0.01)$$
$$R = 1.466e^{0.121T} \quad (r = 0.782,\ p < 0.05)$$
$$b = 1.252e^{-0.112T} \quad (r = -0.915,\ p < 0.05)$$

In these formulas: T = mean water temperature, E = mean euglenoid abundance, S = mean species number, F = mean appearance frequency, P = appearance probability, R = the largest species abundance, b = b value, (all are the same below except T).

Annual rainfall

Water is one of the most important conditions for algal growth. The rainfall in a region affects arid and humid degree of the environment as well as accumulation and distribution of surface water, which directly exerts an influence on algal growth and distribution. The annual rainfall of the regions fundamentally tend to decrease from east to west (Table 4), and it bears conspicuous exponential relationships with other indices of the qualitative and quantitative distribution:

$$E = 1.00e^{0.001W} \quad (r = 0.800,\ p < 0.05)$$
$$S = 0.612e^{0.00109W} \quad (r = 0.813,\ p < 0.05)$$
$$F = 1.229e^{0.000553W} \quad (r = 0.950,\ p < 0.01)$$
$$P = 37.648e^{0.000594W} \quad (r = 0.817,\ p < 0.05)$$
$$R = 1.466e^{0.121W} \quad (r = 0.782,\ p < 0.05)$$
$$b = 0.320e^{-0.000858} \quad (r = 0.815,\ p < 0.05)$$

In these formulas, W = annual rainfall, others are as given above.

Altitude

The altitude of the seven regions progressively increase from east to west, which obviously also brings about progressive change of the qualitative and quantitative distribution of the euglenoids. Most of indices (except b value) in Table 4 decrease with increasing altitude and the relationships between them mostly show distinct logarithmic functions except b value to be exponential and R, linear:

$$E = 2.01 - 2.00 InA \quad (r = -0.941,\ p < 0.01)$$
$$S = 4.827 - 0.447 InA \quad (r = -0.845,\ p < 0.05)$$
$$F = 4.183 - 0.279 InA \quad (r = 0.797,\ p < 0.05)$$
$$P = 127.4 - 9.118 InA \quad (r = -0.817,\ p < 0.05)$$
$$R = 19.345 - 2.585A \quad (r = -0.799,\ p < 0.05)$$
$$b = 0.0973e^{0.0002A} \quad (r = 0.800,\ p \approx 0.05)$$

In these formulas, A = altitude, others are as given above.

The altitude indirectly exerts an influence on the growth and distribution of algae. The difference in altitude results in the change of hydrological conditions and other environmental variables, especially temperature, which undoubtedly has an effect on algae. The vertical distribution of the euglenoids is mainly influenced in quantity of the individual species, but not in the composition of the flora, as the euglenoids are mainly dependent on microclimates and microhabitats. For this reason, their distribution is obviously different from the vertical zonal formation of higher plants.

Discussion

This paper deals with the euglenoid distribution, but its results may also be useful for the study of the distribution of other fresh-water algae. Zonal formations of species composition in higher plants along with latitude, altitude and different natural regions have been studied, but thorough studies and reports concerning the distribution of fresh-water algae are so deficient that the distributing laws have not been fully discovered. The results in the paper show that species compositions in the euglenoids are not distinctly variable in different natural regions, but relationships of quantitative change between them is obvious, which may be a reflection of euglenoid characteristics. Because the euglenoids are unicellular and their growth period may be very short, they can grow and multiply rapidly in very short time as soon as they encounter suitable microhabitats, no matter what the macro-environment is. Many other fresh-water algae show similarities to euglenoids in terms of strategies. Thus it is likely that the distribution and growth of many fresh-water algae may be similar to that of the euglenoids, but further thorough studies are necessary for correct conclusions.

Some of quantitative relationships in the paper can not be analysed owing to insufficient data (e.g. areas and depths of water bodies, vegetation etc.) and difficulties in obtaining relevant data of environmental factors in the field. Although the data of pH recorded in the field are sufficient, a result analysis shows that they bear no relation to euglenoid distribution. It is caused possibly because of the examination by inexact pH paper. The quantitative relationship were analysed only according to recorded factors. However, it is clear that those factors of the macro-environmens have an influence on quantitative changes. The qualitative and quantitative distribution of the fresh-water algae has been discussed by mathematical statistics, which will certainly give more accurate results in keeping with objective reality and will also be indicative for studies in the future. This paper was only a primary attempt, the methods will still have to be replenished, refined and perfected for the future.

Acknowledgements

Grateful acknowledgement is expressed to the National Natural Science Foundation of China for supporting the project. The author wishes to express his sincere thanks to Cao Wen-xuan, Chen Yi-Yu, Chen Jia-You, He Tian-Fu and Wu Zhuo-tian etc. for collecting many algal samples from these regions.

References

Gerrath, J. F. & P. Deey, 1979. Freshwater algae of Sierra Leone, 1. Euglenophyta. Nova Hedwigia 31: 525–365.
Jao Chin-Chih, 1964. Some fresh-water algae from Southern Tibet (in Chinese). Oceanol. Limnol. Sinica 6: 170–189.
Odum, E. P., 1971. Fundamentals of ecology. W. B. Saunders Company, Philadelphia, 574 pp.
Plieou, E. C., 1969. An introduction to mathematical ecology. John Wiley & Sons (Interscience), New York, 286 pp.
Prescott, G. W., 1951. Algae of the Western Great Lakes area. Chanbrook Institute of Science, Michigan, 660 pp.
Shi Zhi-Xin, 1988. A preliminary study on Euglenophyta from Western Hubei, China (in Chinese). Acta Hydrobiol. 12: 169–179.
Shi Zhi-Xin, 1989. Euglenophyta from Suoxiyu nature preserve area, Hunan (in Chinese). In Li Shanhao et al. (ed.), The algal flora and aquatic fauna of Suoxiyu nature preserve area, Huana, China. Science Press, Beijing: 61–67.
Shi Zhi-Xin, 1992a. A preliminary statistical analysis on distribution of Euglenophyta in the Jianghan Plain, Hubei, China (in Chinese). Acta Hydrobiol. 16: 322–330.
Shi Zhi-Xin, 1992b. Euglenophyta of the Xizang Plateau. In Lee Yao-Ying et al. (ed.), The algae of the Xizang Plateau (in Chinese). Science Press, Beijing: 120–192.
Shi Zhi-Xin, 1994a. Study on Euglenophyta of the Wuling Mountain region (in Chinese). In Shi Zhi-Xin et al. (ed.), Compilation of report on survey of algal resources in southwestern China. Science Press, Beijing: 131–282.
Shi Zhi-Xin, 1994b. Study on Euglenophyta of the Hengduasn Mountain region (in Chinese). In Shi Zhi-Xin et al. (ed.), Compilation of report on survey of algal resources in southwestern China. Science Press, Beijing: 245–282.
Sneath, P. H. & R. R. Sokal, 1973. Numerical taxonomy. W. H. Freeman & Company, San Francisco, 573 pp.
Skuja, H., 1939. Beitrag zur Algenflora Lettlands, II. Acta Horti Bot. Univ. Latviensis, 11/12: 41–169.
Yung, Y. K. & P. G. Stokes, 1985. Algae of selected continental and maritime bogs in North America. Can. J. Bot. 64: 1825–1833.

Hydrobiologia **336**: 67–82, 1996.
J. Kristiansen (ed.), Biogeography of Freshwater Algae.

7. Tundra stream macroalgae of North America: composition, distribution and physiological adaptations

Robert G. Sheath[1], Morgan L. Vis[2], Julie A. Hambrook[3] & Kathleen M. Cole[4]
[1] *Dean's office, College of Biological Science, University of Guelph, Guelph, Ontario, N1G 2W1, Canada*
[2] *Department of Environmental and Plant Biology, Porter Hall, Ohio University, Athens, OH 45701, USA*
[3] *Byrd Polar Research Center, Ohio State University, Columbus, OH 43210, USA*
[4] *Department of Botany, University of British Columbia, Vancouver, BC V6T 1Z4, Canada*

Key words: arctic, Chlorophyta, cyanobacteria, North America, stream macroalgae

Abstract

Eighty-three infrageneric taxa of stream macroalgae have been reported from tundra regions of North America, composed of 32 cyanobacteria, 35 Chlorophyta, 10 Chrysophyta and 6 Rhodophyta. There are few if any endemics represented in this flora. The most widespread species are the cyanobacteria *Rivularia minutula, Nostoc commune* and *Tolypothrix tenuis* as well as asexual populations of the chlorophyte genus *Zygnema*. The relative contribution of cyanobacteria compared to that of the Chlorophyta increases from the low to high arctic. Number of species per segment ranges from 0 to 7, with a mean of 2.8, and varies little between the low and high arctic. The percentage of stream bottom covered by macroalgae ranges from 0 to ca. 75%; mean cover values for low and high arctic streams are ca. 12 and 8%, respectively. Tundra macroalgae tend to be more abundant and diverse in less rigorously flowing stream sections. Most species tolerate prolonged freezing by forming resistant vegetative cells with thick walls, plentiful reserves and low molecular weight solutes to lower the freezing point. Many tundra stream macroalgae also produce 'sunscreen' pigments to reduce exposure to damaging radiation in the blue and ultraviolet regions. Nutrients tend to be low and phosphorus is often limiting in these systems. Arctic streams appear to differ from those of Antarctica in having potential grazers of macroalgae, such as the chironomid *Diamesa*, the mayfly *Baetis* and the caddisfly *Brachycentrus*.

Introduction

Stream macroalgae can be defined as those species occurring in flowing freshwaters and having a mature thallus which is benthic and a discrete structure recognizable with the naked eye (Sheath & Cole, 1992). Morphological forms include mats, colonies, gelatinous and free filaments, tissue-like thalli, tufts and crusts. These forms possess various adaptive features to tolerate flow-related drag (Sheath & Hambrook, 1990). Arctic species are subjected to additional stresses associated with a shortened growing season during which time the photoperiod is altered to an extended light phase, low nutrients, and potential desiccation prior to a prolonged freezing period (Prescott, 1963).

The arctic tundra occurs above the latitudinal treeline and comprises about 20% of North America, about 2.5 million km^2 in Canada, 0.2 million km^2 in Greenland and 0.3 million km^2 in Alaska (Bliss, 1988) (Figure 1). It can be divided into the low and high arctic based on a number of environmental characteristics, including length of growing season (3–4 vs. 1.5–2.5 mo), mean July air and soil temperatures at −10 cm (8–12 and 5–8 vs. 3–6 and 2–5°C) and accumulated degree-days above 0°C (600–1400 vs. 150–600) (Bliss, 1988). The low arctic has a more diverse vascular plant flora with woody forms being more common than the high arctic.

This review will examine the hydrology and other characteristics of tundra streams, composition of the macroalgal flora, and some aspects of physiological

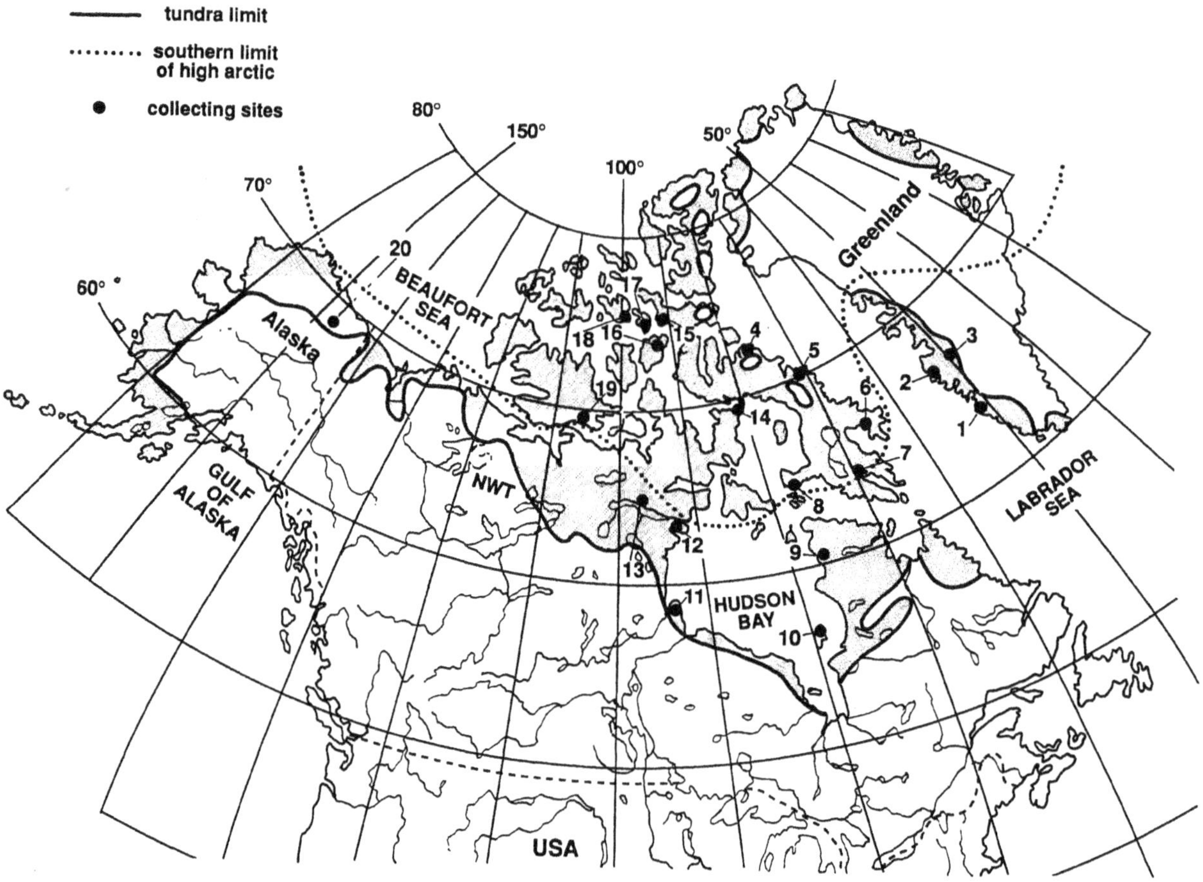

Figure 1. Collecting sites for tundra stream macroalgae of North America, many from the survey of Sheath & Cole (1992). The locations are as follows: 1–3. Greenland. 1. Nuuk, 2. Sisimuit, 3. Kangerlussuaq. 4–8. Baffin Island, N.W.T. 4. Pond Inlet, 5. Clyde River, 6. Pangnirtung, 7. Iqaluit, 8. Cape Dorset. 9. Northern Quebec – Povungnituk. 10. Belcher Islands, N.W.T. – Sanikiluaq. 11. Manitoba – Churchill. 12–13. Keewatin, N.W.T. 12. Rankin Inlet, 13. Baker Lake. 14. Igloolik, N.W.T. 15–18. Queen Elizabeth Islands, N.W.T. 15. Western Devon Island, 16. Cunningham Inlet, Somerset Island, 17. Resolute, Cornwallis Island, 18. Polar Bear Pass, Bathurst Island. 19. Cambridge Bay, Victoria Island, N.W.T. 20. Toolik region, Alaska.

adaptations of arctic species to this harsh environment. The trends will be compared to those of other biomes in North America and stream systems of Antarctica.

Hydrology

There are two basic groups of tundra riverine systems: those that flow solely in the arctic and those that have their origins in the boreal forest biome (Woo, 1991). The majority of stream systems containing macroalgae are of the former type since latter systems are typically quite large and become too deep and turbid to support benthic autotrophs. Most tundra streams have negligible to no flow during the winter, followed by the spring freshet associated with snowmelt (Craig & McCart, 1975; Rydén, 1977; Woo, 1991). The peak flow may occur in late May to early June in low arctic lowland sites while it often takes place from late June to mid July in many parts of the high arctic (Figure 2). The peak typically extends only two to three weeks but accounts for about 80% of the yearly runoff (Rydén, 1981). This is due to the combined effect of high intensities of solar radiation, a decrease in albedo of snow cover and percolation being greatly reduced by permafrost.

In small streams that are strictly nival (snowmelt dominated) in their hydrology, the channel typically becomes dry after the spring flow except for occasional high rainfalls (Woo, 1991). Glacier-fed streams can have a brief low discharge rate after snowmelt due to cold temperatures and then a second increase

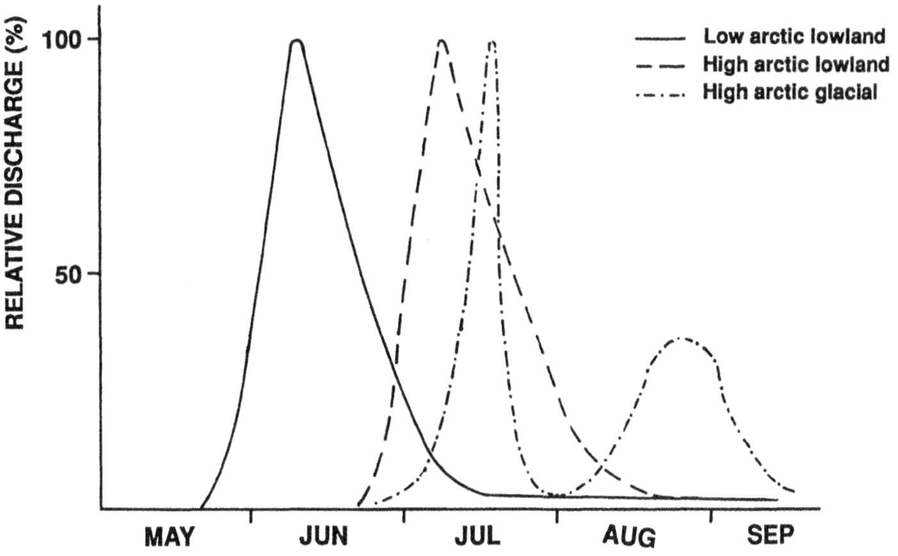

Figure 2. Streamflow regimes of three types of tundra streams based on studies of Craig & McCart (1975), Rydén (1977) and Woo (1991). These regimes typically show additional smaller peaks corresponding to summer rainfall. The base flow after the spring freshet may drop to zero depending on the hydrology of the basin.

from glacier melt (Figure 2). Streams flowing through wetlands have relatively low runoff ratios due to poor drainage and water retention in ponds and peaty soils (Woo, 1988, 1991). Hence, streamflow response to rainfall is attenuated. Lakes also attenuate flow due to water storage. Coastal plain arctic streams often form a series of pools along their course ('beaded streams') and they may have a number of parallel and intertwining channels ('braided streams') (Harper, 1981). Upland-derived streams tend to have relatively large spring flows which may extend longer than those that are strictly lowland (Figure 2) (Craig & McCart, 1975; Rydén, 1977).

In comparison, antarctic stream discharge appears to be less seasonally predictable with high levels of variability at all time levels (e.g. Vincent et al., 1993b).

Other characteristics of tundra streams

Tundra streams containing macroalgae vary considerably in physical and chemical characteristics (Table 1). However, they are typically 1st- to 4th-order reaches which are less than 100 cm in depth and with moderate current velocities. Nival streams tend to be relatively turbid during the snowmelt peak whereas glacial streams may be turbid throughout much of the discharge period (Milner & Petts, 1994). Tundra streams tend to be low in nutrients (Moore, 1979; Peterson et al., 1985; Murray & Svoboda, 1989; Peterson et al., 1985, 1993; Hamilton & Edlund, 1994) and levels of nitrogen and phosphorus do not appear to be related to the nature of bedrock drained (Murray & Svoboda, 1989). Phosphorus largely limits algal production in streams examined on the north slope of Alaska and biological processes at all trophic levels are affected by its addition (Peterson et al., 1985, 1993). Phosphorus limitation in other areas is confirmed by high N:P ratios (e.g. Murray & Svoboda, 1989). However, when tundra streams are phosphorus enriched, nitrogen can become limiting (Hullar & Vestal, 1989). While antarctic streams also tend to have low nutrients, concentrations of nitrogen and phosphorus range considerably with season and time of day associated with freeze-thaw cycles (e.g. Vincent et al., 1993b).

Lowland streams in the low arctic can have high densities of invertebrates ($\bar{x}$ = 1000 – 22000 individuals m^{-2}) compared to mountain streams at the same latitudes ($\bar{x}$ = 293 individuals m^{-2}) (Craig & McCart, 1975). Common invertebrates in these streams include larvae of the chironomid families Diamesinae and Orthocladinae, the mayfly *Baetis laponicus*, the caddisfly *Brachycentrus americanus* and the black flies *Stegopterna ornata* and *Prosimulium martini* (Slack et al., 1979; Hershey & Hiltner, 1988; Peterson et al., 1993). *Baetis* and *Brachycentrus* potentially graze

Table 1. Physical and chemical characteristics of tundra streams sampled in North America (see Sheath & Cole, 1992 for details)

Location[1]	Maximum width (m)	Maximum depth (cm)	Current velocity (cm s^{-1})	Temperature (°C)	pH	Specific conductance (μS cm^{-1})
1	1.2–3.3	14–35	0–17	7–17	7.1–8.3	20–150
2	0.1–15.0	10– > 100	0–75	4–15	8.0–8.3	10–223
3	0.7–4.0	16–67	1–80	9–13	8.1–8.3	20–90
4	0.9–7.0	15–23	1–59	2–10	–	10–70
5	1.5–8.0	16–160	14–105	2–8	6.6–6.9	0–40
6	0.7–3.0	11–34	8–74	1–3	5.9–6.3	0–10
7	1.0–2.3	15– > 100	1–109	0–9	6.2–7.8	0–20
8	0.8–4.0	12–50	0–72	2–7	6.6–7.8	0–90
9	1.3–5.0	19–60	0–58	9–12	6.6–7.5	20–350
10	0.9–8.3	15–30	8–96	4–11	6.1–8.1	60–440
11	3.5–20.0	46– > 100	10–58	12–17	7.2–8.0	290–630
12	1.3–16.0	15–65	0–40	5–14	7.7–8.3	60–1050
13	1.8–20.0	20–45	15–58	4–16	7.4–8.3	10–250
14	1.9–6.0	9–50	12–86	0–5	8.1–8.9	80–230
15	2.2–20.0	5– > 100	10–127	6–10	8.1–8.9	0–440
16	1.0–20.0	10– > 100	30–56	5–11	8.2–8.3	120–170
17	1.5–12.0	20–56	14–49	3–8	8.0–8.6	110–340
18	5.0–20.0	30– > 100	4–16	6–10	8.4–8.8	120–140
19	2.4–20.0	17–62	34–61	4–10	8.1–8.6	58–500
20	1.5–30.0	11– > 100	0–111	5–15	6.7–8.4	10–300

[1] See Figure 1 for specifics.

macroalgae (Hambrook & Sheath, 1987) and their growth rates increase in response to phosphorus addition (Peterson et al., 1993). Lowland streams in the low arctic are also used as spawning and rearing areas by the grayling (*Thymallus arcticus*) (Craig & McCart, 1975; Peterson et al., 1993). Glacier streams support relatively few benthic invertebrates near the glacier snout and these stretches are often dominated by the chironomid genus *Diamesa* which has been found to graze filamentous algae, such as the cyanobacterium *Phormidium* (Milner & Petts, 1994). Downstream the invertebrate fauna becomes more diverse as channels become more stable. Streams flowing from lakes tend to have high populations of filter feeders (Harper, 1981). Much less is published on the animal communities of high arctic streams. Tributaries of a small lake on Cornwallis Island, N.W.T. are dominated by Chironomidae of *Diamesa* and Orthocladinae (Stockner & Hynes, 1976).

Composition of stream macroalgae

Table 2 is a compilation of macroalgal species from our survey of 150 stream segments from 20 locations (Figure 1) (partly given in Sneath & Cole, 1992) plus ten previous reports from other researchers who clearly indicated that they collected in arctic stream habitats from North America. There have been 83 infrageneric taxa reported, composed of 32 cyanobacteria (39%), 35 Chlorophyta (42%), 10 Chrysophyta (12%) and 6 Rhodophyta (7%). However, 12 of these taxa are not identified to the species level. In addition, some of the identifications may need to be re-examined, such as those of *Prasiola* species (Hamilton & Edlund, 1994). The diversity of this flora is similar to other widely sampled biomes in North America, which have 84 to 100 infrageneric taxa (Sheath & Cole, 1992). Thus, the typical increase in species numbers from the arctic to the tropics that is evident among seaweeds (e.g. Alvarez et al., 1988) is not true of stream macroalgae. However, other riverine organisms also show relatively low difference in diversity with latitude (Patrick, 1988). The majority of the tundra lotic taxa are either

Table 2. Distribution of tundra stream macroalgae in North America

Taxon	Location[1]
Cyanophyta	
Chroococcales	
Aphanothece pallida (Kütz.) Rabh.	11 (L)
Chlorogloea microcystoides Geit.	15, 17 (H)
Gloeocapsa sanguinea (C.Ag.) Kütz.	14, 15 (H)
Tychonema bornetii Anag. & Komár.	18 (H)
Oscillatoriales	
Leptolyngbya tenuis (Gom.) Anag. & Komár.	1, 3, 7 (H,L)
Lyngbya aestuarii (Mert.) Lieb.	17 (H)
Microcoleus sociatus W. & G. S. West	18 (H)
Phormidium autumnale (C.Ag.) Gom.	I, VIII (L)
P. corium (C.Ag.) Gom.	3 (L)
P. retzii (C.Ag.) Gom.	1, 17 (H,L)
P. subfuscum Kütz.	3, 4, 7, 8, 12, 13, 17, 18, 20 (H,L)
P. tenue Gom.	I (L)
P. valderiae (Delp.) Geit.	VIII (L)
P. sp.	VII (L)
Pseudophormidium tenue (Thur. ex Gom.) Anag. & Komár.	14 (H)
Schizothrix calcicola (C.Ag.) Gom.	15, 17, V (H)
S. fuscescens Kütz.	1, 7, 8, 20, VIII (H,L)
S. mexicana Gom.	V (H)
S. muelleri Näg.	1, 19, VI (H,L)
Nostocales	
Nostoc commune Vauch.	1, 2, 4, 7, 8, 9, 10, 12, 13, 14, 15, 17, 18, 19, V (H,L)
N. pruniforme C.Ag.	11, 20 (L)
N. verrucosum Vauch.	14 (H)
N. sp.	I (L)
Rivularia haematites (D.C.) C. Ag.	4, 17, 20, VIII (H,L)
R. minutula (Kütz.) Born. & Flah.	1, 3, 6, 7, 8, 9, 10, 11, 12, 13, 14, 15, 17, 18, 19, 20 (H,L)
Scytonema myochrous (Dillw.) C. Ag.	17 (H)
S. tolypothricoides Kütz.	1, 7, 8, 12, 14 (H,L)
Tolypothrix distorta Kütz.	V, IX (H,L)
T. limbata Thuret	VIII (L)
T. tenuis Kütz. emend. J. Schmidt	1, 2, 8, 9, 10, 12, 13, 15, 17, 18, 19, 20 (H,L)
Stigonematales	
Stigonema mamillosum (Lyng.) C. Ag.	2, 7, 8, 14, 20, VI, VIII, IX (H,L)
S. ocellatum (Dillw.) Thuret	8 (H)
Chlorophyta	
Tetrasporales	
Tetraspora sp.	I (L)
Ulotrichales	
Cylindrocapsa conferta W. West	6 (H)
Hormidiopsis ellipsoideum Presc.	7 (H)
Microspora elegans Hansg.	1, 2, 12, 20 (L)
M. loefgrenii (Nordst.) Lagerh.	6, 7, 11 (H,L)

Table 2. (continued)

Taxon	Location[1]
M. tumidula Hazen	2, 3, 4, 6, 7, 9, 12, 14, 20, V (H,L)
M. sp.	I, III (L)
Ulothrix aequalis Kütz.	IX (L)
U. tenerrima Kütz.	17 (H)
U. tenuissima Kütz.	13, 20 (L)
U. variabilis Kütz.	11, 16, 17 (H,L)
U. zonata (Weber & Mohr) Kütz.	17, VIII (H,L)
Chaetophorales	
Chaetophora elegans (Roth) C. Ag.	20 (L)
C. incrassata (Huds.) Hazen	9, 10 (L)
Draparnaldia glomerata (Vauch.) C. Ag.	20 (L)
D. mutabilis (Roth) Cedergren	11 (L)
D. sp.	III (L)
Stigeoclonium flagelliferum Kütz.	20 (L)
S. nanum Kütz.	13 (L)
S. subsecundum Kütz.	7 (H)
Dichotomosiphonales	
Dichtomosiphon tuberosus (Braun) Ernst	14 (H)
Oedogoniales	
Bulbochaete sp.	20 (L)
Oedogonium spp.	2, 7, 8, 18, 20, IX (H,L)
Cladophorales	
Rhizoclonium hieroglyphicum (C.Ag.) Kütz.	10, 19 (H,L)
R. hookeri Kütz.	11 (L)
Prasiolales	
Prasiola fluviatilis (Sommerf.) Aresch.	V, X (H)
P. mexicana J. Ag.	2, 12 (L)
Zygnematales	
Cylindiocystis brebissonii var. *brebissonii* Croads.	1, 14 (H,L)
Mougeotia spp.	10, VI, IX (H,L)
Mougeotiopsis calospora Palla	4 (H)
Spirogyra groenlandica Rosenvinge	2, 4, I (H,L)
S. inflata (Vauch.) Kütz.	III (L)
S. spp.	1, 18, 20, I (H,L)
Zygnema insigne (Hassall) Kütz.	12 (L)
Z. spp.	1, 2, 3, 4, 5, 6, 7, 8, 9, 10, 12, 13, 14, 15, 17, 18, 19, 20, I, III, VII, IX (H,L)
Chrysophyta	
Chromulinales	
Hydrurus foetidus (Vill.) Trev.	20, I, IX (L)
Tribonematales	
Tribonema affine (G. S. West) G. S. West	V (H)
T. utriculosum (Kütz.) Hazen	14 (H)
T. viride Pasch.	12 (L)
T. sp.	I (L)
Fragilariales	
Tabellaria flocculosa (Roth) Kütz.	1, 2, 3, 9 (L)

Table 2. (continued)

Taxon	Location[1]
Meridion circulare (Grev.) C. Ag.	16, VII (H)
Naviculales	
Cymbella cistula (Ehr.) Kirchn.	17, 18 (H)
C. prostrara (Berk.) Cl.	11 (L)
Didymosphenia geminata (Lyngb.) M. Schmidt	20 (L)
Rhodophyta	
Acrochaetiales	
Audouinella hermannii (Roth) Duby	20 (L)
Batrachospermales	
Batrachospermum gelatinosum (L.) DC.	4, 6, 7, 8, 9, 11, 20 (H,L)
B. skujae Geitler	1 (L)
B. spermatoinvolucrum Vis & Sheath	1, 3, 4, 5, 6, 7, 13, 20 (H,L)
B. sp.	1, 3, 6, 10, 11, 19, 20, I (H,L)
Lemanea borealis Atk.	3, 20 (L)

[1] 1–20 stream locations from our survey (see Figure 1); I – Bachmann (1921) – West Greenland; II – Petersen (1924) – north coast of Greenland; III – Yamagishi (1967) – Beaufort Sea coast of Alaska; IV – Kobayashi (1967) – Beaufort Sea coast of Alaska; V – Croasdale (1973) – Lake Hazen, Ellesmere Island, N.W.T.; VI – Moore (1974) – southern Baffin Island, N.W.T.; VII – Stockner & Hynes (1976) – Cornwallis Island, N.W.T.; VIII – Moore (1979) – Beaufort Sea drainage area of N.W.T.; IX – Johansson (1980) – Narssaq area, south Greenland; X – Hamilton & Edlund (1994) – northern Ellesmere Island, N.W.T.; H = high arctic; L = low arctic.

mat-forming filaments (59%) or colonial forms (22%). Antarctic streams also are predominated by cyanobacteria and green algae, many of which are in mats or colonies (Broady, 1982; Vincent & Howard-Williams, 1986; Hawes, 1989; Davey & Clarke, 1992; Vincent et al., 1993a). In antarctic streams, green algae have their greatest abundance in the maritime zone while cyanobacteria typically dominate streams on the Antarctic continent (Vincent et al., 1993b). It is noteworthy that Rhodophyta have not been collected in antarctic streams, whereas they are common in the arctic.

The most widespread species in our survey of 20 arctic locations were the cyanobacteria: *Rivularia minutula* (15), *Nostoc commune* (13) and *Tolypothrix tenuis* (12) (Table 2). In addition, asexual populations of the chlorophyte genus *Zygnema* were present in all but two locations. Some of the widely distributed species are presented in Figures 3–17. *Nostoc commune* and *Zygnema* spp. are also common in antarctic streams (Hawes, 1989; Vincent et al., 1993a).

Of the 83 infrageneric taxa in tundra streams of North America, 34 were found only in the low arctic, 22 only in the high arctic and 27 were collected in both regions (Table 2). The proportion of cyanobacteria and Chlorophyta is largely reversed between the low arctic (34 and 46%, respectively) and high arctic (48 and 38%, respectively). At least 21% of the high arctic taxa are capable of nitrogen fixation, based on the presence of heterocysts, but additional species of cyanobacteria probably undergo this process (e.g. Pandey et al., 1992).

The number of species per stream segment (minimum area examined 100 m^2) in our survey of tundra streams varies from zero to seven with a mean of 2.8 (Figure 18). The high arctic streams tend to have a slightly lower number (most frequently two per segment, $\bar{x} = 2.7$) than those of the low arctic (most frequently three per segment, $\bar{x} = 2.9$). These numbers compare to mean numbers of 2.9–3.6 per stream segment in other biomes in North America (Sheath & Cole, 1992).

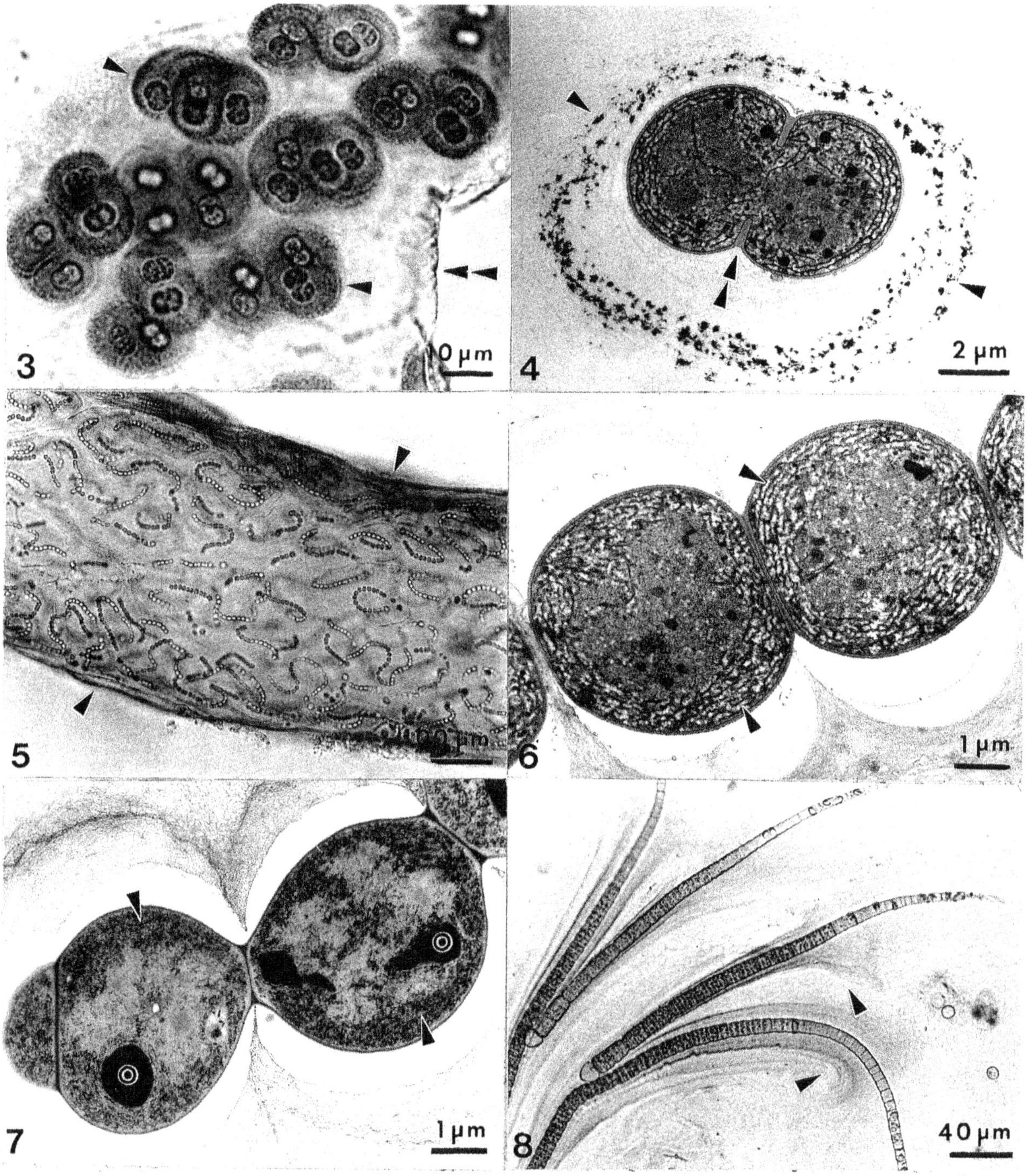

Figures 3–8. Common tundra stream cyanobacteria: 3–4 *Gloeocapsa sanguinea* collected at snowmelt from Igloolik, N.W.T. (81): 3. Light micrograph showing crimson-colored sheath surrounding small packets of cells (single arrowheads) all enclosed in a common colorless colonial sheath (double arrowhead). 4. Electron micrograph showing granulate, layered sheath (single arrowheads) surrounding a cell undergoing division (double arrowhead). The peripheral thylakoids surround the central cytoplasm which has several, small electron dense globules. 3–7. *Nostoc commune* from various locations and conditions: 5. Light micrograph of a cross-section of a colony showing brown-colored, leathery sheath (arrowheads) enclosing many filaments (N.W.T. 83 – Igloolik at snowmelt). 6. Electron micrograph of two cells fixed at the time of snowmelt (N.W.T. 81, Igloolik) with typical cell structure, including concentric, peripheral thylakoids (arrowheads) and small, electron-dense globules. 7. Electron micrograph of two cells from a highly desiccated colony collected over one month after the spring freshet (N.W.T. 97, Bathurst Island). The peripheral thylakoids are not as obvious (arrowheads), the central cytoplasm is less granular and large osmiophilic granules (O) are present in each cell. 8. *Rivularia minutula* collected at snowmelt at Igloolik, N.W.T. (84). Light micrograph showing tapering filaments surrounded by lamellated, brownish, flared sheath (arrowheads).

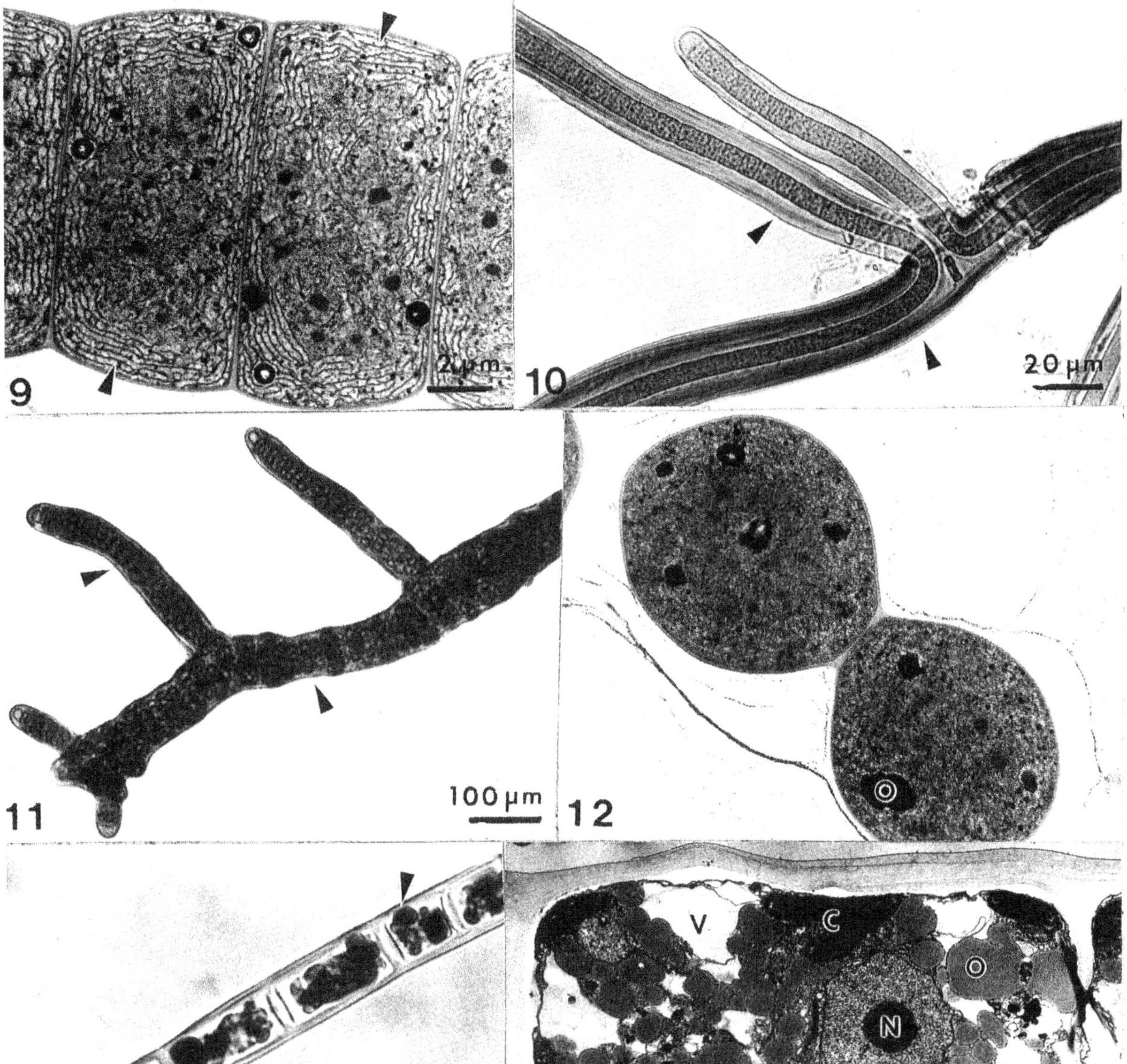

Figures 9–14. Common cyanobacteria (9–12) and Tribophyceae (13–14) from tundra streams: 9. *Rivularia minutula* collected at snowmelt from Igloolik, N.W.T. (84). Electron micrograph of two cells with typical concentric, peripheral thylakoids (arrowheads) and numerous, small, osmiophilic globules. 10. *Scytonema tolypothricoides* collected at snowmelt from Igloolik, N.W.T. (85). Light micrograph showing the lamellated sheath which is yellow-to brown-colored (arrowheads). 11–12. *Stigonema mamillosum* collected from different locations: 11. Light micrograph of multiseriate filament with brown-colored sheath collected after spring freshet (GLD 10). 12. Electron micrograph of two cells from a filament collected at snowmelt (N.W.T. 86, Igloolik) with typical cell structure and large, osmiophilic globules in most cells (O). 13–14. *Tribonema utriculosum* collected and fixed while still frozen at the time of snowmelt (N.W.T. 83, Igloolik): 13. Light micrograph of a filament stained with sudan Black B, showing positively-stained lipid bodies (arrowheads). 14. Electron micrograph of a longitudinal section through a cell with many osmiophilic globules (o), small vacuoles (v), a central nucleus (n), and peripheral chloroplasts (c).

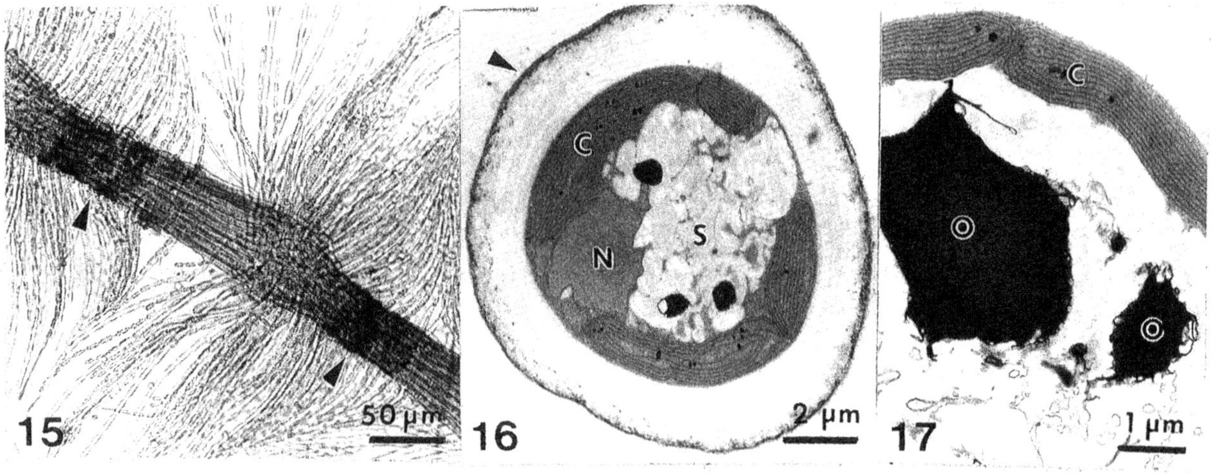

Figures 15–17. Batrachospermum gelatinosum collected from different locations and conditions: 15. Light micrograph of a filament collected after spring freshet (AK 58, Toolik) showing orange-brown-pigmented cortication surrounding main axis (arrowheads). 16–17. Electron micrographs of cells from a filament collected at snowmelt (N.W.T. 90, Iqaluit). 16. Fascicle cell transverse section, showing very thick cell wall (arrowhead), peripheral chloroplasts (c), nucleus (n), plentiful starch (s) and some osmiophilic globules. 17. Periphery of fascicle cell containing chloroplasts (c) with intact thylakoids and large osmiophilic globules (o).

There appear to be relatively few endemics in the tundra freshwater algal flora, either in lotic or lentic systems (e.g. Sheath & Steinman, 1982; Sheath & Cole, 1992). However, Douglas & Smol (1993) note that in parts of the arctic, the taxa are poorly described and some may represent new species. In terms of stream macroalgae, from our survey of 1000 stream segments throughout North America, it was found there were 13 out of 259 infrageneric taxa identified that were collected only in the tundra (Sheath & Cole, 1992). However, most of these taxa have been reported for different biomes by other researchers.

The diatom flora has changed significantly in some high arctic ponds on Cape Herschel, Ellesmere Island, N.W.T. beginning in the 19th century in response to anthropogenic impacts on the environment (Douglas et al., 1994). Douglas et al. (1994) propose that it may already be too late to catalogue natural assemblages in the arctic because of these changes. Such an analysis is not possible for stream macroalgae since most taxa are soft-bodied and not preserved in sediments.

Effects of environmental factors and physiological adaptations

As in other biomes, the distribution of tundra stream macroalgae is quite variable within and among streams. The percentage of stream bottom covered by macroalgae varies from 0 to ca. 75% with the majority of tundra streams having 1–10% (Figure 19). The mean cover values for high and low arctic streams are ca. 8 and 12%, respectively, which compare to a range of 12–21% for other biomes (Sheath & Cole, 1992). Even within patches of macroalgae, the biomass and net productivity vary considerably in the same stream (Table 3). In addition, the contribution to total net productivity by macroalgae can range from ca. 4 to 82%.

Table 3. Comparison of macroalgal and periphytic biomass and productivity in a stream near Toolik, Alaska[1]

Date (1990)	Macroalgal species[2]	Macroalgal biomass (g fw m^{-2})	Net productivity (mg O_2 h^{-1} m^{-2} ± SE) macroalgae	periphyton
June 22	Rm, Bg	558	35.8 ± 10.1	8.0 ± 2.6
	Rm	68	0.7 ± 0.4	10.6 ± 5.5
July 10	Rm, Bg	897	60.9 ± 4.7	16.8 ± 3.1
July 23	Rm, Zs, Bg	217	0.6 ± 0.0	14.3 ± 4.7
	Rm, Zs, Bg	398	24.5 ± 2.3	14.1 ± 4.9

[1] Measurements taken from patches of macroalgae in Okstrukuyik Creek. See Bowden et al. (1992) for details of fertilization and productivity measurement.
[2] Rm = *Rivularia minutula*, Bg = *Batrachospermum gelatinosum*, Zs = *Zygnema* sp.

Streams are often considered to be physically controlled environments in which flooding, droughts and rapid temperature changes can act as major sources of density-independent mortality (Hart, 1983). This is particularly true of tundra streams. In a mapping study

Table 4. Effect of stream slope on macroalgal abundance in a high arctic stream at Resolute, N.W.T. (NWT 104)

Distance from source[1] (m)	Slope (cm/m)	Channel width (m)	Mean current velocity (cm s^{-1})	Stream macroalgae[2]			
				Mean Number of entities	Frequency (% of quadrats)	Mean cover (%)	Species number
1. Gradual slope region:							
0	9	24.5	10	440	100	5	3
60	9	14.0	15	330	100	1	2
180	11	14.5	13	540	100	3	2
420	11	17.5	14	300	100	0.9	3
2. Steep slope region:							
540	18	5.0	41	150	100	0.2	2
600	18	4.0	48	20	25	0.1	1
720	18	7.0	37	0	0	0	0
960	22	1.2	64	0	0	0	0

[1] Mapped in m^2 quadrats across the stream segment at these distances from the outflow of a tundra pond.
[2] In order of frequency: *Rivularia minutula*, *Zygnema* sp., *Scytonema myochrous* and *Nostoc commune*; entities are either discernable colonies or mats.

of a high arctic stream in Iqaluit, N.W.T., we found that the domninant species, the rhodophyte *Batrachospermum gelatinosum*, was essentially an avoider of high current velocities by occurring only along the banks where the substratum is stable and the flow is moderate (Figure 20). Similarly, in a relatively large tundra stream in Resolute, N.W.T., the two predominant macroalgae, the cyanobacterium *Rivularia minutula* and chlorophyte *Zygnema* sp., were present in high numbers in all quadrats which had a gently slope and low current velocity ($\bar{x} = 0 - 14$ cm s^{-1}). However, they were virtually absent when the slope increased, the channel narrowed and the current velocity was relatively high ($\bar{x} = 37 - 64$ cm s^{-1}) (Table 4). It is likely that the spate which occurs during spring snowmelt in tundra streams partially creates these distribution patterns by removing the larger macroalgae in high flow parts of the channel more readily than microscopic forms (e.g. Peterson & Stevenson, 1992). Substrata movement due to strong currents will also reduce macroalgal density (Hambrook & Sheath, 1991). In antarctic streams expanded sheets of *Prasiola* (Broady, 1989) and mats of *Zygnema* (Hawes, 1989) are also restricted to less rigorous flowing regions.

The fact that so many taxa tolerate an eight-to ten-month period of freezing is somewhat surprising because few tundra species undergo sexual reproduction to form a resistant spore for over-wintering (Prescott, 1963; Sheath & Steinman, 1982). Hence, vegetative cells must be physiologically adapted to withstand prolonged freezing. This trend is also true of antarctic stream macroalgae (Vincent & Howard-Williams, 1986; Davey, 1989; Hawes, 1990). The common cyanobacteria species have a thick extracellular mucilaginous layer and an accumulation of reserve products within the cytoplasm, but otherwise the over-wintering cell is typical (Figures 3–12). The reserve products are osmiophilic, indicating that they are either cyanophycin granules or polyphosphate bodies (Jensen, 1985). While the chrysophyte *Tribonema utriculosum* is still encased in ice at the end of winter it has thick walls, plentiful lipid globules, numerous small vacuoles, but the nuclei and chloroplasts are in good shape (Figures 13–14). At snowmelt the rhodophyte *Batrachospermum gelatinosum* also has very thick walls (up to 2 μm thick) and the major reserve product is floridean starch (Figure 16). There are some osmiophilic globules in these cells which appear to be lipoidal (Figure 17) based on previous studies of this genus (Sheath & Whittick, 1995). The common condition among all of these algae in the over-wintering state, a thick extracellular covering and plentiful reserve products, is typical of many algae under prolonged stress (Morison & Sheath, 1985). The ability of these resistant vegetative cells to survive the long winter is due to prevention of intracellular ice crystallization (Hawes, 1990). Crystallization of water inside the cell may occur by internal nucleation or by penetra-

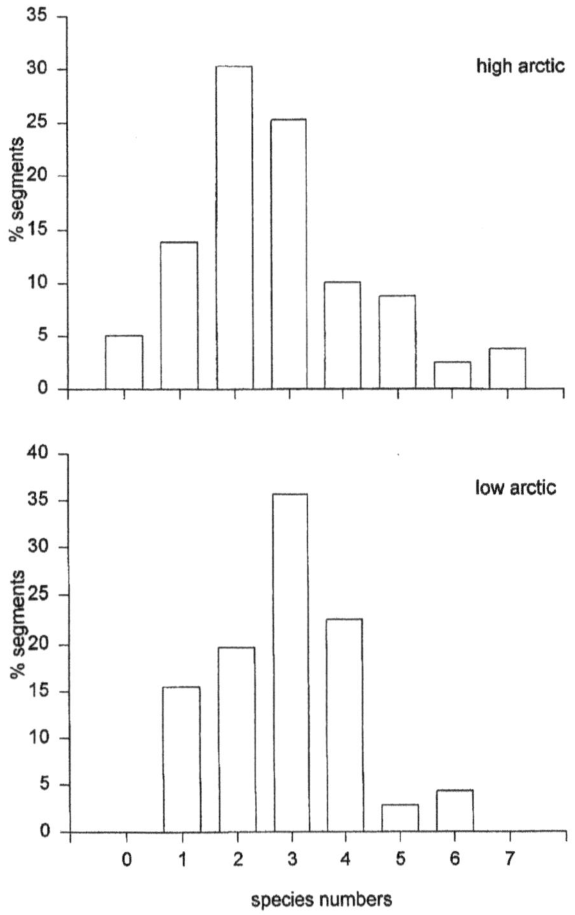

Figure 18. Diversity of stream macroalgae in high arctic (n = 79) and low arctic stream segments (n = 71) from our survey of 150 segments from 20 locations (see Figure 1).

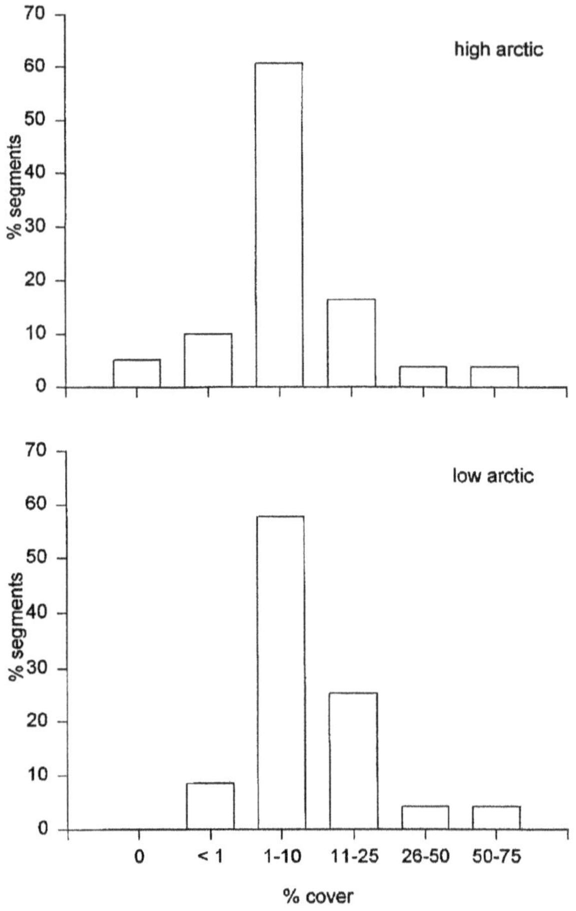

Figure 19. Frequency of cover values in each stream segment from the high arctic (n = 79) and low arctic (n = 71).

tion into the cell by external ice crystals (Guy, 1990). Many algae produce low-molecular solutes to lower the freezing point, thereby functioning as cryoprotectants. In addition, restructuring of lipid components in membranes is likely. These phenomena are in turn affected by rate of freezing, number of freeze-thaw cycles, prevailing light conditions and amount of available free water (Davey, 1989; Hawes, 1990). The chrysophyte *Tribonema viride*, collected at Igloolik, N.W.T., exhibits a slight but significantly greater frequency of cell viability after freezing if the cells are previously partially desiccated (Table 5). Cyanobacterial mats in antarctic meltwater streams retain a high metabolic capacity while they are dry and frozen and they respond rapidly to rehydration (Vincent & Howard-Williams, 1986). Nonetheless, severe desiccation causes some cellular disruption in *Nostoc commune* collected on Bathurst Island, N.W.T. (Figure 7).

Pools may act as refugia for macroalgae in 'beaded' arctic streams because they retain their water much longer after snowmelt than shallow riverine channels. In our survey of 150 tundra streams, pools constituted the most common habitat of certain taxa, such as species of *Microspora* and *Batrachospermum* (Sheath et al., unpubl.). It may be that a more gradual desiccation allows sufficient time for these taxa to produce cryoprotectant compounds that can be used for tolerance of both drying and freezing (Morison & Sheath, 1985). Distribution of filamentous green algae in streams of Signy Island, Antarctica is also related to depth (Hawes, 1989).

Photoperiod affects tundra stream macroalgae in a variety of ways. The paucity of resting spores among

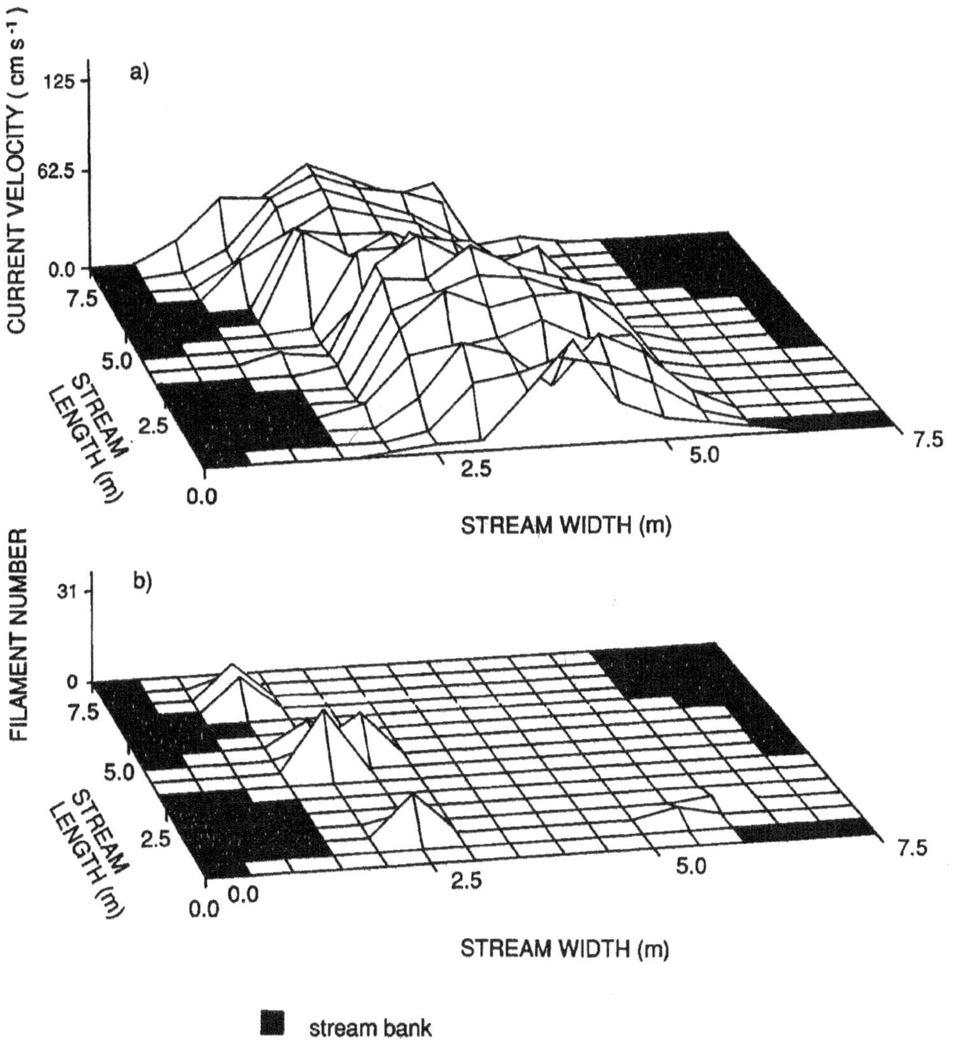

Figure 20. Distribution of the dominant macroalga, *Batrachospermum gelatinosum*, in relation to current velocity within a mid-sized stream segment in Iqaluit, N.W.T. The stream was mapped in 0.25 m^2 quadrats in which current velocity and filament number were determined on July 2–3, 1988 (near the end of spring freshet).

these species is partly due to a lack of diurnal light-dark alternation during the growing season which prevents those taxa requiring daylength induction from undergoing sporulation or gametogenesis (Prescott, 1963; Sheath & Steinman, 1982). This situation contrasts with ecotypes of arctic vascular plants which have physiological responses to daylengths greater than 20 hours (Vankat, 1979).

The long photoperiod also results in potentially high exposure to damaging radiation in the blue and ultraviolet regions (Stonehouse, 1989). This is compounded by a recent rise in ultraviolet-B flux in parts of the North American arctic due to ozone depletion (Vincent & Roy, 1993). A number of tundra macroalgae produce 'sunscreen' pigments to filter out these wavelengths. The cyanobacteria frequently have dark-colored sheaths, such as the crimson one of *Gloeocapsa sanguinea* (Figures 3–4) or the yellowish to brown-colored ones of *Rivularia minutula* (Figure 8), *Scytonema tolypothricoides* (Figure 10), and *Stigonema mamillosum* (Figure 11). The reddish color has been ascribed to gloeocapsin and the yellow-brown color to scytonemin (Whitton, 1992). Antarctic cyanobacterial mats also are enriched with scytonemin (Vincent et al., 1991a). Scytonemin is not unique to polar cyanobacteria but is common to more than 30 species

Table 5. Viability counts of *Tribonema viride* (Tribophyceae, Chrysophyta) collected from Igloolik, N.W.T. and subjected to different moisture and temperature treatments in culture (number of live cells per 100 cells)

Temperature treatment	Moisture treatment	
	BBM (control)	BBM (desiccated)
10°C (control)	97.5 ± 1.3[1]	88.0 ± 1.6[2]
−15°	89.0 ± 1.9[3]	98.3 ± 2.4[1,2]

[1] Values are means of four petri dishes ± SD. Viability determined using 0.05% Evan's blue (Crippen & Perrier, 1974). Moisture regimes maintained for 28 days prior to subjecting them to the temperature conditions for an additional 28 days.
[2] Denotes a significant difference from moisture control at $P < 0.05$.
[3] Denotes a significant difference from temperature control at $P < 0.05$.

Table 6. Frequency of orange-brown cortication surrounding main axis in *Batrachospermum gelatinosum* (Batrachospermales, Rhodophyta) from different biomes of North America

Biome	Frequency of orange-brown cortication (%)	Sample size
tundra	67	15
boreal forest	50	14
western coniferous forest	29	7
hemlock hardwood forest	21	14
deciduous forest	17	12
coastal plain	20	7

which have been exposed to intense solar radiation (Garcia-Pichel & Castenholz, 1991). The pigment is lipid soluble and has a prominent absorption maximum in the near ultraviolet range of the spectrum. Tundra populations of the red alga *Batrachospermum gelatinosum* often have orange-brown pigmented cortication surrounding the main axis which may be due to an increase in carotenoid pigments (Figure 15). The axial cells, which become surrounded by the colored cortical filaments, are typically colorless and hence they do not have carotenoids to act as an internal 'sunscreen' for protection of nucleic acids. Colored cortication is not unique to tundra populations of *Batrachospermum gelatinosum* but its incidence is considerably higher in the arctic compared to that for other biomes in which this taxon occurs (67 vs 17–50%, respectively) (Table 6). A second advantage to having dark pigmentation in the outer layers of thalli of many tundra stream macroalgae is a slight but possibly important rise in thallus temperature (Hebert & McWalter, 1983). This increase in temperature may allow them to complete the life history within the cold waters and short growing season. Arctic species probably employ other mechanisms for UV-protection noted by Vincent & Roy (1993), including production of mycosporine-like amino acids for additional 'sunscreening', quenching of toxic products of UV and oxygen with carotenoid pigments and superoxide dismutase, and efficient DNA repair mechanisms.

The typically low concentrations of nutrients present during the brief growing season are an additional source of potential limitation to the production of tundra stream macroalgae. For example, when the Kuparuk River on the north slope of Alaska is fertilized with phosphate and ammonium, epilithic chlorophyll *a* and photosynthetic rates are greatly increased, particularly in pools (Bowden et al., 1992). Much of this production is due to periphytic diatom communities (Miller et al., 1992) but macroalgae are also enhanced (Hambrook, unpubl.). When these tundra streams are fertilized for a long period (seven years) certain mosses, such as *Fontinalis neomexicana* and *Hygrohypnum* spp. become abundant (Bowden et al., 1994), and compete for suitable substrata with macroalgae. In contrast, nutrient enrichment experiments on glacier-fed ephemeral streams of southern Victoria Land, Antarctica, do not support the concept of N or P limitation to pigment content or photosynthetic rates of cyanobacterial mats (Howard-Williams & Vincent, 1989).

Little has been published on grazing of tundra stream macroalgae or other trophic dynamics involving these taxa. Hawes (1989) noted that herbivory of green algal mats in streams on Signy Island, Antarctica was nil due to the absence of invertebrate grazers. Rather, sloughing was the major mechanism of biomass reduction. Thus, the presence of potential grazers in arctic streams, as noted previously, is one key difference from those of Antarctica.

Conclusions

The tundra stream macroalgal flora at North America constitutes a northern extension of a portion of the temperate community rather than an assemblage of unique species. These taxa appear to be well adapted to this harsh environment by producing resistant vegetative cells for over-wintering, 'sunscreen' pigments to reduce high energy ultra-violet and blue radiation and the ability to grow in a short period of time at low temperatures and nutrients. It is clear that there is much to study before we fully understand the biology of stream macroalgae of the tundra. Because few of these taxa are endemic to the arctic, molecular studies comparing tundra populations with those from other biomes and other polar regions would be useful in determining biogeographic patterns. Biochemical mechanisms of cellular cryoprotection should be examined in terms of the types of low molecular solutes formed for freezing point depression. Population dynamics of recruitment, growth, reproduction and dispersal would be useful to study. In addition, trophic dynamics of tundra stream macroalgae should be related to epilithic periphyton and to localized physical factors.

Acknowledgements

This work was funded by various sources, including NSF grant BRS 8906986, NSERC OGP 0105629 and Polar Continental Shelf Project 636-94 to RGS; NSERC Grant 0645 to KMC; NSF grant DPP 9010190 to JAH. Research centre support at Igloolik and Iqaluit from the Science Institute of the Northwest Territories is also gratefully acknowledged. Assistance in field work from William Bowden, J. Finlay and Mary Koske and in preparation of parts of this manuscript from Carolyn Emerson, Christine Everson, Roy and Sylvia Ficken, Kelly Morris and Kirsten Müller is appreciated. A helpful review from Warwick F. Vincent is also acknowledged.

References

Alvarez, M., T. Gallardo, M. A. Ribera & Gómez Garetta, 1988. A reassessment of North Atlantic seaweed biogeography. Phycologia 27: 221–233.

Bachmann, H., 1921. Beträge zur Algenflora des Süßwassers von Westgrönland. Mitt. Naturforsch. Ges. Luzern 8: 1–181.

Bliss, L. C., 1988. Arctic tundra and polar desert biome. In M. G. Barbour & W. D. Billings (eds), North American Terrestrial Vegetation. Cambridge University Press, Cambridge: 1–32.

Bowden, W. B., B. J. Peterson, J. C. Finlay & J. Tucker, 1992. Epilithic chlorophyll *a*, photosynthesis, and respiration in control and fertilized reaches of a tundra stream. Hydrobiologia 240: 121–131.

Bowden, W. B., J. C. Finlay & P. E. Maloney, 1994. Long-term effects of PO_4 fertilization on the distribution of bryophytes in an arctic river. Freshwat. Biol. 32: 445–454.

Broady, P. A., 1982. Taxonomy and ecology of algae in a freshwater stream in Taylor Valley, Victoria Land, Antarctica. Arch. Hydrobiol. Suppl. 63, 3: 331–349.

Broady, P. A., 1989. The distribution of *Prasiola calophylla* (Carmich.) Menegh. (Chlorophyta) in Antarctic freshwater and terrestrial habitats. Antarctic Sci. 1: 109–118.

Craig, P. C. & P. J. McCart, 1975. Classification of stream types in Beaufort Sea drainages between Prudhoe Bay, Alaska, and the Mackenzie Delta, N.W.T., Canada. Arctic Alpine Res. 7: 183–198.

Crippen, R. W. & J. L. Perrier, 1974. The use of neutral red and Evan's blue for live-dead determinations of marine plankton. Stain Technol. 49: 97–104.

Croasdale, H., 1973. Freshwater algae of Ellesmere Island, N.W.T. National Museums of Canada. Publications in Botany No. 3, Ottawa, 131 pp.

Davey, M. C., 1989. The effects of freezing and desiccation on photosynthesis and survival of terrestrial Antarctic algae and cyanobacteria. Polar Biol. 10: 29–36.

Davey, M. C. & K. J. Clarke, 1992. Fine structure of a terrestrial cyanobacterial mat from Antarctica. J. Phycol. 28: 199–202.

Douglas, M. S. V. & J. P. Smol, 1993. Freshwater diatoms from high arctic ponds (Cape Hershel, Ellesmere Island, N.W.T.). Nova Hedwigia 57: 511–552.

Douglas, M. S. V., J. P. Smol & W. Blake Jr., 1994. Marked post – 18th century environmental change in high-arctic ecosystems. Science 266: 416–419.

Garcia-Pichel, F. & R. W. Castenholz, 1991. Characterization and biochemical implications of scytonemin, a cyanobacteria sheath pigment. J. Phycol. 27: 395–409.

Guy, C. L., 1990. Cold acclimation and freezing tolerance: role of protein metabolism. Annu. Rev. Plant Physiol. Plant Mol. Biol. 41: 187–223.

Hambrook, J. A. & R. G. Sheath, 1987. Grazing of freshwater Rhodophyta. J. Phycol. 23: 656–662.

Hambrook, J. A. & R. G. Sheath, 1991. Reproductive ecology of the freshwater red alga *Batrachospermum boryanum* Sirodot in a temperate headwater stream. Hydrobiologia 218: 233–246.

Hamilton, P. B. & S. A. Edlund, 1994. Occurrence of *Prasiola fluviatilis* (Chlorophyta) on Ellesmere Island in the Canadian arctic. J. Phycol. 30: 217–221.

Harper, P. P., 1981. Ecology of streams at high latitudes. In M. A. Lock & D. D. Williams (eds), Perspectives in Running Water Ecology, Plenum Press, New York: 313–337.

Hart, D. D., 1983. The importance of competitive interactions within stream populations and communities. In J. R. Barnes & G. W. Minshall (eds), Stream Ecology, Application and Testing of General Ecological Theories, Plenum Press, New York: 99-136.

Hawes, I., 1989. Filamentous green algae in freshwater streams in Signy Islands, Antarctica. Hydrobiologia 172: 1–18.

Hawes, I., 1990. Effects of freezing and thawing on a species of *Zygnema* (Chlorophyta) from the Antarctic. Phycologia 29: 326–331.

Hebert, P. D. N. & D. B. McWalter, 1983. Cuticular pigmentation in arctic *Daphnia*: adaptive diversification of asexual lineages? Am. Nat. 122: 286–291.

Hershey, A. E. & A. L. Hiltner, 1988. Effect of a caddisfly on black fly density: interspecific interactions limit black flies in an arctic river. J. N. Am. Benthol. Soc. 7: 188–196.

Howard-Williams, C. & W. F. Vincent, 1989. Microbial communities in southern Victoria Land streams (Antarctica) I. Photosynthesis. Hydrobiologia 172: 27–38.

Hullar, M. A. & R. Vestal, 1989. The effects of nutrient limitation and stream discharge on the epilithic microbial community in an oligotrophic Arctic stream. Hydrobiologia 172: 19–26.

Jensen, T. E., 1985. Cell inclusions in the cyanobacteria. Arch. Hydrobiol. Suppl. 21(1/2): 33–73.

Johansson, C., 1980. Attached algal vegetation in some streams from the Narssaq area South Greenland. Acta Phytogeogr. Suec. 68: 89–96.

Kobayashi, Y., 1967. *Prasiola crispa* and its allies in the Alaskan arctic and Antarctica. Bull. Nat. Sci. Mus. Tokyo 10: 211–220, 2 pls.

Miller, M. C., P. DeOliveira & G. G. Gibeau, 1992. Epilithic diatom community response to years of PO_4 fertilization: Kuparuk River, Alaska (68 N Lat.). Hydrobiologia 240: 103–119.

Milner, A. M. & G. E. Petts, 1994. Glacial rivers: physical habitat and ecology. Freshwat. Biol. 32: 295–307.

Moore, J. W., 1974. Benthic algae of southern Baffin Island III. Epilithic and epiphytic communities. J. Phycol. 10: 456–462.

Moore, J. W., 1979. Distribution and abundance of attached, littoral algae in 21 lakes and streams in the Northwest Territories. Can. J. Bot. 57: 568–577.

Morison, M. O. & R. G. Sheath, 1985. Responses to desiccation stress by *Klebsormidium rivulare* (Ulotrichales, Chlorophyta) from a Rhode Island stream. Phycologia 24: 129–145.

Murray, J. L. & J. Svoboda, 1989. Available nitrogen and phosphorus content in streams feeding wet sedge meadows at Sverdrup Pass, Ellesemere Island, N.W.T. Musk-Ox 37: 54–59.

Pandey, K. D., A. K. Kashyap & R. K. Gupta, 1992. Nitrogen fixation by cyanobacteria associated with moss communities in Schirmacher Oasis, Antarctica. Israel J. Bot. 41: 187–198.

Patrick, R. A., 1988. Importance of diversity in the functioning and structure of riverine communities. Limnol. Oceanogr. 33: 1304–1307.

Petersen, J. B., 1924. Freshwater-algae from the north-coast of Greenland collected by the late Dr. Th. Wulff. Den. II. Thule Ekspedition til Grønlands Nordkyst 1916–18. Nr. 13. XIII. Medd. Grønland 64: 305–319.

Peterson, B. J., L. Deegan, J. Helfrich, J. E. Hobbie, T. E. Ford, A. Hershey, A. Hiltner, G. Kipphut, M. A. Lock, D. M. Fiebig, M. C. Miller, J. R. Vestal, R. Ventullo & G. Volk, 1993. Biological responses of a tundra river to fertilization. Ecology 74: 653–672.

Peterson, B. J., J. E. Hobbie, A. E. Hershey, M. A. Lock, T. E. Ford, J. R. Vestal, V. L. McKinley, M. A. J. Hullar, M. C. Miller, R. M. Ventullo & G. S. Volk, 1985. Transformation of a tundra river from heterotrophy to autotrophy by addition of phosphorus. Science 229: 1383–1386.

Peterson, C. G. & R. J. Stevenson, 1992. Resistance and resilience of lotic algal communities: importance of disturbance and current. Ecology 73: 1445–1461.

Prescott, G. W., 1963. Ecology of Alaskan freshwater algae. II. Introduction: general considerations. Trans. Am. Microsc. Soc. 82: 83–142.

Rydén, B. E., 1977. Hydrology of Truelove Lowland. In L. C. Bliss (ed.), Truelove Lowland, Devon Island, Canada. University of Alberta Press, Edmonton: 107–136.

Rydén, B. E., 1981. Hydrology of northern tundra. In L. C. Bliss, O. W. Heal & J. J. Moore (eds), Tundra ecosystems: a comparative analysis. Cambridge University Press, Cambridge: 115–137.

Sheath, R. G. & K. M. Cole, 1992. Biogeography of stream macroalgae in North America. J. Phycol. 28: 448–460.

Sheath, R. G. & J. A. Hambrook, 1990. Freshwater ecology. In K. M. Cole & R. G. Sheath (eds), Biology of the Red Algae. Cambridge University Press, Cambridge: 423–453.

Sheath, R. G. & A. D. Steinman, 1982. A checklist of freshwater algae of the Northwest Territories, Canada. Can. J. Bot. 60: 1964–1997.

Sheath, R. G. & A. Whittick, 1995. The unique gonimoblast propagules of *Batrachospermum breutelii* (Batrachospermales, Rhodophyta). Phycologia 34: 33–38.

Slack, K. V., J. W. Nauman & L. J. Tilley, 1979. Benthic invertebrates in a north-flowing stream and a south-flowing stream, Brooks Range, Alaska. Wat. Res. Bull. 15: 108–135.

Stockner, Z. S. J. & H. B. N. Hynes, 1976. Studies on the tributaries of Char Lake, Cornwallis Island, Canada. Hydrobiologia 49: 97–102.

Stonehouse, B., 1989. Polar Ecology. Chapman & Hall, New York, 222 pp.

Vankat, J. L., 1979. The Natural Vegetation of North America. An Introduction. John Wiley & Sons, New York, 261 pp.

Vincent, W. F., M. T. Downes, R. W. Castenholz & C. Howard-Williams, 1993a. Community structure and pigment organization of cyanobacteria-dominated microbial mats in Antarctica. Eur. J. Phycol. 28: 213–221.

Vincent, W. F. & C. Howard-Williams, 1986. Antarctic stream ecosystems: physiological ecology of a blue-green algal epilithon. Freshwat. Biol. 16: 219–233.

Vincent, W. F., C. Howard-Williams & P. A. Broady, 1993b. Microbial communities and processes in Antarctic flowing waters. In Antarctic Microbiology, Wiley-Liss: 543–569.

Vincent, W. F. & S. Roy, 1993. Solar ultraviolet-B radiation and aquatic primary production: damage, protection, and recover. Envir. Rev. 1: 1–12.

Whitton, B. A., 1992. Diversity, ecology, and taxonomy of the cyanobacteria. In N. H. Mann & N. G. Carr (eds), Photosynthetic Prokaryotes. Plenum Press, New York: 1–51.

Woo, M.-K., 1988. Wetland runoff regime in northern Canada. In K. Senneset (ed.), Permafrost. Fifth International Conference Proceedings Vol. 1, Tapir Publishers, Trondheim: 644–649.

Woo, M.-K., 1991. Arctic streamflow. In M.-K. Woo & D. J. Gregor (eds), Arctic Environment: Past, Present & Future. Dept. of Geography, McMaster University, Hamilton: 105–111.

Yamagishi, T., 1967. Some filamentous Chlorophyceae in the Alaskan arctic. Bull. Nat. Mus. Tokyo 10: 201–206, 2 pl.

Hydrobiologia **336**: 83–91, 1996.
J. Kristiansen (ed.), Biogeography of Freshwater Algae.

8. Biogeography of *Vaucheria* species from European freshwater/soil habitats: implications from chloroplast genomes

Karl-Heinz Linne von Berg[1] & Klaus V. Kowallik[2]
[1]*Botanisches Institut der Universität Köln, Gyrhofstr. 15, D 50931 Köln, Germany*
[2]*Botanisches Institut der Universität Düsseldorf, Universitätsstr. 1, D 40225 Düsseldorf, Germany (corresponding author)*

Key words: chloroplast genomes, biogeography, species concept, *Vaucheria*

Abstract

Infraspecific heterogeneity of chloroplast genomes was found in four *Vaucheria* species (*V. bursata, V. cruciata, V. geminata, V. prolifera*) collected from six European countries. The degree of sequence variability among strains of each of the four species, as demonstrated by restriction site analysis, exceeds that of higher plant species or even genera. Mainly single base substitutions and, to a much lesser extent, minor insertions/deletions account for such differences, whereas the linear gene arrangement remains unaffected. Chloroplast genotypes found to be identical among strains collected from different geographical localities are considered the common genotype of a given species. These findings are discussed with respect to evolution, biogeographical distribution and the species concept of this genus.

Introduction

The genus *Vaucheria* comprises some 70 species of worldwide distribution. In Europe about 40 species are known from mostly freshwater and terrestrial habitats (Rieth, 1980). Ultrastructural as well as biochemical characters relate this genus to the Xanthophyceae. However, both the morphology of the thallus and the mode of sexual and vegetative propagation, which are unique within the Xanthophyceae, render this genus monotypic, even in comparison to other chlorophyll $a + c$-containing algae.

The taxonomy of this genus is primarily based upon the morphology of antheridia and, to a much lesser extent, upon oogonia, the mode of vegetative reproduction, and characters of the siphonaceous filaments (Heering, 1921; Venkataraman, 1961; Rieth, 1980; Christensen, 1987). As pointed out by Rieth (1980), species identification may be difficult due to overlapping characters among closely related species. Nevertheless, numerous subspecies and/or varieties have been described when morphological characters indicated evolutionary divergence within a given species. As culture experiments which can exclude the potential influence of external factors upon the morphology of a species are rare, it is obvious that subspecies as well as varieties described for this genus are of questionable taxonomic value (cf. Rieth, 1980). On the other hand, strains of a given species that are indistinguishable by morphological characters are not necessarily genotypically homogeneous. In particular, our knowledge about the molecular variability of this genus at different taxonomic levels is meagre, as is the knowledge about ecology and biogeographic distribution of its species. This lack of information is mainly due to the fact that *Vaucheria* samples from natural habitats – though easily recognized at the genus level – rarely exhibit sexual organs and are usually treated as unidentified species in sampling lists. Differences within populations of *Vaucheria* species that must be taken into account, due to inherent effects of selection pressure, drift, migration, and mutation, are entirely unknown on the molecular level, as such knowledge requires a sampling strategy focussing upon strains, combined with extensive culture studies.

Mating type analyses (Coleman, 1977), isozyme banding patterns (Gallagher, 1982; Sako et al., 1991, and others), different nutritional requirements (Post et al., 1994, and others), restriction site polymorphisms (Gross et al., 1988; Spanier et al., 1992; Stabile et al., 1992; Scholin and Anderson, 1994), as well as nucleotide sequences of rbcL genes (Scholin et al., 1994) and ITS regions of Rubisco (Destombe and Douglas, 1991) and nuclear ribosomal RNA genes (Bakker et al., 1992; Kooistra et al., 1992; Coleman et al., 1994) have documented that the genetic diversity in algal species exceeds by far that of land plants, even on the infraspecific level. To date, most of these investigations were based on marine algae, as genotypic differences are more easily to explain by hydrogeographic events responsible for the biogeographic distribution of a given species. Fresh water species from ponds and lakes, however, are subject to complex migration strategies that are difficult to evaluate. Yet such investigations are required to understand evolutionary processes in fresh water algae with respect to population dynamics and biogeographic distribution.

In general, molecular and genetic data indicating genotypic variability among strains of a species frequently conflict with characters defining a morphological species. They are controversely interpreted with respect to their taxonomic value. Nevertheless, molecular as well as biochemical data usually resemble the only available markers to manifest genetic diversity on a lower taxonomic level. They may be used to document genotypic variability in morphological and even biological species. They are especially affirmative in elucidating evolutionary traits in populations as well as in establishing the biogeography of a given species, rather than to serve for taxonomic revisions in cases that are not substantiated by morphological characters.

Approaches to study the biogeographic distribution of *Vaucheria* species

In order to assess the genotypic status of different strains of four *Vaucheria* species (*V. bursata* (Müller) Agardh, *V. cruciata* (Vaucher) de Candolle, *V. geminata* (Vaucher) de Candolle, *V. prolifera* Dangeard), and to delineate affiliations among species and strains relative to their geographic distribution, we analyzed chloroplast genomes by restriction site and gene mapping experiments. It was already shown that chloroplast genomes characterized in such a way may be used as molecular markers (Linne von Berg & Kowallik, 1988).

Restriction fragment length polymorphisms (RFLPs) are either caused by point mutations (single base exchanges), insertions/deletions, and inversions, thereby resulting in the loss of an existing, or the gain of a new, restriction site. Any restriction fragment whose position on the chloroplast genome has been mapped, represents a binary character (present or absent) and may therefore be used to create a matrix appropriate for cladistic analyses. Fragments with identical sizes (same Rf values) may be treated as homologs, if their homology has been proven by Southern hybridization experiments using isolated or cloned restriction fragments of a given strain as probes.

In order to establish the genotypic variability of chloroplast genomes in *Vaucheria*, strains were collected from various locations in central Europe, with the exception of *V. bursata* 2 which was obtained from the Culture Collection of Göttingen (Table 1, Figure 1). The samples, mostly from terrestrial habitats, were flooded with distilled water and incubated for four to six weeks at 15 °C and 14:10 hours light:dark at 120 μE PAR. During this period antheridia and oogonia developed which allowed us to identify the species. For further studies, fertile filaments were dissected, rinsed, and transferred to Waris medium (Waris, 1953) enriched with vitamins (von Stosch & Drebes, 1964) and trace elements (Braune et al., 1976). Methods to cultivate the algae in 25 l aerated vessels for chloroplast isolation, to isolate and purify chloroplast DNA, and to characterize the chloroplast genomes by restriction site and gene mapping experiments were the same as already described (Linne von Berg & Kowallik, 1988).

Chloroplast genomes and biogeography of Vaucheria bursata

The chloroplast DNA of *Vaucheria bursata* (Sectio Corniculatae) reveals a G+C-contents of 31%, comparable to other chromophytic chloroplast genomes (K. V. Kowallik, unpubl. results). Consequently, endonucleases like Bam HI or Sal I that recognize sites rich in G+C-concent create a moderate number of restriction fragments (up to 10). Thus, differences among strains of a given species may be obscured due to a low number of restriction fragments. On the other hand, restriction endonucleases like Eco RI that cut the DNA at sites enriched in A and T yield a higher number of restriction fragments (in *Vaucheria* species up to 30), thereby

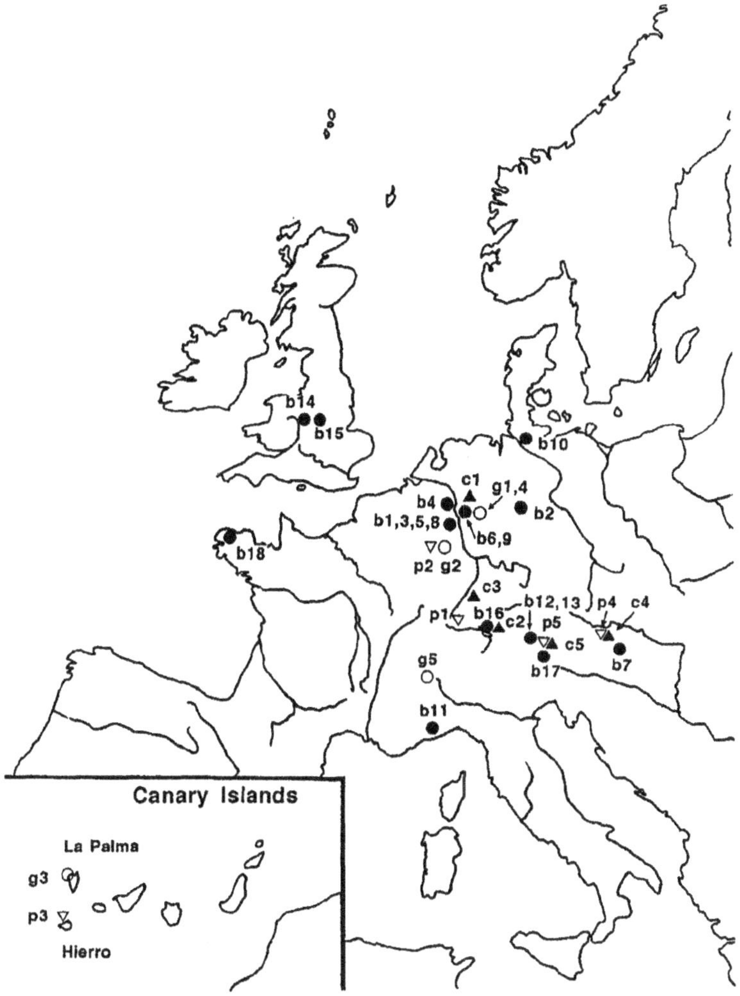

Figure 1. Geographic distribution of sampling sites of strains from *V. bursata* (b1–b18), *V. cruciata* (c1–c5), *V. geminata* (g1–g5), and *V. prolifera* (p1–p5). Inset: Canary Islands, drawn at a different scale.

allowing the construction of detailed restriction site maps necessary to identify gene locations.

Eco RI restriction fragment patterns from 18 *V. bursata* strains are shown in Figure 2. Apparently, the fragment patterns may be grouped into two major assemblages (group I: strains 1–7; group II: strains 8–15). Three strains (16–18, Table 1) have only little in common with each other or with either of the two major groups. Only strain 17 reflects the pattern of group I chloroplast genomes to some extent. Strain 6 differs from the remaining group I types by a single additional restriction site (Figure 2), whereas five different genomes may be encountered in group II strains due to the extinction/appearance of individual recognition sites. This grouping of the genomes is confirmed by chloroplast DNA patterns additionally generated by Sal I, Bam HI, and Pvu II (data not shown, cf. Linne von Berg & Kowallik, 1988).

Surprisingly, strains isolated from a distinct sampling site may exhibit different restriction profiles (e.g., strains 1, 3, 5, when compared with strain 8). Vice versa, chloroplast genomes from strains collected at different localities may be indistinguishable when cut with all four restriction enzymes used that together recognize some 300 nucleotides (strains 1, 3, 5, when compared with strains 2 and 6). Strains 14 and 15 from a disjunct locality appear to be genotypically identical for all four restriction enzymes, but slightly differ from

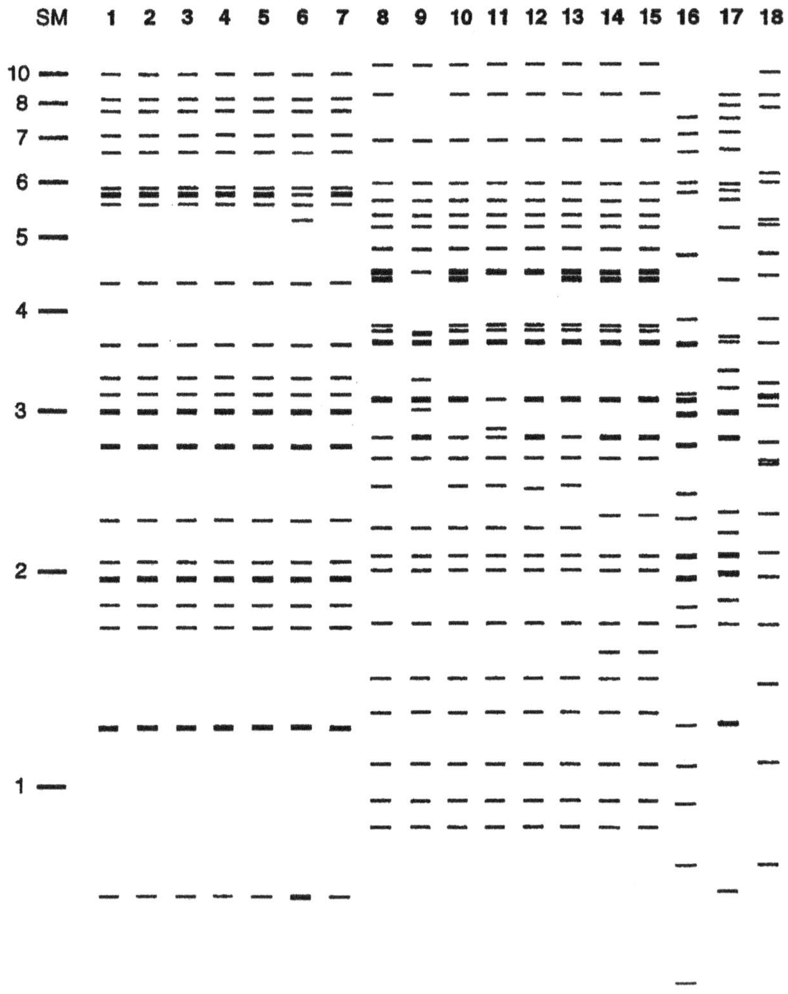

Figure 2. Diagrammatic representation of Eco RI restriction fragment patterns of chloroplast DNAs isolated from 18 strains of *V. bursata*. The DNA fragments were separated on a 0.8% agarose gel. Restriction fragments occurring as duplicates are indicated by bold faced lines. A size marker (kilobases) is shown at the left hand side.

other group II members with respect to their Eco RI profiles (Figure 2).

Differences of chloroplast genomes among strains of a single species that are used as genotypic markers must be corroborated by morphological data. Mean values of at least 100 measurements of thallus diameter, shape of zoosporangia (coefficient length:breadth), and grouping of antheridia and oogonia relative to each other revealed a clearcut picture: the two groups of *V. bursata* (1 to 7 and 8 to 15) differ in all these respects, indicating a divergence of this species into two subspecies. As already indicated by restriction fragment analysis, strain 17 resembles the group I type, whereas morphologically strains 16 and 18 are more closely related to the group II strains.

Chloroplast genomes and biogeography of Vaucheria cruciata, V. geminata, *and* V. prolifera

Figure 3 illustrates the Eco RI restriction fragment patterns from each of five strains of *V. cruciata* (Sectio Anomalae), *V. geminata* (Sectio Corniculatae), and *V. prolifera* (Sectio Globiferae). With the addition of restriction fragment profiles generated by Pvu II (not illustrated) similarities as well as differences found among the strains of the three species generally resem-

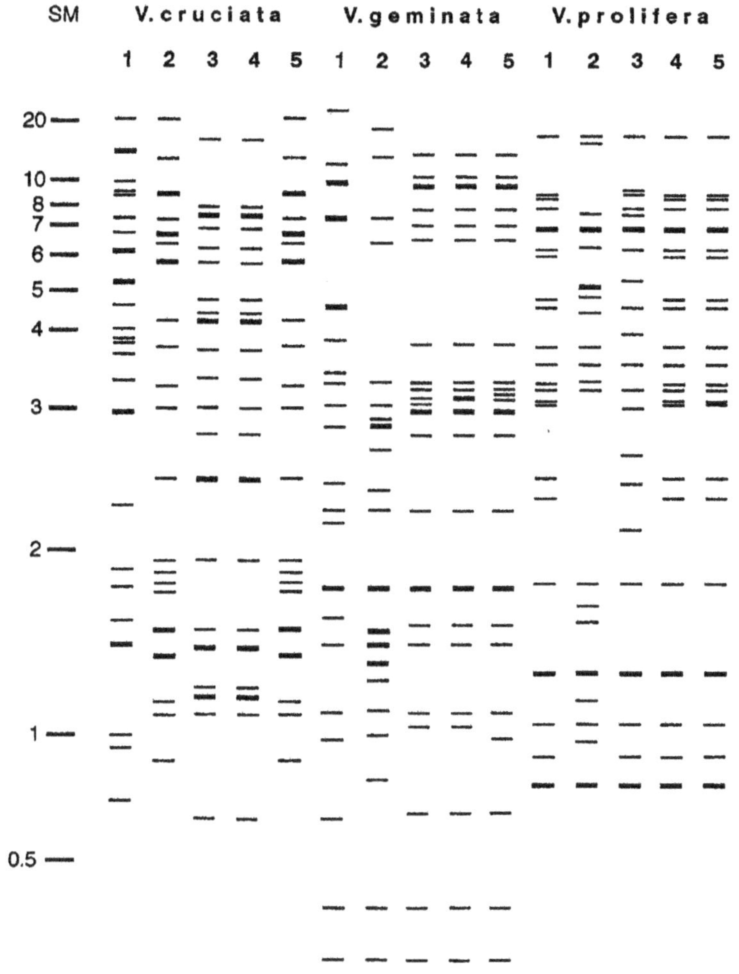

Figure 3. Diagrammatic representation of Eco RI restriction fragment patterns of chloroplast DNAs isolated from each of five strains of *V. prolifera, V. curciata,* and *V. geminata.* The DNA fragments were separated on a 1% agarose gel. Restriction fragments occurring as duplicates are indicated by bold faced lines. A size marker (kilobases) is shown at the left hand side.

ble the degree of restriction site polymorphism in *V. bursata* strains.

V. cruciata reveals four different chloroplast genome types. Only strains 3 and 4, though from different localities, are indistinguishable from each other. Strain 1 exhibits some similarity to either strains 3 and 4, or 2 and 5, respectively. However, it was not possible to attribute specific morphological characters to either of these strains.

The chloroplast genomes of five strains of *V. geminata* may all be discriminated by their restriction fragment patterns. Although from widely distant geographic regions, strains 1, 3, 4, and 5 appear to be remarkably similar to each other, but differ clearly from strain 2. Culture experiments revealed that strain 2 exclusively developed branches with two oogonia, which is typical for this species. Strains 3, 4, and 5 sometimes (up to 5%) showed sexual branches characteristic for *V. verticillata* Meneghini, whereas strain 1 generated fruiting branches similar to those of *V. taylorii* Blum when cultivated below 10 °C, but resembled *V. verticillata* when grown above 10 °C.

In *V. prolifera* strains 1, 4, and 5 appear to be genotypically identical, whereas strains 2 and 3 differ from each other as well as from the remaining strains. A detailed microscopic investigation of these five strains did not reveal any morphological differences among strains 1, 3, 4, and 5 which clearly resemble *V. prolifera*

Table 1. Collection sites of different strains of *Vaucheria* species

V. bursata (O.F. Müller) Agardh 1812	(Corniculatae)
strain 1	Neuss, Germany
strain 2	Culture Collection Göttingen
strain 3	Neuss, Germany
strain 4	Kamp Lintfort, Germany
strain 5	Neuss, Germany
strain 6	Düsseldorf, Germany
strain 7	Neusiedler See, Austria
strain 8	Neuss, Germany
strain 9	Düsseldorf, Germany
strain 10	Geversdorf, Germany
strain 11	San Bernardo, Italy
strain 12	Berchtesgaden, Germany
strain 13	Berchtesgaden, Germany
strain 14	Brewood, England
strain 15	Tyrley, England
strain 16	Bodensee, Germany
strain 17	Sauerfeld, Austria
strain 18	Britany, France
V. cruciata (Vauch.) DC. 1805	(Anomalae)
strain 1	Erkrath, Germany
strain 2	Mindelsee, Germany
strain 3	Ottenheim, Germany
strain 4	Steinwandklamm, Austria
strain 5	Schneibstein, Germany
V. geminata (Vauch.) DC. 1805	(Corniculatae)
strain 1	Düsseldorf, Germany
strain 2	Monschau, Germany
strain 3	La Palma, Canary Islands
strain 4	Düsseldorft, Germany
strain 5	San Bernardo, Italy
V. prolifera Dangeard 1939	(Globiferea)
strain 1 (fa. *prolifera*)	Freiburg, Germany
strain 2 (fa. *corniculata*)	Monschau, Germany
strain 3 (fa. *prolifera*)	Hierro, Canary Islands
strain 4 (fa. *prolifera*)	Steinwandklamm, Austria
strain 5 (fa. *prolifera*)	Schneibstein, Germany

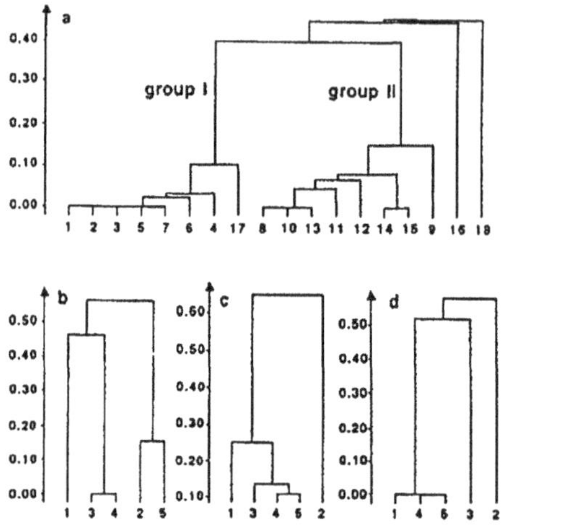

Figure 4. Cladistic analysis of restriction fragment patterns, revealing the heterogeneity of chloroplast genomes in strains of *V. bursata* (a), *V. cruciata* (b), *V. geminata* (c), and *V. prolifera* (d). For all species restriction fragment patterns generated by Eco RI and Pvu II were used to construct the cladograms. In addition, Sal I and Bam HI patterns were included for *V. bursata*. The scale is proportional to the degree of similarity/dissimilarity between the strains.

var. *prolifera* f. *corniculata* (Rieth, 1978). Strain 2 differs from the remaining ones by its bulb-like antheridium and therefore resembles *V. prolifera* var. *prolifera* Dangeard (Rieth, 1980).

A cladistic analysis using each restriction fragment generated by the four enzymes used as an independent character confirmed the results of the restriction fragment analysis. In *V. bursata* (Figure 4a) the two subgroups clearly form two subclades, with strains 16 and 18 being most distantly related to all other strains. Strain 17 clusters among group I individuals, though separated from the remaining group I strains. Altogether, strains 1, 2, 3, 5, 7 are indistinguishable from one another in group I. Strains 8, 10, 13, and strains 14 and 15, respectively, do not differ among each other in group II.

In *V. cruciata* (Figure 4b) and *V. geminata* (Figure 4c) two out of five strains appear to be identical (strains 3 and 4) or nearly identical (strains 4 and 5), whereas in *V. prolifera* (Figure 4c) three strains (1, 4, 5) are indistinguishable upon restriction fragment patterns for Eco RI and Pvu II. Several such congeneric identical strains are from different geographical sites.

Gene mapping experiments clarified the nature of the RFLPs in this genus (Figure 5). Despite the differences in restriction fragment profiles all chloroplast genomes are 120–125 kb in size. Differences in size probably resulted from minor insertion/deletion events, whereas differences in restriction fragment patterns are caused by changes in the nucleotide sequences. Inversion/transposition events which appear to be typical for chromophytic plastid genomes (Kowallik, 1989) did not affect the linear arrangement of genes mapped to chloroplast genomes of *Vaucheria* species and strains (Figure 5).

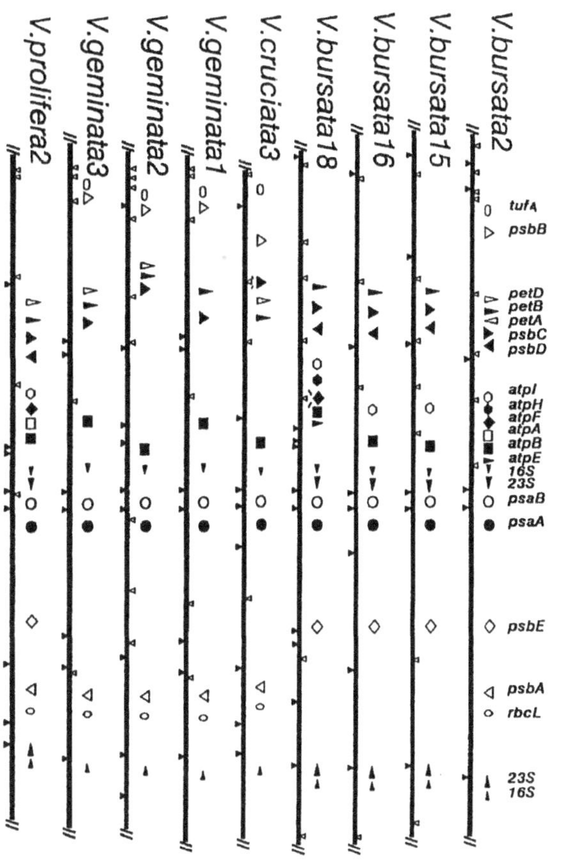

Figure 5. Linearized restriction site maps (Sal I: empty traingels, Bam HI: filled triangles) of the circular chloroplast genomes of *V. bursata* (strains 2, 15, 16, 18), *V. cruciata* (strain 3), *V. geminata* (strains 1, 2, 3,) and *V. prolifera* (strain 2). Sites used to linearize the genomes are marked by double obliques. Positions of chloroplast genes identified by Southern hybridizations are indicated by symbols above the maps. The maps were aligned along the BamHI restriction sites between the psaA and psaB genes, as these sites are homologous in all chloroplast genomes investigated.

Discussion

Although genotypic variability is known from many algae which are characterized by only few morphological characters, the degree of sequence divergence among chloroplast genomes of *Vaucheria* strains that appear to be identical on a morphological basis is striking. This finding raises questions as to the significance of such characters in terms of infraspecific evolution, biogeographic distribution, and taxonomy.

In land plant chloroplast genomes RFLPs have been reported for different taxonomic levels. Comparisons among congeneric or closely related species indicated little, if any, infraspecific variability (Clegg & Zurawski, 1986). Even on the intergenus level, as e.g. in *Nicotiana* (Salts et al., 1984), *Lycopersicon* and *Solanum* (Palmer & Zamir, 1982), and *Epilobium* (Schmitz & Kowallik, 1986) chloroplast genomes did not reveal a degree of sequence divergence comparable to that found among strains of a single *Vaucheria* species. Moreover, restriction fragment patterns of angiosperm species from different continents or even from different genera (v. Westtstein, 1983) may have more restriction fragments in common than those of *Vaucheria* strains isolated from a single sampling site.

Nevertheless, there are few cases known in land plants where considerable sequence variation exists on the populational level. In *Trifolium pratense* infrapopulational as well as interpopulational diversity seems to be a common feature (Milligan, 1991), whereas screening of *Lupinus texensis* strains from 21 different populations only revealed 3% variant chloroplast genome types (Banks & Birky, 1985).

A much higher degree of genotypic variability was found among season-dependent strains of the marine centric diatom *Skeletonema costatum* using isozyme patterns as markers to discriminate between strains (Gallagher, 1982). These data clearly demonstrate that the degree of sequence divergence in algal species exceeds by far that of most higher plants. In addition, both the *Skeletonema* and *Vaucheria* findings imply that biogeographic analysis of species distribution based upon morphological markers will only provide a crude and poorly graded insight into what had really occurred during evolution of a given species or genus.

The diversity among *Vaucheria* strains documented here does not reflect sequence variation within a population, as only individual strains from distant localities have been considered. Generally, a population is characterized by a common genotype and many rare ones. In a figurative sense this holds also for chloroplast genomes which have co-evolved over time with the corresponding nuclear genomes. This implies that RFLPs of chloroplast genomes do reflect underlying genomic differences. If such differences are not expressed morphologically, genotypical diversity may be obscured. As a consequence, strains that are indistinguishable by traditional, i.e. nuclear inherited taxonomic characters, may be mistakenly considered as being genotypically homogeneous.

Although the data from 18 *V. bursata* strains represent the most comprehensive research on the lowest taxonomic level in algae to date, they do not clearly reveal a chloroplast genotype typical for each of the two subgroups. It may be possible that strains 1, 2, 3, 5, and 7, which are indistinguishable on the basis of four restriction enzymes, represent the common genotype in group I, as these strains are from disjunct locations in both Germany and Austria. Likewise, the chloroplast genomes of strains 8, 10, and 13 may be regarded as the common group II genotype. The remaining genotypes would then accompany the two common genotypes. They altogether reflect a degree of sequence variability which – though understood on the structural level – remains unexplained with respect to the underlying motives. However, this kind of genome diversity perfectly agrees with findings from RNA populations generated by random mutations *in vitro* (Nowak & Schuster, 1989).

Taken into consideration the high degree of ploidy of most chloroplast types, including those of *Vaucheria*, and the high number of chloroplasts transmitted maternally to the zygote in this genus, thereby eliminating possible recombination events, then this high degree of sequence variability becomes even more difficult to understand. Probably, one has to consider accelerated molecular clocks and modes of evolution in chloroplasts of certain chromophyte algae, when compared with the rate of evolution in cyanobacteria (Castenholz, 1992), red algal (Goff & Coleman, 1988) or metaphyte chloroplasts (Palmer et al., 1985). An assessment of mutation rates, using the equation of Brown et al. (1979), revealed molecular differences among the 10 *V. bursata* strains that are 20 times higher than those among different species of *Pisum* (Palmer et al., 1985). The mutation rates even exceed those found between Hordeum & Zea (Zurawski et al., 1984), and are about half as much as among mitochondrial DNAs from different primates (Brown et al., 1979).

Acknowledgements

We gratefully acknowledge financial support from DFG grants Ko 439/6–1 and 439/6–3.

References

Bakker, F. T., J. L. Olsen, W. T. Stam & C. van den Hoek, 1992. Nuclear ribosomal DNA internal transcribed spacer regions (ITS1 and ITS2) define discrete biogeographic groups in *Cladophora albida* (Chlorophyta). J. Phycol. 28: 839–845.

Banks, J. A. & C. W. Birky, Jr., 1985. Chloroplast DNA diversity is low in a wild plant, *Lupinus texensis*. Proc. Natl. Acad. Sci. USA 82: 6950–6954.

Braune, W., A. Leman & H. Taubert, 1976. Praktikum zur Morphologie und Entwicklungsgeschichte der Pflanzen. G. Fischer Verlag Stuttgart, 448 pp.

Brown, W. M., M. George Jr. & A. C. Wilson, 1979. Rapid evolution of animal mitochondrial DNA. Proc. Natl. Acad. Sci. USA 76: 1967–1971.

Castenholz, R. W., 1992. Species usage, concept, and evolution in the cyanobacteria (blue green algae)? J. Phycol. 28: 737–745.

Christensen, T., 1987. Seaweeds of the British Isles, Vol. 4: Tribophyceae (Xanthophyceae). British Museum (Natural History), 36 pp.

Clegg, M. T. & G. Zurawski, 1986. Chloroplast DNA and the study of plant phylogeny: present status and future prospects. In: Molecular Systematics of Plants, P. S. Soltis, D. E. Soltis & J. J. Doyle (eds), Chapman and Hall, New York: 1–13.

Coleman, A. W., 1977. Sexual and genetic isolation in the cosmoplitan algal species *Pandorina morum*. Am. J. Bot. 64: 361–68.

Coleman, A. W., A. Suarez & L. J. Goff, 1994. Molecular delineation of species and syngens in volvocacean green algae (Chlorophyta). J. Phycol. 30: 80–90.

Destombe, C. & S. E. Douglas, 1991. Rubisco spacer sequence divergence in the rhodophyte alga *Gracilaria verrucosa* and closely related species. Curr. Genet. 19: 395–398.

Gallagher, J. C., 1982. Physiological variation and electrophoretic banding patterns of genetically different seasonal populations of *Skeletonema costatum* (Bacillariophyceae). J. Phycol. 18: 148–162.

Goff, L. J. & A. W. Coleman, 1988. The use of plastid DNA restriction endonuclease patterns in delineating red algal species and populations. J. Phycol. 24: 357–368.

Gross, C. H., L. P. W. Ranum & P. A. Lefebvre, 1988. Extensive restriction fragment length polymorphisms in a new isolate of *Chlamydomonas reinhardtii*. Curr. Genet. 13: 503–508.

Heering, W., 1921. Siphonales. In: Die Süßwasserflora Deutschlands, Österreichs und der Schweiz, Heft 7, A. Pascher (ed.), G. Fischer Verlag Jena, 31 pp.

Kooistra, W. H. C. F., W. T. Stam, J. L. Olsen & C. van den Hoek, 1992. Biogeography of *Cladophoropsis membranacea* (Chlorophyta) based on comparisons of nuclear rDNA ITS sequences. J. Phycol. 28: 660–668.

Kowallik, K. V., 1989. Molecular aspects and phylogenetic implications of plastid genomes of certain chromophytes. In: The Chromophyte Algae: Problems and Perspectives, J. C. Green, B. S. C. Leadbeater & W. L. Diver (eds), The Systematics Association, Clarendon Press, Oxford 38: 101–124.

Linne von Berg, K.-H. & K. V. Kowallik, 1988. Structural organization and evolution of the plastid genome of *Vaucheria sessilis* (Xanthophyceae). BioSystems 21: 239–247.

Milligan, B. G., 1991. Chloroplast DNA diversity within and among populations of *Trifolium pratense*. Curr. Genet. 19: 411–416.

Nowak, M. & P. Schuster, 1989. Error thresholds of replication in finite populations. Mutation frequencies and the onset of Muller's Ratchet. J. Theor. Biol. 137: 375–395.

Palmer, J. D. & D. Zamir, 1982. Chloroplast DNA evolution and phylogenetic relationships in *Lycopersicon*. Proc. Natl. Acad. Sci. USA 79: 5006–5010.

Palmer, J. D., R. A. Jorgensen & W. F. Thompson, 1985. Chloroplast DNA variation and evolution in *Pisum*: patterns of change and phylogenetic analysis. Genetics 109: 195–213.

Post, A. F., I. Cohen & E. Romem, 1994. Characterization of two *Chlorella vulgaris* (Chlorophyceae) strains isolated from waste-water oxidation ponds. J. Phycol. 30: 950–954.

Rieth, A., 1978. Beiträge zur Kenntnis der Vaucheriaceae. XIX. Der Formenkreis von *Vaucheria prolifera* Dangeard, 1939. Arch. Protistenkde. 120: 278–286.

Reith, A., 1980. Xanthophyceae, 2. Teil. In: Süßwasserflora von Mitteleuropa, Vol. 4, H. Ettl, J. Gerloff & H. Heynig (eds), G. Fischer Verlag Stuttgart, New York: 147 pp.

Sako, Y. S., K. Shrestha, A. Uchida & Y Ishida, 1991. Isozyme analysis of mating populations of *Gonium pectorale* (Chlorophyta). J. Phycol. 27: 309–315.

Salts, Y., R. G. Herrman, N. Peleg, U. Lavi, S. Izhar, R. Frankel & J. S. Beckmann, 1984. Physical mapping of plastid DNA variation among eleven *Nicotiana* species. Theor. Appl. Genet. 69: 1–14.

Schmitz, U. K. & K. V. Kowallik, 1986. Polymorphism and gene arrangement among plastomes of ten *Epilobium* species. Plant Mol. Biol. 7: 115–127.

Scholin, C. A. & D. M. Anderson, 1994. Identification of group- and strain-specific generic markers for globally distributed *Alexandrium* (Dinophyceae). I. RFLP analysis of SSU rRNA genes. J. Phycol. 30: 744–754.

Scholin, C. A., M. Herzog, M. Sogin & D. M. Anderson, 1994. Identification of group- and strain-specific genetic markers for globally distributed *Alexandrium* (Dinophyceae). II. Sequence analysis of a fragment of the LSU rRNA gene. J. Phycol. 30: 999–1011.

Spanier, J. G., J. E. Graham & J. W. Jarvik, 1992. Isolation and preliminary characterization of three *Chlamydomonas* strains interfertile with *Chlamydonomas reinhardtii* (Chlorophyta). J. Phycol. 28: 822–828.

Stabile, J. E., E. T. Wurtzel & J. C. Gallagher, 1992. Comparison of chloroplast DNA and allozyme variation in winter strains of the marine diatom *Skeletonema costatum* (Bacillariophyta). J. Phycol. 28: 90–94.

Stosch, H. A. von & G. Drebes, 1964. Entwicklungsgeschichtliche Untersuchungen an zentrischen Diatomeen. IV. Die Planktonalge *Stephanopyxis turris* – ihre Behandlung und Entwicklungsgeschichte. Helgol. Wiss. Meeresunters. 11: 209–257.

Venkataraman, G. S., 1961. Vaucheriaceae. Indian Council of Agricultural Research, Vol. IX, New Delhi, 112 pp.

Waris, H., 1953. The significance for algae of chelating substances in the nutrient solutions. Physiol. Pl. 6: 538–543.

Wettstein, D. von, 1983. Genetic engineering in the adaptation of plants to evolving human needs. Experientia 39: 687–713.

Zurawski, G., M. T. Clegg & A. H. D. Brown, 1984. The nature of nucleotide sequence divergence between barley and maize chloroplast DNA. Genetics 106: 735–749.

Hydrobiologia **336:** 93–97, 1996.
J. Kristiansen (ed.), Biogeography of Freshwater Algae.

9. The algal flora of the Primorsky Region, Russian Far East

A. A. Gontcharov
Russian Academy of Sciences, Far Eastern Branch, Institute of Biology and Pedology, Vladivostok, Russia

Key words: Biodiversity, locality types, regional distribution types

Abstract

A survey of the freshwater algal flora of the Primorsky Region, Russian Far East, is presented. During more than 70 years of investigations 1659 species (including varieties and forms 2276 taxa) belonging to 346 genera of different groups of algae have been recorded from this territory. The Primorsky Region is rich in habitats attractive to algae-rivers, lakes, reservoirs, and ponds. The details of the distribution of algae in different types of waterbodies have been examined. The algal flora of the region is comparable to the areas, and its taxonomical structure is typical for Eurasia. From this territory some new species and varieties have been described – *Achnanthes chankensis, Cymbella ventricosa* var. *hankensis, Surirella alisoviana*, and others. The flora includes some rare species – *Fragilaria magocsyi, Perone dimorpha*, and *Heterothrix monochloron*, and some endemics of Asia – *Navicula scutelloides* var. *baicalensis, N. globulifera* var. *nipponica, Gomphonema sphaerophorum*, and *Surirella tientsinensis*.

The Primorsky Region is located in the south-eastern part of the Russian Far East. It has an area of 165 900 km^2. The main part of the region consists of the Sikhote-Aline and the East-Manchurian Mountains. The vegetation is dominated by forests of conifers and deciduous trees. The region is rich in water bodies such as rivers, lakes, swamps, reservoirs, and ponds. The river systems are concentrated in three basins. (1) The Sea of Japan basin – small and medium-sized rivers on the eastern slope of the Sikhote-Aline and East-Manchurian Mountains, (2) the upper part of the Ussury River Basin, with the rivers Bolshay Ussurca and Bikin, and (3) The Hanka Lake Basin.

The lakes are mainly situated in the Western Primorsky Plain. Their total area is 4321 km^2, 4070 km^2 of which constitute the Hanka Lake. 99% of the lakes are smaller than 1 km^2, they are mainly meander-shaped. There are a few large (5–10 km^2) saline and brackish lagoons in the southern part of the Primorsky Region. Further there are 119 reservoirs, most of them small.

The area comprises six nature conservation areas (four forest areas, one swamp-lake area, and one marine area).

During 70 years the species composition in 79 rivers, 68 smaller streams, 26 lakes and 7 reservoirs has been studied. A few detailed reviews of the history of these studies have been published (Kukharenko, 1974, 1989; Medvedeva & Barinova, 1990).

At present the algal flora of the Primorsky Region amounts to 1659 species (including varieties and forms 2276 taxa) from 9 divisions, 16 classes, 40 orders, 48 families and 346 genera. It should of course be emphasized that these numbers certainly are far from being exact or exhaustive and that further studies may change or increase them considerably.

The highest number of species in the flora is reached by the green algae: 659 species (814 taxa), which account for 39,8% of the total species number. Desmidiales – 297 species, and Chlorococcales – 200 species, are the most numerous groups in this division. The diatoms occupy the second place, 509 species, 30.7%, but they surpass the green algae in the number of taxa, viz. 810. The main part of the diatoms belong to the Pennales: Araphidineae 42 species and Raphidineae 434 species; the Centrales comprise only 33 species. The cyanophytes are in the third place – 234 species (300 taxa), amounting to 14.1%, among

Table 1. Systematic composition of the algal floras of the Primorsky Region and some other territories (sp\%)

Department	Primorya	Komi A. R.	W. Siberia	Yakutia
Cyanophyta	234\14.1	127\15.1	244\14.0	335\15.2
Euglenophyta	108\6.5	35\4.2	232\13.3	193\8.8
Pyrrophyta	22\1.3	9\1.0	28\1.6	550\2.3
Chrysophyta	61\3.8	36\4.3	46\2.7	140\6.4
Bacillariophyta	509\30.7	382\45.3	463\26.5	628\28.5
Xanthophyta	57\3.4	14\1.7	40\2.3	185\8.4
Rhodophyta	5\0.3	1\0.1	3\0.2	2\0.1
Chlorophyta	659\39.7	237\28.1	678\38.8	657\29.8
Charophyta	4\0.2	2\0.2	15\0.9	10\0.5
Total	1659\100	843\100	1749\100	2200\100

Table 2. Genera most rich in number of species in the algal flora of the Primorsky Region.

Genera	Number of species
Cosmarium	95
Navicula	77
Closterium	55
Nitzschia	52
Trachelomonas	43
Pinnularia	43
Cymbella	41
Eunotia	38
Oscillatoria	37
Staurastrum	36

these the Oscillatoriales and the Nostocales are the most numerous groups. The euglenophytes account for 108 species (141 taxa), 6.5% of the total number of species.

The freshwater algal flora of the Primorsky Region is comparable to the floras of other, adjacent, regions in the temperate zone as regards the number of species, in some cases the number of species is even still higher (Table 1). Also the taxonomic structure is similar and typical for Eurasia (Getzen, 1978; Muzafarov, 1981; Safonova, 1984; Vassilieva, 1989).

Some of the species, varieties and forms found in water bodies in the region are rare and have only been found once or twice outside. A number of species, described by Skvortzov mainly from Lake Hanka, have been recorded from many places in our region but there are very few records outside: *Fragilaria magocsyi* Skv., *Eunotia rostellata* Skv., *Perone dimorpha* Skv., and *Heterothrix monochloron* Skv. The following species

Table 3. Widely distributed species in Primorya

Amphora ovalis Kütz.
A.ovalis var. *pediculus* Kütz.
Cymbella ventricosa Kütz.
C. sinuata Greg.
C. turgidula Grun.
C. affinis Kütz.
C. cistula (Hemp.) Grun.
Gomphonema angustatum (Kütz.) Rabenh.
G. anqustatum var. *productum* Grun.
G. olivaceum (Lyngb.) Kütz.
Rhopalodia gibba (Ehr.) O. Müll.
R. gibba var. *ventricosa* (Ehr.) Grun.
Hantzschia amphioxys (Ehr.) Grun.
Nitzschia palea (Kütz.) W. Sm.
Surirella angustata Kütz.
S. robusta var. *splendida* Ehr.
S. ovata Kütz.
S. ovata var. *pinnata* (W. Sm.) Hust.
Tribonema minus Hazen
Tribonema viride Pascher
Trachelomonas volvocina Ehr.
Scenedesmus quadricauda (Turp.) Breb.
Ulothrix zonata (Web. & Mohr.) Kütz.
Stigeoclonium tenue (Ag.) Kütz.

Table 4. Systematic composition of the freshwater algal flora in different types of water bodies in the Primorsky Region.

	Lakes	Reservoirs	Rivers	Streams
Cyanophyta	149\15.3	113\13.4	97\15.6	38\26.8
Euglenophyta	51\5.2	63\7.5	14\2.2	7\4.9
Pyrrophyta	17\1.7	10\1.2	4\0.6	
Chrysophyta	23\2.3	52\6.5	5\0.8	1\0.7
Bacillariophyta	290\29.8	237\28.1	307\28.1	71\50.0
Xanthophyta	20\0.2	34\4.0	19\3.0	3\2.1
Rhodophyta			4\0.6	2\1.4
Chlorophyta	419\46.0	334\39.6	177\28.5	20\14.4
Total	971\100	843\100	620\100	142\100

can be considered endemics of Asia: *Achnanthes hankensis* Skv., *Surirella alisoviana* Skv., *S. robusta* var. *hankensis* Skv., and also *Surirella pantocsekii* Skv., *S. tentsinensis* Skv., *Navicula globulifera* var. *nipponica* Skv., *N. scutelloides* var. *baicalensis* Skv., and *Gomphonema sphaerophorum* Ehr., which occur in this region too.

In general the freshwater algal flora of the Primorsky Region consists of widely distributed species,

Table 5. Genera most rich in number of species in lakes

Genera	Number of species
Cosmarium	66
Closterium	37
Nitzschia	36
Navicula	33
Staurastrum	32
Cymbella	31
Eunotia	29
Pinnularia	27
Trachelomonas	24
Anabaena	20

Table 6. Genera most rich in number of species in reservoirs

Genera	Number of species
Cosmarium	43
Closterium	32
Navicula	30
Trachelomonas	30
Cymbella	29
Nitzschia	26
Pinnularia	18
Eunotia	17
Oscillatoria	17
Staurastrum	15

but it also includes boreal, arcto-alpine, and arctic elements; these are, however, less numerous. There have also been identified a few species which are believed to have their main distribution in the Tropics: *Actinella brasiliensis* Grun. *Melosira cataractarum* Hust., *Cymbella turgidula* Grun., and *Staurastrum urinator* var. *brasiliensis* Grönbl. Genera with the highest number of species are shown in Table 2. These are mainly desmids and diatoms. These genera account for 31.4% of the whole species list. In temperate algal floras, these genera are normally among the leading as regards number of species. 141 of the genera (40.8 %) are monotypic.

Table 7. Genera most rich in number of species in rivers

Genera	Number of species
Navicula	50
Nitzschia	40
Cymbella	40
Closterium	29
Pinnularia	25
Eunotia	24
Achnanthes	21
Gomphonema	21
Surirella	14
Synedra	12

Table 8. Genera most rich in number of species in streams

Genera	Number of species
Cymbella	17
Gomphonema	12
Nitzschia	12
Phormidium	11
Eunotia	8
Navicula	7
Closterium	7
Lynqbya	7
Cylindrospermum	6
Ulothrix	5

43% of the taxa were found in only one of the 180 studied water bodies. 10.5% (218) of the taxa in more than 5% of them, and only 2.7% (55 taxa) in more than 10%. Widely distributed in this region, occurring in more than 20% of the water bodies, are only 24 taxa (1.1%). These are shown in Table 3. They are mainly diatoms, but also one euglenophyte, two xanthophytes and three chlorophytes. They are cosmopolitans or at least widely distributed in the world. Most of them prefer oligohalobic conditions. As regards saprobility, they are mainly oligo-saprobic, but alpha- and beta-mesosaprobic are also present. All the benthic and periphytic species belong to the widely distributed group of species.

A comparison of the species compositions of the different types of water bodies in the Primorsky Region (Table 4) shows that the highest species number is found in the lakes – 971 species (1318 taxa), which account for 57.9% of the total species number. Chlorophytes and diatoms constitute the major part of the species list, with 419 species (46%) and 290 species (29%), respectively. The leading genera are the same as for the whole flora (Table 5). Our lakes are very different in morphometry and other characteristics of importance for differences in species composition. Many of them are swampy and thus responsible for abundant development of desmids, others are rich in nutrients and warm in summer, thus favouring the development of Chlorococcales. In the northern part of the region, the lakes are connected with the lower parts of mountain rivers; they are cold and oligotrophic, and the dominant algae here are the diatoms.

The species composition of Lake Hanka is the most diverse, 300 taxa, 224 of which are diatoms growing in masses on silt and sand substrates. In the southern swampy part of the lake, desmids are abundant too. It should be mentioned that the species composition of this lake – one of the largest lakes in Asia – is still not well studied, and its species list should be expected to number 800–1000 species.

32% of the lake species only occur in one lake. A wide occurrence, in more than 25% of the lakes, have only 5.1% (70 taxa). Chlorophytes have a somewhat higher occurrence rate, mainly caused by the genera *Scenedesmus (S. bijugatus* Kütz., *S. falcatus* Chod., *S. obtusus* Meyen, *S. acuminatus* (Lagerh.) Chod.) and *Coelastrum microporum* Näg., *Pediastrum tetras* (Ehr.) Ralfs occurred in almost half of the lakes and had the highest score. Among the cyanophytes, *Merismopedia tenuissima* Lemm., *Snowella lacustris* (Chod.) Kom.& Hindák, and often species of *Microcystis* and *Gloeocapsa* also occur. The occurrence of the Nostocales is more specific for each lake, only *Aphanizomenon flos-aquae* (L.) Ralfs was occurring in every third lake.

Species composition of the reservoirs in the Primorsky Region was also rather diverse, with altogether 843 species (1083 taxa), 50.9% (Table 4). The highest number of species was noted in the Artemovsky Reservoir: 537 species (Barinova, 1990), and in the water-cooling reservoir of the Primorsky power station: 482 species (Kukharenko & Gontcharov, 1991). As in the lakes, chlorophytes dominate above the diatoms – 334 species (39.4%) and 222 species (26.3%), respectively. Cyanophytes in reservoirs account for 113 species (13.4%). Euglenophytes, xanthophytes, chrysophytes, and dinophytes are more numerous here than in the lakes. The list of genera with highest species number (Table 6) is very similar to that of the lakes and of the whole flora, but there are many smaller species, e.g. in *Trachelomonas.*

Similarity in species composition between the reservoirs is higher than between the lakes. 16% of the species occur in four of the seven reservoirs, and the percentage of species found in only one reservoir is lower, 23.9%. Besides taxa widely distributed in all water bodies, a few are especially common in reservoirs: *Eudorina elegans* Ehr., *Cyclotella stelligera* Cleve & Grun., and *Synedra ulna* (Nitzsch) Ehr., *Pediastrum tetras* (Ehr.) Ralfs, *P. duplex* Meyen, *Closteriopsis acicularis* (G. M. Smith) Belcher & Swale, *Kirchneriella obesa* (W. West) Schmidle, *Gonatozygon monotaenium* De Bary, and *Hyalotheca dissiliens* (J. E. Smith) Bréb. The chlorophytes have more species widely distributed in reservoirs than the diatoms.

Rivers amount to almost half of the number of water bodies studied in the region. The list of algae found in rivers contains 620 species, 23.3%. Here the diatoms are dominating 307 species, 49.5%. Chlorophytes amount to only 177 species, 28.6%. Also the diversity of cyanophytes is lower – 97 species (109 taxa), 15.6%. The role of the other groups is small. 9 out of 10 leading genera are diatoms, only the desmid *Closterium* is number 4 by number of species (Table 7).

More than half of the taxa found have narrow distribution, occurring only in one river, but among diatoms this group is significantly smaller – 37.5%. Only 37 taxa have frequent occurrence, found in more 20% of the rivers studied. In addition to some diatoms, these are *Ulothrix zonata* (Web. & Mohr) Kütz., *Hydrurus foetidus* (Villars) Trevisan, *Stigeoclonium tenue* (Ag.) Kütz., and some species of *Phormidium.*

As true phytoplankton is almost absent in the Primorsky rivers, benthos and periphyton are the main types of algal communities here. By type of habitat, it is possible to distinguish three types of these communities: (1) communities on rocks in the central part of the rivers, (2) litoral communities, and (3) communities on mosses (Hakhina, 1934). Typical rheophilic species dominate in the first group. It can be filamentous chlorophytes and chrysophytes, on which katharobic and oligosaprobic diatoms obtain mass development, also covers of diatoms (*Ceratoneis arcus* (Ehr.) Kütz., *Meridion circulare* Ag., species of *Diatoma, Synedra, Cymbella*) and crusts of cyanophytes (*Phormidium autumnale* (Ag) Gom., *P. uncinatum* (Ag.) Gom., *Chamaesiphon incrustans* Grun., *Homoeothrix simplex* Woron).

In places with slow current, species of the genera *Spirogyra, Zygnema, Tribonema* and *Oscillatoria* are present in masses, just as *Rhopalodia gibba* (Ehr.) O. Müll., *Aulacoseira italica* (Ehr.) Sim., *Tabellaria fenestrata* (Lyngb.) Kütz. and *T. flocculosa* (Roth) Kütz., and species of *Eunotia.*

Algal communities on mosses are the most diverse. Several taxa of diatoms usually grow here: *Didymosphenia geminata* (Lyngb.) M. Schmidt, *Navicula mutica* Kütz., *Achnanthes lanceolata* (Bréb.) Grun., *A. kryophila* Peters., *Stauroneis anceps* Ehr., *Eunotia polydentula* Brun, *Frustulia rhomboides* (Ehr.) De Toni, and *Surirella pantocsekii* Skv.

In the middle and lower parts of the rivers algal diversity usually increases, and at the same time x-o-saprobic communities change to mesosaprobic.

In comparison, the flora of algae in smaller streams is poor – 142 species (8.6%). Diatoms with 71 species account for 50%. The second place is occupied by the cyanophytes with 38 species (42 taxa), 26.8%. Chlorophytes amount to 20 species (24 taxa), 14.1%. The leading genera are less rich in species than in rivers and other water bodies (Table 8). Besides diatoms there are chlorophytes, such as *Ulothrix* and the desmid *Closterium*, and cyanophytes – *Lyngbya* and *Cylindrospermum*.

The present survey shows the richness of the algal flora in the Primorsky Region in comparison to the neighbouring regions. It also shows the differences within the region and between the algal floras of the different types of water bodies. This implies different geographic distribution types among the algae.

Acknowledgments

My special thanks are due to Dr Jørgen Kristiansen, University of Copenhagen, for critically reading the manuscript and correcting the language.

References

Barinova, S., 1990. Analysis of the algal flora of the Artemovsky Reservoir (Primorsky Region). In L. N. Vassilieva (ed.), Cryptogamic investigations in the Far East. Far Eastern Branch Acad. Sci. USSR, Vladivostok: 29–44. (In Russ.)

Getzen, M. V., 1978. Peculiarities of the flora of the eastern part of the Bolshesemelskay tundra: Algae. In M. V. Getzen (ed.), Flora and fauna of the waterbodies of North Europe. Nauka, Leningrad: 11-20. (In Russ.)

Hakhina, A. G., 1934. Algae of the upper tributaries of the Suifun River. Materials to the algal flora of the Far East Region. Vestnik Far Eastern Branch Acad. Sci. USSR. 10: 77–88. (In Russ.)

Kukharenko, L. A., 1974. Review of the algological investigations in the Soviet Far East. In L. N. Vassilieva (ed.), Cryptogamic plants of the Soviet Far East. Far East Science Centre Acad. Sci. USSR, Vladivostok: 5–16. (In Russ.)

Kukharenko, L. A., 1989. Algae of the freshwater bodies of the Primorsky Region. Far Eastern Branch Acad. Sci. USSR, Vladivostok. 143 pp. (In Russ.)

Kukharenko, L. A. & A. A. Gontcharov, 1991. Algae of the fish ponds and the water-cooling reservoir of a power station (Primorsky Region). Far Eastern Branch Acad. Sci. USSR, Vladivostok. 42 pp. (In Russ.)

Medvedeva, A. & S. S. Barinova, 1990. Algological investigations in the Far East (Review 1971–1986). In L. N. Vassilieva (ed.), Cryptogamic investigations in the Far East. Far East Branch Acad. Sci. USSR, Vladivostok: 4–22. (In Russ.)

Muzafarov, A. M., 1981. Geographical distribution of Algae. Izd. Fan. Taskent. 238 pp. (In Russ.).

Safonova, T. A., 1984. The algal flora, its peculiarities and role in biological productivity of the waterbodies of Western Siberia. In L. S. Berdichev, G. I. Galasey & G. B. Iogansen (eds), Biological resources of inland waterbodies of Sibiria and the Far East. Nauka, Moscow: 108–117. (In Russ.)

Vassilieva, I. I., 1989. Analysis of species composition and dynamics of the development of algae in waterbodies of Yakutia. Yakut Sci. Centre Sib. Branch Acad. Sci. USSR, Yakutsk. 48 pp. (In Russ.)

Hydrobiologia **336**: 99–106, 1996.
J. Kristiansen (ed.), Biogeography of Freshwater Algae.

10. Snow algae of the Windmill Islands region, Antarctica

H. U. Ling
Antarctic Division, Channel Highway, Kingston, Tasmania, Australia 7050

Key words: snow algae, taxonomy, ultrastructure, new species

Abstract

A list of the 24 species of snow algae identified from the region, a resume of what is currently known about the major species, and avenues for further research are provided. New species discovered include 2 *Desmotetra* spp., one *Chlorosarcina* sp., 2 *Chloromonas* spp. and a *Palmellopsis* sp. Several of these are from genera whose members have previously been found only in the soil flora. Not only was it necessary to elucidate the life cycle of these species, but it was also essential to examine them ultrastructurally to determine their taxonomic positions.

Introduction

The Windmill Islands region, Wilkes Land, consists of a group of islands and peninsulas centred around the Australian Antarctic station of Casey. In summer, semi-permanent snow banks and permanent ice tinted grey, red, orange and green are a common feature of the landscape in this region. The colours are usually caused by blooms of algae which thrive in the increased temperature and day length. Most of the algae appear to be identifiable with previously described species. However, on closer examination, many of the dominant species were found to be new and of uncertain generic status. Not only was it necessary to culture the algae to elucidate their life cycles, but it was also essential in some cases to investigate the ultrastructure, especially of the flagellar apparatus, of the cells to determine their taxonomic positions.

This paper summarises what is known about the region's snow algae and suggests what further research is required.

Materials and methods

Collection, examination and culture methods were as described by Ling & Seppelt (1993). Samples for scanning (SEM) electron microscopy were prepared according to Marchant & Thomas (1983). For transmission electron microscopy (TEM) the growth medium was used for diluting fixatives and for rinses. Cells were fixed in 2% glutaraldehyde for 1 hour followed by 1% osmium tetroxide. After dehydration in acetone, cells were embedded in Spurr's resin. Ultra-thin sections were stained in aqueous uranyl acetate for 5 minutes, then lead citrate for 10 minutes.

Results

To date twenty four species of snow algae have been observed (Table 1). All of the major species have been isolated into culture and are currently maintained at the Australian Antarctic Division. Most of the cultures are clonal, or at least unialgal. The majority of the cultures, especially species with mucilage envelopes, are contaminated with bacteria.

The algae can be conveniently divided into groups according to the colour they impart to the snow, although some algae may colour snow differently depending on the stage of their life cycle.

Mesotaenium berggrenii *(Wittr.) Lagerheim forma*

Grey snow is dominated by a form of *Mesotaenium berggrenii*. This form has a reddish-brown cell sap instead of the bluish-purple found in the type. It is the most abundant and widespread of the region's snow

Table 1. Snow algae recorded from the Windmill Islands region. A (aquatic) and T (terrestrial) indicates other habitats of these algae. and C (culture) that these algae are currently maintained in culture at the Australian Antarctic Division.

CHLOROPHYTA			
Chlamydomonas pseudopulsatilla Gerloff			
Chlamydomonas sp.			C
Chloromonas brevispina (Fritsch) Hoham, Roemer & Mullet			
Chloromonas polyptera (*Scotiella polyptera* Fritsch)			C
Chloromonas rubroleosa Ling & Seppelt			C
Chloromonas sp. 1			C
Chloromonas sp. 2			
Chlorosarcina sp.			C
Desmotetra sp. 1			C
Desmotetra sp. 2			C
Mesotaenium berggrenii (Wittr.) Lagerheim forma			C
Monoraphidium sp.			
Palmellopsis sp.			C
Prasiococcus calcarius (Boye-Pet.) Vischer	A	T	C
Prasiola crispa (Lightf.) Meneghini	A	T	C
Raphidonema helvetica Kol			
Raphidonema nivale Lagerheim			
Raphidonema tatrae Kol			
Stichococcus bacillaris Näigeli		T	C
Stichococcus minutus Grintzesco & Peterfi			
Ulothrix sp. 1			C
Ulothrix sp. 2			
XANTHOPHYTA			
Ellipsoidion sp.?			
CHRYSOPHYTA			
unidentified chrysophyte			

algal species although its occurrence as dust-like particles masks its presence. Results of research on this species has already been published (Ling & Seppelt, 1990).

Chloromonas rubroleosa *Ling & Seppelt*

The most common of the red snow species was previously identified as *Protococcus nivalis* (Bauer) Agardh (Llano, 1965) which is synonymous with *Chlamydomonas nivalis* (Bauer) Wille. Research has established it as a new *Chloromonas rubroleosa* (Ling & Seppelt, 1993) whose green chloroplasts are masked by fine, red-pigmented oil droplets.

Chlorosarcina *sp.*

A second red snow species proved to be even more unusual. It was usually found as clumps of red spores (Figure 1) in pink snow collected from the islands rather than from the peninsulas. The spores are very similar to cells previously described as *Chlamydomonas antarcticus* Wille (Gain, 1912; Kol, 1968) but lack pyrenoids. Successful germination of the spores (Figure 2) produced cultures containing loose-cell, green, cubical packages held together by mucilage. The cells, about 7–13 μm in diameter, are oblong to spherical when solitary but becoming angular from rapid division. Each cell has a parietal, cup-shaped chloroplast which may be bi-lobed in older cells. Neither pyrenoids nor contractile vacuoles have been observed. As the cultures approach the stationary phase, the cell packages develop into colonies with dichotomously branched mucilaginous stalks characteristic of the genus *Hormotila*. Biflagellate, elongate zoospores without any stigmata are often produced when cultures are transferred into fresh media.

The loose-cell, semi-sarcinoid nature of the vegetative cells (Figure 3), their mode of division, absence of a pyrenoid and production of *Hormotila*-like stages make classification of this alga difficult. There are comparable species in the *Chlorosarcinopsis* but the cells in this genus possess pyrenoids. The plant has tentatively been placed in the genus *Chlorosarcina* because it has various characteristics in common with those of *Chlorosarcina stigmatica* Deason (1959) but it must be borne in mind that there is considerable controversy surrounding *C. stigmatica* (Sluiman & Blommers, 1990) and the genus *Chlorosarcina* itself. Confirmation that the alga belongs in the Chlorophyceae would require a study of the ultrastructure, especially of the flagellar basal apparatus of the zoospores. A preliminary study suggests a chlorophycean flagellar apparatus, but further research is required.

An investigation into the life history and ecology of this species is now complete but the results cannot be published in the absence of flagellar apparatus data.

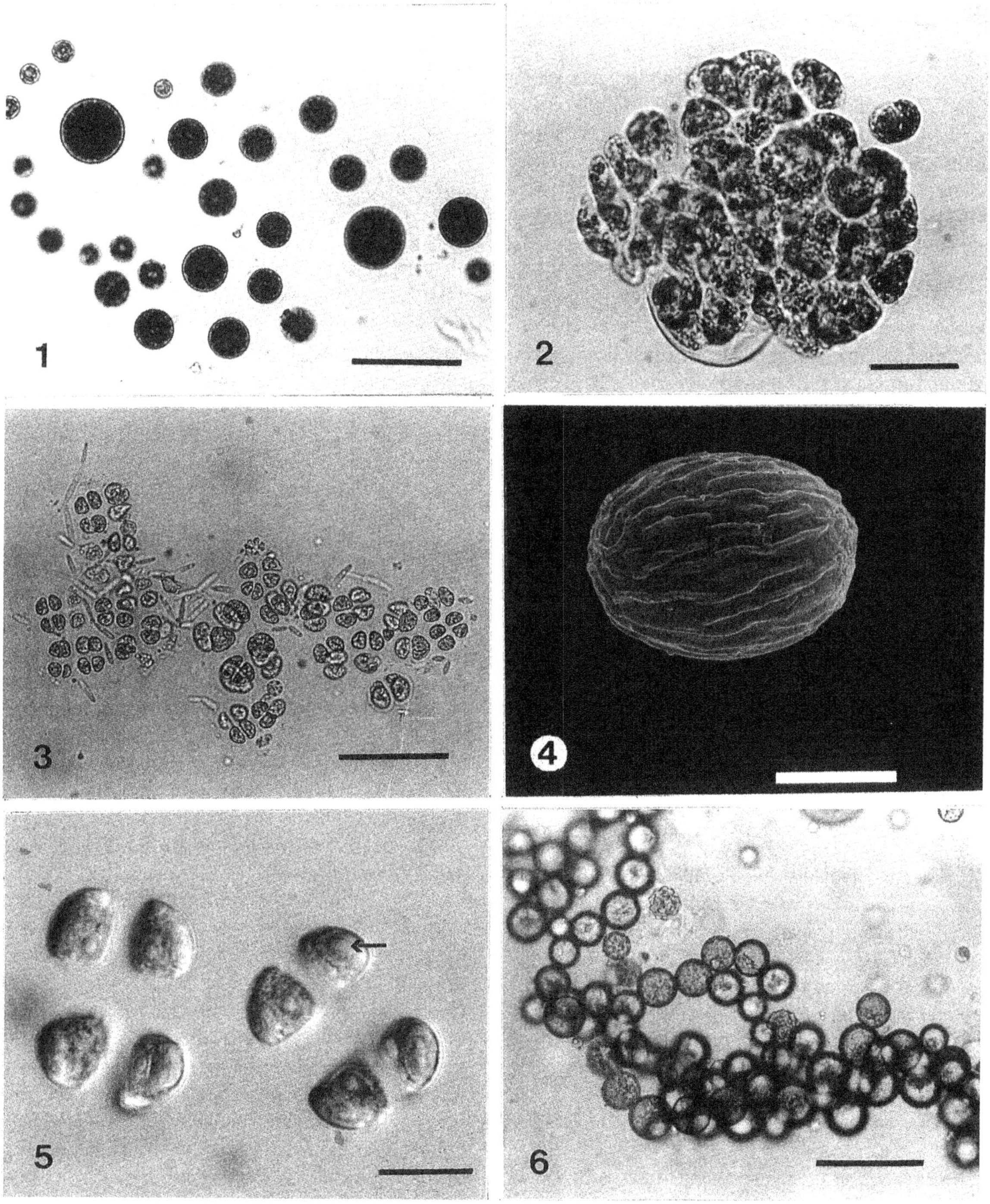

Figures 1–6. Figures 1–3. *Chlorosarcina* sp. Figure 1. Clump of spores. Note wide size range and mucilage envelope around spores. Scale bar = 50 μm. Figure 2. Spore germination. Release of multicellular spore contents. Scale bar = 20 μm. Figure 3. Vegetative cells from field collection. Rod-shaped cells are yeast cells. Scale bar = 50 μm.

Figure 4. *Chloromonas polyptera.* SEM of spore. Scale bar = 10 μm.

Figures 5–6. *Desmotetra* sp. 1. Figure 5. Vegetative cells from field collection preserved in acetic Lugol. Note nuclei and pyrenoid (arrow). Scale bar = 10 μm. Figure 6. Spores. Scale bar = 50 μm.

Chloromonas polyptera (Scotiella polyptera *Fritsch)*

Orange snow is dominated by reddish-orange spores (Figure 4) that are identical to cells previously described as *Scotiella polyptera* Fritsch by various authors (Fritsch, 1912; Kol, 1971; Akiyama, 1979). Hoham et al. (1983) found what they believed to be *Scotiella polyptera* from snow in the United States. Their investigation of the life history of this alga led them to transfer *S. polyptera* to the genus *Chloromonas*. The life history of the Windmill Islands' *S. polyptera* has also been investigated. Although it is also a *Chloromonas* it differs from the American one in spore shape and in the size of the motile green cells. Also, the spores appear to be asexual in contrast to the sexual spores of the American alga. The spores germinate to produce four daughter cells. Although the zoospores from the two isolated cultures do not have stigmata, some zoospores from a mixed culture derived from a large number of spores do have a stigma. In this respect the zoospores are similar to the American ones.

There is little doubt that the Windmill Islands' *Chloromonas* is the same as the original described by Fritsch, however, it is believed that the American *Chloromonas* described by Hoham et al. (1983) may be a separate species. The results are being prepared for publication.

Green snow is caused by a multiplicity of species whose life cycle stages present a bewildering array of vegetative cells, zoospores, gametes, planozygotes, zygospores and akinetes which would be impossible to classify without intensive research and culture. Adding to this confusion, vegetative stages of *Chlorosarcina* sp. and *Chloromonas polyptera* are also frequent components of green snow. All the dominant species, *Desmotetra* sp. 1 and 2, *Palmellopsis* sp. 1 and *Ulothrix* sp. 1 and 2 are unusual and or problematical. The two *Desmotetra* spp. have been the most difficult.

Desmotetra *spp.*

Desmotetra is a relatively new genus erected by Deason & Floyd (1987) principally to accommodate their finding of a new pyrenoid type (lacking starch sheath) in *Chlorosarcina stigmatica*. Additional reasons given are the retention of the flagellar apparatus, a stigma and contractile vacuoles in the vegetative cells; however, these were not included in the genus diagnosis.

Individual cells of *Desmotetra* sp. 1 are oval to globular (although they are often hemispherical or quarter of a sphere in shape as a result of recent division) with a single parietal, bowl-shaped chloroplast with an indistinct pyrenoid (Figure 5). The pyrenoid is perceived only in occasional cells, especially recently divided cells stained with acetic Lugol. One to three contractile vacuoles may usually be observed in each cell. The cells divide to form loose-cell cubical packets of from 2–16 cells held together in a common gelatinous matrix.

The cells produce, separately, zoospores of two different sizes both with stigmata. The larger zoospores are oblong in shape measuring 4–8 by 9–11.5 μm. The smaller zoospores are initially spindle-shaped later becoming pyriform to spherical, measuring 3.56 μm in diameter. Only these small zoospores, or gametes, were observed to fuse in pairs to form smooth-walled golden spores (Figure 6). Mature spores can be germinated with comparative ease, each spore releasing a large number of cells.

An investigation of the ultrastructure of the cells revealed a pyrenoid traversed by parallel bands of thylakoids (Figure 7). Starch grains are scattered through the chloroplast and do not form a distinct sheath around the pyrenoid as they do in many green algae. The flagellar apparatus of the zoospores have non-overlapping basal bodies suggesting a clockwise absolute orientation (Figure 8). Figure 8 is comparable with the figure of a chlorophycean flagellar apparatus depicted in Figure 4 in Deason (1989) but more ultrastructural work is required.

The results on *Desmotetra* sp. 1 are also being prepared for publication

The morphology of the vegetative cells and also the processes of vegetative reproduction and zoosporogenesis in *Desmotetra* sp. 2 cannot be distinguished from those of *Desmotetra* sp. 1. In fact, it is very difficult to distinguish between field collections of vegetative cells of *Desmotetra* sp. 1, *Desmotetra* sp. 2 and *Chlorosarcina* sp. and from the presence of mixed spores there is little doubt that the three species do grow together.

Sexual reproduction has also been observed in field collections of *Desmotetra* sp. 2. No morphologically distinct gametes were observed, the zoospores normally associated with vegetative propagation fusing in pairs to form spores with a thick wall ornamented with ridges in the form of 5 to 7-sided polygons (Figures 9, 10). Attempts to germinate individually isolated spores have so far been unsuccessful although a few odd spore germinations have been observed in old field samples irrigated with fresh media.

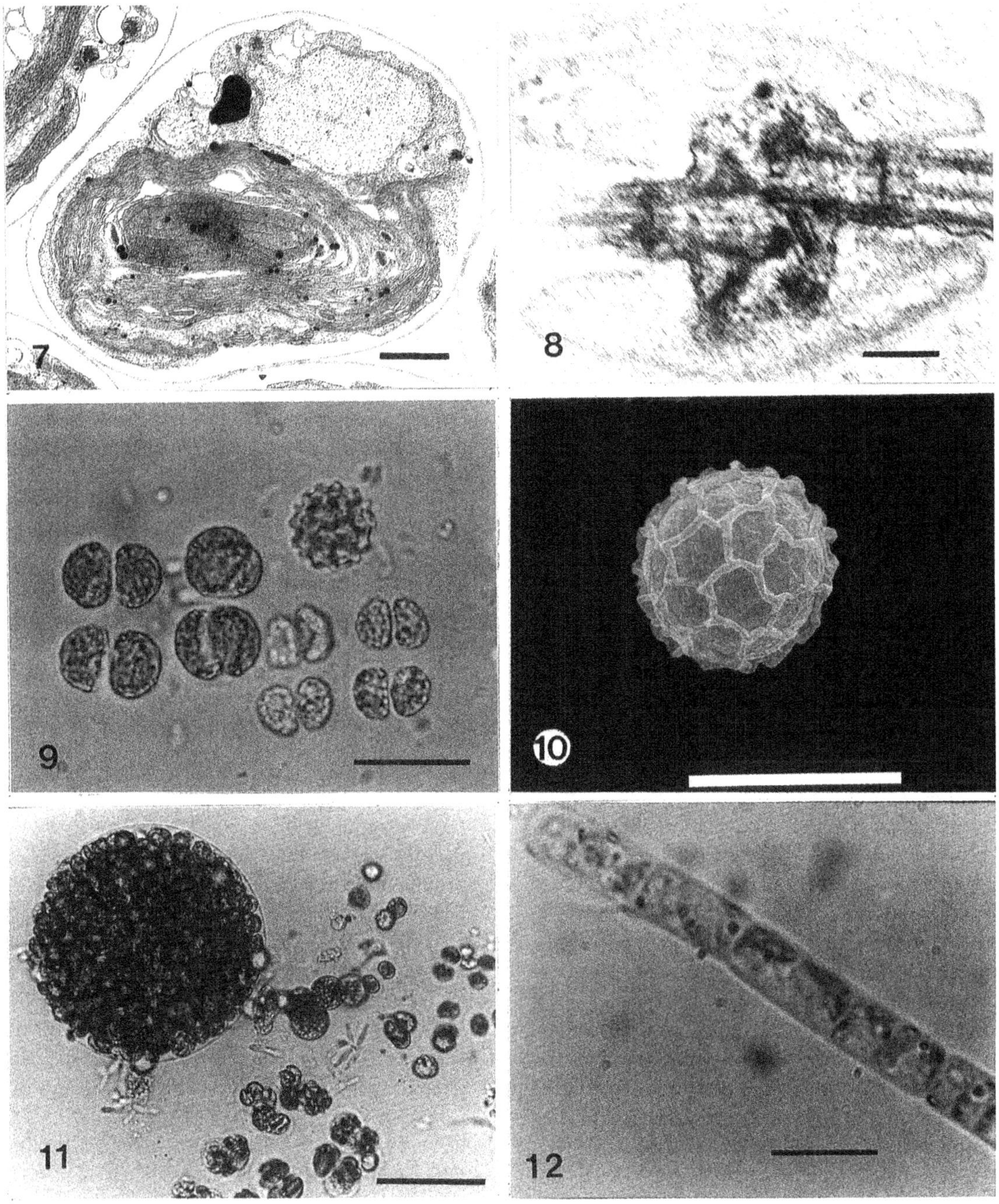

Figures 7–12. Desmotetra sp. 1. Figure 7. TEM of vegetative cell showing pyrenoid traversed by parallel bands of thylakoids. Scale bar = 1.0 μm. Figure 8. TEM of the flagellar basal apparatus of a zoospore as seen from the cell posterior. Scale bar = 200 nm.
Desmotetra sp. 2. Figure 9. Vegetative cells and a single spore. Scale bar = 20 μm. Figure 10. SEM of spore. Scale bar = 10 μm.
Figure 11. *Palmellopsis* sp. Upper left, cell that has divided into a large number of zoospores. Vegetative cells of *Desmotetra* sp. on right. Scale bar = 50 μm.
Figure 12. *Ulothrix* sp. 1. Single filament. Scale bar = 10 μm.

The spores agree with cells found in yellow snow, South Orkneys and described as *Trochiscia antarctica* Fritsch (1912, p. 116; Pl. 1, Figure 30). On present evidence the alga has been placed in *Desmotetra* because of its resemblance to *Desmotetra* sp. 1. Research on the ultrastructure of the cells is essential to confirm this.

Palmellopsis *sp.*

Another green snow alga has tentatively been identified as a *Palmellopsis*. Individual cells are spherical with an urn-shaped chloroplast and up to four contractile vacuoles in the portion of the cell with clear cytoplasm. This alga is similar to *Palmellopsis muralis* Bold & King (Bold et al., 1981) but lacks a pyrenoid. Asexual reproduction is by biflagellate, pyriform to ovoid zoospores, averaging 10×13 μm in size, of which from a few up to about a hundred may be produced by each cell (Figure 11). Two anterior contractile vacuoles were observed in some zoospores and there is a stigma situated about a quarter down the side of the cell and a median-posterior nucleus. Further research, especially at the ultrastructural level, is required.

Ulothrix *spp.*

Uniseriate green filaments (Figure 12) are often a component of the snow algal flora. They are similar to plants described previously as *Klebsormidium* sp. A by Broady (1981). There appear to be two species in the Windmill Islands region. The larger filaments have been observed to produce zoospores each with a single stigma and either two or four flagella. Pre-emergent and settled zoospores of the smaller filaments usually have two stigmata each though some may have only one stigma. The number of flagella on the zoospores have yet to be determined.

From the shape of the zoospores and the attachment of the flagella it appears the plants belong to the *Ulothrix* genus. *Klebsormidium* zoospores lack stigmata and are dorsiventral in shape with subapically and asymmetrically inserted flagella (Lokhorst, 1991).

Of the minor species, spores of *Chloromonas brevispina* are often found with spores of *Chlorosarcina* sp. The life history and ecology of *C. brevispina* has been well documented by Hoham et al. (1979). *Chloromonas* sp. 1 cells each have a pair of flattened flagella and forms spores with fine granules on the walls. *Prasiola crispa, Prasiococcus calcarius* and *Stichococcus bacillaris* are typical soil algae and may have been deposited from soil onto the snow/ice by melt water, though it is believed the latter two species are capable of multiplying in snow. The true identity of *Ellipsoidion* sp. and the unidentified chrysophyte has yet to be established. The chrysophyte consists of thick-walled pyriform to spherical cyst-like cells with a plug at one end. Two golden brown discoid chloroplasts are clearly visible in young cells while some of the old cells may have an aperture and be devoid of contents.

All the snow algae cultures grow well at the temperature of 3 °C, where they are currently maintained. *Chloromonas rubroleosa* and *Chloromonas polyptera* cultures died when grown at 10 °C. *Chlorosarcina* sp. cultures were in poor condition with most of the cells dead at 10 °C. *Palmellopsis* sp. and *Desmotetra* sp. 1 cultures were still growing at 10 °C but died at 15 °C while *Desmotetra* sp. 2 and *Chloromonas* sp. 1 cultures were barely alive at 15 °C.

A survey of the non-marine algae of the region is nearing completion. Though some spores of the various species have been observed in soil samples collected near snowdrifts, vegetative cells of the various snow algae have not been observed growing in soil or lake sites.

Discussion

From the species list (Table 1) it is evident that only very few of the species have been identified with certainty. In a taxonomic survey it is reasonable to expect a small number of species to be problematical and also an even smaller number, usually the rarer ones, to be new. However, when more than half of the species, and especially when practically all of the dominant species, are problematical and/or new then the task of identification becomes difficult. The normal procedure is to determine the life cycle of a difficult species from frequent field collections or to isolate it into culture. The former has not been possible with the chlorosarcinalean species because of the similarities of the vegetative cells as well as the zoospores of the various species. Even the latter procedure has been inadequate. When the task necessitates the additional examination of the ultrastructure of the flagellar basal apparatus of the zoospores just to determine what class (Chlorophyceae or Pleurastrophyceae) the alga belongs to then the task becomes onerous. It becomes nearly impossible when one is on a limited employment contract to identify, not just the snow algae, but all the non-marine algae of the Windmill Islands' region. No apology is offered for the incomplete species list. On the contrary, as a paper belonging to a collection of biogeographical papers, it serves as a reminder that unless one is certain of the identity of the algae under discussion, it can be misleading to draw conclusions on their biogeography.

Snow algal flora is thought to be dominated generally by members of the Chlamydomonadaceae (Kol, 1968; Hoham, 1980). Though the Chlamydomonadaceae are an important part of the snow algal flora of the Windmill Islands region, the dominant alga is a member of the Mesotaeniaceae. Even more remarkable is the fact that three of the major species are identifiable with members of the Chlorosarcinales of which the genera and species so far described are, almost without exception, members of the soil algal flora (Bold & Wynne, 1985). However, recent ultrastructural studies by various workers indicate that many of the formerly chlorosarcinalean algae actually belong to three different classes of the Chlorophyta and that the order is artificial and should be abandoned (Deason, 1989). With the systematics of the Chlorosarcinales in a state of flux, research on the ultrastructure, especially of the flagellar apparatus of these snow algae would help clarify the taxonomic position of members of this order. *Chlorosarcina, Desmotetra* and *Palmellopsis* all contain only a small number of known species and the flagellar basal apparatus of only *Desmotetra stigmatica* (Deason) Deason & Floyd and *Chlorosarcina stigmatica* strain T105 are known.

The origins of these species are unclear. Though their closest relatives are members of the soil algal flora, their absence from the soil flora, their ability to complete their entire life cycle in the snow and form blooms in melting snow and ice where the temperature remains at 0–1 °C and their susceptibility to temperatures above 10–15 °C leave little doubt that they are not only true snow algae but also obligately cold adapted. The only other known obligately cold-adapted taxa belong to the Chlamydomonadaceae and Ulotrichaceae (Seaburg et al., 1981). *Chloromonas pichinchae* (Lagerh.) Wille from Washington (USA) snow grew best at 1 °C, almost as well at 5 °C (Hoham, 1975) while *Chloromonas rubroleosa* have optimum growth temperatures of about 1–4 °C (Ling & Seppelt, 1993). Other closely related algae from Antarctica such as *Chlorosarcinopsis* sp. isolated from an algal mat and *Tetracystis* sp. isolated from a glacial meltstream have temperature growth ranges of 2–20 °C and 2–25 °C for different clones of the former and an additional temperature range of 2–30 °C for different clones of the latter (Seaburg et al., 1981).

The results on *Desmotetra* sp. 1 also throw light on the controversy regarding the identity of the original isolate of *C. stigmatica* versus *C. stigmatica* strain T 105. *C. stigmatica* was initially described (Deason, 1959) as lacking pyrenoids. However, in a recent ultrastructural study by Deason & Floyd (1987) pyrenoids were unexpectedly found in the original isolate. The authors suggested that either pyrenoids were present in the original material but had been missed or pyrenoids have developed since then. Strain T 105 is a sarcinoid alga from the Innsbruck (Austria) Algal Culture collection. It possesses all the characteristic features mentioned in the original light microscopic description of *C. stigmatica* by Deason (1959), including the all-important lack of pyrenoids. This last feature was confirmed in ultrastructural studies by both Gärtner et al. (1988) and Sluiman & Blommers (1990). From the results presented here it is very likely that the presence of the pyrenoid was missed when the species was first described (Deason, 1959). (The pyrenoid in *Desmotetra* sp. 1 was detected only when the cells were examined ultrastructurally.) The oversight was remedied when Deason & Floyd (1987) created the new genus *Desmotetra* and renamed *C. stigmatica* as *Desmotetra stigmatica*. Strain T 105 should perhaps be described as a new species. Its production of gibbous akinetes (Gärtner et al., 1988) may be distinctive.

The type species of *Mesotaenium berggrenii* has previously been reported in the Antarctic region from Wiencke Island (Gain, 1912; Wille, 1924) and Signy Island (Kol, 1972) and elsewhere from equatorial glaciers of New Guinea (Kol & Peterson, 1976) and alpine regions of Ecuador, North America and Europe, and Greenland (Kol, 1968).

Acknowledgments

This work was conducted while the author was an expeditioner (wintering biologist) with the Australian Antarctic Division. The author would like to thank Dr Rod Seppelt for advice during the course of the investigation, Anna McEldowney and Gerry Nash of the Antarctic Division for technical assistance, and the Centre for Aquatic Resources Utilization and Management, Deakin University for financial support in the completion of this manuscript.

References

Akiyama, M., 1979. Some ecological and taxonomic observations on the colored snow algae found in Rumpa and Skarvsnes, Antarctica. Mem. Natl Inst. Polar Res., Spec. Issue 11: 27–34.

Bold, H. C., R. D. Slocum, J. M. King & F. J. MacEntee, 1981. *Palmellopsis muralis* sp. nov. (Chlorophyceae, Tetrasporales, Palmellaceae). J. Phycol. 17: 9–14.

Bold, H. C. & M. J. Wynne, 1985. Introduction to the algae. 2nd edn. Prentice–Hall, Englewood Cliffs, New Jersey, 720 pp.

Broady, P. A., 1981. Non-marine algae of Cape Bird, Ross Island and Taylor Valley, Victoria Land, Antarctica. MUPAS Report No. 37. University of Melbourne.

Deason, T. R., 1959. Three Chlorophyceae from Alabama soil. Am. J. Bot. 46: 572–8.

Deason, T. R., 1989. A re-examination of the green algal taxon Chlorosarcinales – an ultrastructural approach. Critical Reviews in Plant Sciences 8: 259–272.

Deason, T. R. & G. L. Floyd, 1987. Comparative ultrastructure of three species of *Chlorosarcina* (Chlorosarcinaceae, Chlorophyta). J. Phycol. 23: 187–195.

Fritsch, F. E., 1912. Freshwater algae collected in the South Orkneys by Mr R. N. Rudmose Brown, B. Sc., of the Scottish National Antarctic Expedition, 1902-04. J. linn. Soc., Bot. 40: 293–338.

Gain, L., 1912. La flore algologique des régions antarctiques et subantarctiques. Deuxième Expédition Antarctique Française, (1908–1910), commandée par le Dr Jean Charcot, Sciences Naturelle Documents Scientifiques-Paris, 156–218.

Gärtner, G., A. Hofer & E. Ingolic, 1988. Morphological and taxonomical observations on some strains of *Chlorosarcina*, *Chlorosarcinopsis* and *Planophila* (Chlorophyta, Chlorosarcinales), with special reference to 'vegetative cell division'. Arch. Protistenkd. 135: 119–131.

Hoham, R. W., 1975. Optimum temperatures and temperature ranges for growth of snow algae. Arctic and Alpine Research 7: 13–24.

Hoham, R. W., 1980. Unicellular chlorophytes – snow algae. In E. R. Cox (ed.), Phytoflagellates. Elsevier/North Holland, New York: 61–84.

Hoham, R. W., J. E. Mullet & S. C. Roemer, 1983. The life history and ecology of the snow alga *Chloromonas polyptera* comb. nov. (Chlorophyta, Volvocales). Can. J. Bot. 61: 2416–2429.

Hoham, R. W., S. C. Roemer & J. E. Mullet, 1979. The life history and ecology of the snow alga *Chloromonas brevispina* comb. nov. (Chlorophyta, Volvocales). Phycologia 18: 55–70.

Kol, E., 1968. Kryobiologie. Biologie und Limnologie des Schnees und Eises. I. Kryovegetation. Die Binnengewässer 24: 1–216. E. Schweizerbart'sche Verlagsbuchhandlung, Stuttgart.

Kol, E., 1971. Green snow and ice from the Antarctica. Ann. Hist. Nat. Mu. Nat. Hung. 63: 51–56.

Kol, E., 1972. Snow algae from Signy Island (South Orkney Islands, Antarctica). Ann. Hist. Nat. Mu. Nat. Hung. 64: 63–70.

Kol, E., & J. A. Peterson, 1976. Cryobiology. In G. S. Hope, J. A. Peterson, I. Allison & U. Radok (eds), The Equatorial Glaciers of New Guinea. Balkema, Rotterdam: 81–91.

Ling, H. U. & R. D. Seppelt, 1990. Snow algae of the Windmill Islands, continental Antarctica. *Mesotaenium berggrenii* (Zygnematales, Chlorophyta) the alga of grey snow. Antarctic Science 2: 143–148.

Ling, H. U. & R. D. Seppelt, 1993. Snow algae of the Windmill Islands, continental Antarctica. 2. *Chloromonas rubroleosa* sp. nov. (Volvocales, Chlorophyta) an alga of red snow. Eur. J. Phycol. 28: 77–84.

Llano, G. A., 1965. The flora of Antarctica. In T. Hatherton (ed.), Antarctica. Reed, Wellington: 331–351.

Lokhorst, G. M., 1991. Synopsis of Genera of Klebsormidiales and Ulotrichales. Crypt. Bot. 2/3: 274–288.

Marchant, H. J. & D. P. Thomas, 1983. Polylysine as an adhesive for the attachment of nanoplankton to substrates for electron microscopy. J. Microsc. 131: 127–129.

Seaburg, K. G., B. C. Parker, R. A. Wharton, Jr. & G. M. Simmons, Jr., 1981. Temperature-growth responses of algal isolates from antarctic oases. J. Phycol. 17: 353–360.

Sluiman, H. J. & P. C. J. Blommers, 1990. Ultrastructure and taxonomic position of *Chlorosarcina stigmatica* Deason (Chlorophyceae, Chlorophyta). Arch. Protistenkd. 138: 181–190.

Wille, N., 1924. Susswasseralgen von der Deutschen Sudpolar-Expedition auf dem Schiff 'Gauss'. In E. v. Drygalski (ed.), Deutsche Sudpolar-Expedition 1901–1903, 8: 394–402.

Hydrobiologia **336:** 107–120, 1996.
J. Kristiansen (ed.), Biogeography of Freshwater Algae.

11. The Indo-Malaysian North-Australian phycogeographical region revised

Wim Vyverman
Laboratory of Botany, University of Ghent, K.L. Ledeganckstraat 35, B-9000 Gent, Belgium

Key words: tropics, freshwater algae, Indo-Malaysia North Australia, geographic distribution

Abstract

The present knowledge of the freshwater algal flora in the Indo-Malaysian North Australian region is reviewed. More than 4700 taxa have been recorded from this region. Desmids, diatoms and cyanobacteria are among the better studied groups of organisms, while phytoflagellates have received very little attention. Phytoplankton communities in tropical lakes are generally similar to summer communities of temperate lakes. In addition, there is a large number of typical tropical taxa, including pantropical and regional endemic elements. Local endemism occurs in the old Tertiary lakes of the region. The composition of algal communities changes markedly along an altitudinal gradient, and tropical taxa are gradually replaced by taxa characteristic for cool climatic conditions. Biogeographical distribution patterns are exemplified for the desmids and diatoms. Among the more than 2680 desmids recorded from the Indo-Malaysian North Australian region, about 800 have never been found elsewhere. Dispersal by migratory birds and past climatic changes might explain distribution patterns. Because of incomplete taxonomy and a general lack of understanding of the autoecology, distribution and speciation of freshwater algae, however, there remain serious obstacles for detailed biogeographical analyses.

Introduction

Biogeography has never been very popular among freshwater phycologists. That groups which are distributed by passive dispersal make poor subjects for biogeographical enquiry (Ball, 1975), was a statement which has been vigorously opposed to by workers on freshwater zooplankton, especially by students of rotifers and copepods (e.g. Bayly & Morton, 1978; Dumont, 1980, 1983). Recent studies have shown that among zooplankton clear patterns emerge in the geographic distribution of taxa, which can be related to historical factors such as past climatic conditions, continental drift etc. (e.g. Maly & Bayly, 1991). Why has there not been any recent response from freshwater phycologists?

In the following, I will use the example of the Indo-Malaysian North Australian region to discuss some features of the geographic distribution of tropical freshwater algae and to illustrate some problems related to the interpretation of tropical algal floras. As early as 1902, West & West recognised that the region encompassing the Indian subcontinent, Sri Lanka, tropical south-east Asia and northern Australia shared a number of characteristic desmid species. This was further confirmed by Krieger (1932) who defined the Indo-Malaysian North Australian phycogeographical region, along with nine other phycogeographical provinces, based on the composition of the desmid flora.

Location and general characteristics of the Indo-Malaysian North Australian Region

The Indo-Malaysian North Australian region (IMNAR) (Figure 1) is an extremely diversified region in terms of geology, topography, climate, vegetation, and fauna. It comprises the land masses of the Indian subcontinent, the south-eastern part of the Asian continent, the Indonesian Archipelago, New Guinea and North Australia. The northern margin of the province roughly coincides with the Tropic of Cancer, to the south it is restricted to the more humid regions in the northern part of the Australian continent. The climate ranges from permanently humid and hot in New Guinea and a large part of Indonesia, to the distinct monsoonal climatic conditions at higher latitudes. Temperate and cold climates prevail in mountainous regions such as the central cordillera of New Guinea and Kalimantan. The Indonesian archipelago and New Guinea are important biogeographical boundary zones for the terrestrial fauna and flora, reflecting to a large extent past tectonic and climatic events.

Limnologically, the IMNAR is no less diverse than its terrestrial habitats. There is a wide range of aquatic habitats as regards origin and physical-chemical characteristics, including the numerous tectonic and volcanic lakes of Tertiary and more recent origin, extensive floodplain systems, alpine lakes and man-made lakes and reservoirs. Recent reviews of the limnology in the IMNAR have been published by Chambers (1987), Costa (1994), Giesen (1994), Ho (1994), Khondker (1994), Nontji (1994), and Pearson (1994).

History and nature of phycological research

As for most parts of the world, freshwater algal identification manuals do not exist for the Indo-Malaysian North Australian region. The widely scattered literature in various, often hardly accessible journals, makes it difficult to obtain precise estimates of the amount of work done on this region. We have made a compilation of 330 publications dealing with the systematics and ecology of freshwater algae in the IMNAR (Vyverman, unpubl.). This list is probably not complete but it reflects the amount and nature of the studies done. Despite the considerable number of publications, research in this region still lags considerably behind the better-studied northern temperate regions.

Taxonomic studies of the freshwater algae in the IMNAR started in the middle of last century (Grant, 1842 *cit. in* Suxena & Venkateswarlu, 1966), approximately at the same time when the earliest studies were undertaken in tropical South America and Africa. However, a rapid increase in the number of published studies did not start until the end of the nineteenth century. The large expeditions in the first half of the present century such as the German Sunda expedition (1928–1929) in Indonesia, the Wallacea expedition (1932) in the Philippines, the Sunda islands and Hawaii, have made a major contribution to the knowledge of the ecology and distribution of the freshwater algae in this region. One third of all publications are from India, illustrating the great phycological activity in this region. Indonesia (especially Java, Bali and Sumatra), Bangladesh, Malaysia and Korea are the next best studied regions, in terms of the amount and size of publications. However, in recent years, there have been very few phycological studies in Indonesia, while it was intensively studied earlier in this century. Since the late seventies, there is an increasing number of papers on the algae from Papua New Guinea.

A prerequisite for any biogeographical study is to record species distributions. There have only been few attempts to study the aquatic habitats of the IMNAR in a systematic way (e.g. Ruttner, 1952; Behre, 1956; Hustedt, 1937–1939, 1942; Ling & Tyler, 1986; Vyverman, 1991a, 1991b, 1992a, 1992b, 1992c, 1996). Many of the earlier publications and numerous present studies, rely on samples collected by non-phycologists. Although these publications contain very valuable information, the biogeographically most interesting but often hardly accessible aquatic habitats have only been scarcely studied. In a recent review of the major Indonesian lakes (surface area larger than 1 ha), Giesen (1994) states that among 35 lakes for which limnological data are known, only 14 have been comprehensively studied. The remaining 470 lakes in the archipelago are virtually unknown, illustrating the lack of even basic knowledge of their physical and chemical characteristics, let alone the biological parameters. In Papua New Guinea, the distribution of desmids and diatoms in only 150 of the more than 5000 large and small lakes (surface area larger than 0.1 ha) have been studied (Vyverman, 1992a, 1992b, 1992c); in only about 40 of them complete phytplankton communities have been investigated (Vyverman, subm., unpubl.). These observations are equally applicable for almost any other part of the IMNAR. Consequently, distribution patterns are very incomplete and have to be interpreted with great care as they reflect the distribution of studied localities rather than the geographic distribution of species.

Species diversity

Among the studies in our data-base, general accounts on the phytoplankton composition and ecology prevail (38%). The remaining 62% mainly concern systematic accounts of Chlorophyta (34%), especially desmids (22%), followed by Bacillariophyta (12%), Cyanobacteria (6%) and Rhodophyta (5%). Many of the general systematic accounts mainly concern the desmids (e.g., West & West, 1902; Ling & Tyler, 1986), illustrating

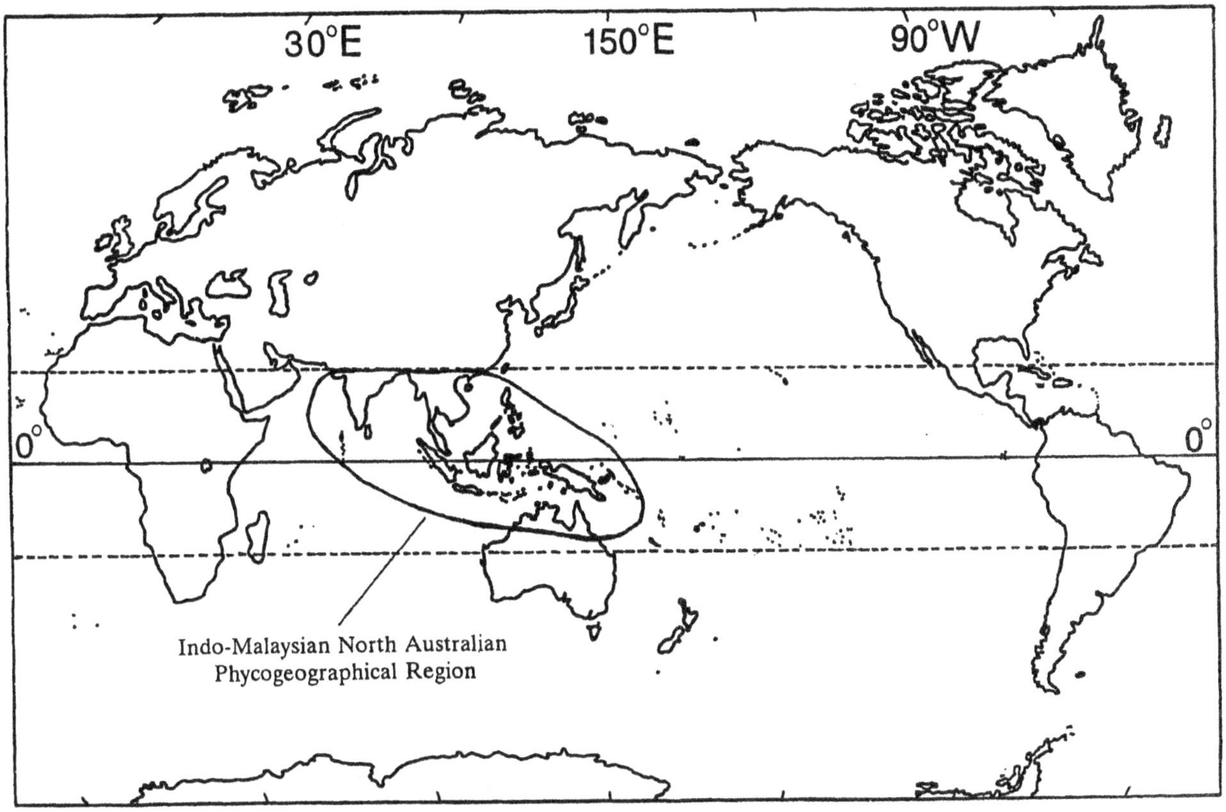

Figure 1. Location of the Indo-Malaysian North Australian phycogeographical region.

the great diversity of this algal group in the INMAR and the scientific preferences of the authors. Phytoflagellates are only poorly studied; studies in Euglenophyta, Cryptophyta, Dinophyta and Chrysophyta comprise only 5% of all studies. The recent discovery in the tropics of a high diversity of silica-scaled chrysophytes (e.g., Cronberg, 1989), formerly believed to be a group of cold-water organisms, confirms that the knowledge of phytoflagellates in tropical waters is still very limited.

It is difficult to estimate the actual number of taxa reported from this region. Not only the literature is widely scattered, but in many cases species lists need taxonomic and nomenclatural revision. With the exception perhaps of the diatoms and desmids, the figures given in Table 1 are only rough approximations of the total number of taxa per major taxonomic group described from the INMAR. We estimate that at least 4700 taxa have been recorded from this region. The overwhelming majority of them are chlorophytes (67%), especially desmids (57%); diatoms (19%) and blue-green algae (6%); whereas there is only a small number of phytoflagellates. New taxa are constantly being described, even in the better known groups such as desmids and diatoms, illustrating the insufficiently known species composition and diversity.

The vast number of taxa and their often incompletely known taxonomy is another great barrier towards a biogeographical analysis of the freshwater algal flora.

Comparison of tropical and temperate algal flora's

Reynolds (1984) gives a general scheme of the composition of phytoplankton assemblages and their seasonal succession in temperate lakes of different trophic status. Tropical lakes represent an equally diverse limnological spectrum as their temperate counterparts; the major factors controlling the wax and wane of algal communities are grossly the same as in temper-

Table 1. The dominant phytoplankton composition in low- and high-altitude lakes of different size and origin in the Indo-Malaysian North Australian region.

Lake	Surface km^2	Altitude m.a.s.	Phytoplankton composition (major taxa)	References
Aman	0.02	?	Eugleno, Chloro, Cyano, *Euglena, Trachelomonas, Ankistrodesmus, Crucigenia, Chlorella, Scenedesmus, Microcystis, Anabaena, Oscillatoria*	Yusoff & Patimah, 1994
Angii	4.5	1780	Chloro, Bacill, Cyano, *Oocystis, Staurastrum, Coelastrum, Surirella, Aulacoseira, Chroococcus*	Vyverman, unpubl. Irian Jaya
Batur	15.9	1031	Bacill, Cyano, Dino, *Nitzschia, Cyclotella, Dactylococcopsis, Anabaena, Lyngbya, Peridinium*	Ruttner, 1952.
Bune	0.1	2200	Chloro, Bacill, Chryso, *Staurastrum, Cosmarium, Aulacoseira, Mallomonas*	Vyverman, unpubl.
Govermas	9.8	40	Cyano, Chloro, Bacill, Dino, *Microcystis, Staurastrum, Cosmarium, Peridinium, Aulacoseira, Rhizosolenia, Eunotia*	Vyverman, subm.
Habbema	0.7	3325	Chryso, Dino, Chloro, *Dinobryon, Peridinium, Ceratium, Cosmarium*	Vyverman, unpubl., Irian Jaya
Imbuando	4.1	20	Eugleno, Cyano, Bacill, Chloro, Chryso, *Euglena, Trachelomonas, Oscillatoria, Aulacoseira, Dictyosphaerium, Mallomonas*	Vyverman, subm.
Keloet	?	1121	Cyano, Bacill, *Anabaena, Surirella*	Huber-Pestalozzi, 1932
Klindungan	1.9	10	Cyano, Dino, *Dactylococcopsis, Anabaenopsis, Lyngbya, Peridinium, Peridiniopsis*	Ruttner, 1952
Kopawunum	0.72	40	Chloro, Cyano, Bacill, Eugleno, Chryso, *Volvox, Dictyosphaerium, Kirchneriella, Oscillatoria, Aulacoseira, Rhizosolenia, Euglena, Trachelomonas, Mallomonas, Dinobryon*	Vijverman, subm.
Kutubu	4.99	808	Chloro, Cyano, *Botryococcus, Staurastrum, Cosmarium, Anabaena*	Bayly et al., 1970, Vyverman, unpubl.
Lamongan	0.34	240	Cyano, Bacill, *Anabaenopsis, Lyngbya, Nitzschia, Melosira, Oocystis, Tetraedron, Scenedesmus*	Ruttner, 1952
Lanao	357	700	Chloro, Cyano, Eugleno, Bacill, Dino, Crypto, *Oocystis, Closterium, Nitzschia, Melosira, Lyngbya, Anabaena, Ceratium, Trachelomonas*	Lewis, 1978
Lau	0.05	2800	Chloro, *Oocystis, Staurastrum*	Vyverman, unpubl.
Mainit	141	27	Cyano, Dino, Crypto, Chloro, *Anabaena, Lyngbya, Peridinium, Gymnodinium, Oocystis, Tetraedron, Rhodomonas*	Lewis, 1973
Merdada	2	2000	Chloro, *Staurastrum, Cosmarium, Scenedesmus*	Huber-Pestalozzi, 1935
Ohoitiel	?	?	Dino, Chloro, *Staurastrum, Arthrodesmus, Peridinium*	Nygaard, 1926
Pansoon Reservoir	130	?	Chloro, Crypto, Bacill, Dino, Chryso, *Euastrum, Staurastrum, Rhodomonas, Cryptomonas, Mallomonas, Dinobryon, Cyclotella, Gonyaulax*	Anton, 1994
Parago	0.9	2350	Dino, Bacill, *Ceratium, Peridinium, Aulacoseira, Cyclotella*	Vyverman, unpubl.
Patengang	?	1600	Chloro, Dino, *Staurastrum, Cosmarium, Ceratium*	Huber-Pestalozzi, 1935
Piunde	0.15	3650	Dino, Chloro, Bacill, *Peridinium, Staurastrum, Rhizosolenia*	Thomasson, 1967, Vyverman, unpubl.
P. Frederik Hendrik	0.2	3	Chloro, Bacill, Cyano, Chryso, Dino, *Staurastrum, Cosmarium, Xanthidium*	Vyverman, in prep.

ate lakes. One major difference, however, is the fairly constant temperature and solar irradiance in tropical lakes.

Pollingher & Berman (1991) state that major differences in the phytoplankton assemblages between lakes of the warm belt and those of other regions are mainly to be seen in large and deep lakes. Previous com-

Table 1. Continued.

Lake	Surface km²	Altitude m.a.s.	Phytoplankton composition (major taxa)	References
(unnamed lake)			*Euastrum, Coelastrum, Scenedesmus, Dictyosphaerium, Kirchneriella Aulacoseira, Eunotia, Peridinium, Dinobryon, Mallomonas*	
Ranau	126	540	Bacill, Chloro, *Synedra, Nitzschia, Oocystis, Botryococcus, Tetraedron, Schroederia, Staurastrum*	Ruttner, 1952
Sindanglaja	0.02	1050	Dino, Bacill, Eugleno, *Ceratium, Peridinium, Cyclotella, Melosira, Synedra, Nitzschia, Trachelomonas*	Ruttner, 1952
Tasek Dayang Bunting	0.24	20	Dino, Chloro, Bacill, *Glenodinium, Staurastrum, Eunotia*	Yadav & Ratnasbapathy, 1974
Telaga Ngebel	1.48	730	Cyano, Chloro, Bacill, *Anabaena, Lyngbya, Anabaenopsis, Botryococcus, Oocystis, Cosmarium, Synedra, Cyclotella*	Ruttner, 1952
Titiwangsa	0.46	?	Cyano, Chloro, Dino, Chryso, *Merismopedia, Microcystis, Staurastrum, Eudorina, Peridinium, Dinobryon, Uroglenopsis*	Yusoff & Patimah, 1994
Tjigombong	0.3	500	Dino, Chloro, *Peridinium, Ceratium, Mallomonas, Dinobryon, Dactylococcopsis, Oocystis, Botryococcus*	Ruttner, 1952
Toba	1130	905	Bacill, Chloro, Cyano, *Denticula, Synedra, Melosira, Planctonema, Closterium, Aphanothece, Peridinium*	Ruttner, 1952

parisons of phytoplankton of tropical and temperate lakes (e.g., Lewis, 1978; Pollingher & Berman *op. cit.*) used the complete phytoplankton composition including spring, summer, autumn, and winter assemblages. A comparison of typical summer assemblages in temperate lakes (e.g., Reynolds, 1984) with phytoplankton assemblages from tropical lakes however indicates a greater overlap than when the complete phytoplankton composition is considered. Table 2 summarizes the dominant phytoplankton composition in some lowland and highland lakes in the Indo-Malaysian North Australian region. Although different sampling techniques were used (net plankton, quantitative samples), some general trends can be seen among the larger phytoplankton species. Typically, phytoplankton communities in tropical lakes are dominated by 'cosmopolitan' taxa, along with pantropical taxa in lowland lakes and 'northern-montane' species in highland lakes; while species with a more restricted geographic distribution usually occur in low densities.

Table 2. Approximate numbers of species per major algal group in the Indo-Malaysian North Australian region, based on a survey of 330 phycological and limnological publications. Taxonomic and nomenclatural revision is needed for most groups.

Taxonomic group	Taxa	%
Cyanobacteria	300	6.3
Bacilliariophyta	900	19.0
Euglenophyta	150	3.2
Chlorophyta	3200	67.3
Desmidiales	2700	56.7
Others	500	10.6
Chrysophyta	90	1.9
Cryptophyta	10	0.2
Dinophyta	50	1.1
Prymnesiophyta	?	?
Xanthophyta	50	1.1
Total	4750	100.0

In lowland lakes, chlorophytes and cyanobacteria, sometimes together with diatoms, are dominating, a feature which was also observed by Lewis (1978) and Pollingher & Berman (1991). There is no evidence for a decrease in species diversity towards the equator. The diatoms *Rhizosolenia* spp., *Aulacoseira granulata* (Ehr.) Simonsen and especially its var. *angustissima* (O.M.) Simonsen, and many genera of the chlorophytes and cyanobacteria typically also develop during the summer months in temperate lakes. Chrysophytes do not seem to play an important role in large tropical lakes, although Hecky & Kling (1987) found that in Lake Tanganyika a considerable fraction of the phytoplankton biomass is attributable to chrysophytes. In smaller lowland lakes such as the floodplain lakes in

New Guinea, chrysophytes are common (Vyverman & Cronberg, 1993) and quantitatively important, especially the genera *Dinobryon, Synura*, and *Mallomonas*. Other phytoflagellates, including euglenoids *(Euglena, Trachelomonas, Strombomonas)*, Volvocales *(Pandorina, Eudorina, Volvox)*, dinoflagellates *(Peridinium, Glenodinium)*, and unidentified green flagellates, constitute a significant fraction of the phytoplankton biomass in these lakes, especially during periods of mixing (Vyverman, 1996). A similar assemblage, dominated by euglenoids, was reported from Amazonian floodplain lakes (Schmidt & Uherkovich, 1973). This assemblage is also reminiscent of the summer phytoplankton of brown-water lakes in temperate regions, where phytoflagellates play an important role (e.g., Ilmavirta, 1988).

With increasing altitude, the phytoplankton composition changes considerably. Cyanobacteria are less important, and chlorophytes, dinoflagellates, diatoms and chrysophytes dominate. This altitudinal zonation in algal communities in tropical lakes is also obvious when benthic assemblages are considered. In Papua New Guinea, a highly significant relation was found between the composition of littoral diatom assemblages and altitude (Vyverman, 1992b; Vyverman & Sabbe, 1995), suggesting a strong effect of temperature and related factors on the composition of algal communities.

Despite the overall similarities in the structure of tropical and temperate phytoplankton communities, tropical lakes show striking some floristic differences with high-latitude lakes. To our knowledge, *Asterionella formosa* Hassal has never been reported from the IMNAR, but it is a common and often dominant species in the spring and summer phytoplankton communities in temperate lakes in the northern and southern hemisphere. In some tropical lakes, the genus *Surirella* is abundant in the plankton (Hustedt 1937–1939), while most temperate species of this genus are benthic. In other, often more eutrophic lowland lakes, *Synedra* and *Nitzschia* spp. may dominate the phytoplankton. The dinoflagellate genus *Peridinium* seems to replace the genus *Ceratium* towards the equator while *Ceratium hirundinella* (O.F. Müller) Dujardin, a very common and dominant species in temperate lakes, is rare; to our knowledge it has never been reported in typical lowland lakes of the INMAR. Whereas we have never found this species in the highland lakes of New Guinea, there are several reports from mountain lakes in Indonesia. In the New Guinea highland lakes the closely related species *C. brachyceros* Daday and *C. cornutum* (Ehr.) Claparède & Lachmann are common (Vyverman & Compère, 1991; Vyverman, unpubl.). It would be interesting to see whether the former reports of *C. hirundinella* agree with the typical forms from the temperate lakes or with *C. brachyceros*, or represent other species of this genus.

In the tropics, phytoplankton studies have mainly focussed on large lakes, whereas there is very little information on phytoplankton communities and species succession in smaller lakes. Other discrepancies between tropical and temperate lakes may be found after careful analysis and further studies of the algal communities in tropical lakes.

Geographic distribution patterns of freshwater algae

A large portion of the algal taxa in tropical lakes appears to have a world-wide distribution. Estimations of the cosmopolitan element in tropical phytoplankton communities invariably quote 50–80% of taxa with a world-wide distribution (e.g., Compère & Iltis, 1983). Only very few freshwater algal genera have a restricted geographic distribution. Among the desmids, the genera *Amscottia, Phymatodocis, Streptonema, Apriscottia*, and *Allorgeia* are confined to tropical regions. The diatom genera *Terpsinoë* and *Hydrosera* have their main distribution in the tropics, although there are some records from temperate regions (Round et al., 1990).

Although the geographic distribution of most algal species is badly known, there is little doubt that the high degree in cosmopolitanism to a large extent is true. However there are some difficulties related to the term 'cosmopolitan'. First, identification of tropical phytoplankton is mostly done using identification manuals or other publications from temperate regions, and the often slight differences in the morphological characteristics are neglected or interpreted as intraspecific variation. There are very few studies dealing with the morphological variation of these cosmopolitan species, using material from both tropical and temperate locations. In addition, there may be ecological or genetic differences between tropical and temperate populations of the same species, but this has not yet been studied in much detail. Finally, species are often referred to as cosmopolitan, but examination of their distribution patterns in the tropics shows that they mainly occur in high-altitude lakes.

The remaining 20 to 50% of the species in phytoplankton assemblages from tropical regions have more restricted distributions. They can be classified as pantropical 'northern-montane or northern hemispheric'. 'southern-montane or southern hemispheric', or 'local and regional endemic' taxa.

A large number of the non-cosmopolitan species can be considered as pantropical and occur widespread over the tropics. Pantropical species are found among all major taxonomic groups (Table 3); their distribution suggests a strict dependence on high and more or less constant temperature regimes (polythermic or warm stenothermic taxa). Some species seem to be more polythermic than others, the latter also occur in subtropical regions. Typical summer species in temperate lakes, such as *Aulacoseira granulata* var. *angustissima*, might also be regarded as warm-water taxa. The present knowledge of the autecology of most species is however too scant to allow more detailed conclusions about the influence of temperature on their occurrence and distribution. We do not know how and to which extent the growth of polythermic species is controlled by temperature, including the frequency and duration of 'cold-water' periods; or if other factors such as the ability to form resting spores, changing competitional strength, or niche availability also play a role. These observations also apply to the numerous 'northern-montane' species and to the increasing number of species described from cool climatic regions in the southern hemisphere, which can be regarded as 'southern-montane' taxa. In tropical high-altitude waters, there is a considerable number of species considered to be oligothermic (= cold stenothermic, in the literature often referred to as temperate- or northern-montane taxa). Again there appears to be a more or less continuous range from cold-stenothermic to poikilothermic species. Round (1981) argues that many of these oligothermic species have much wider distributions but, because of habitat loss, are becoming restricted to the generally less polluted mountainous and high-latitude waters. This may be true to some extent, but the striking similarity of diatom communities in tropical high-altitude lakes (e.g., Servant-Vildary et al., 1990; Vyverman, 1992a) with similar temperate habitats indicates that there is a large number of species confined to cool climatic conditions.

In addition to the numerous species which have a distinct latitudinal distribution there is a number of clear-cut species restricted to a certain geographic region. It is possible to distinguish between 'local endemics' which are only known from one locality such as geologically old lakes or from a particular ecological habitat, and 'regional endemics' which are confined to a larger geographic region. Regional endemic species can be found in most geographic regions. In Lake Tanganyika for example, 8% (30 taxa) of the diatom flora can be considered as local endemic species, while there are only few known endemic species in Lake Malawi and Lake Victoria (Cocquyt et al., 1993; Cocquyt & Vyverman, 1994). Regional endemic species, confined to the African continent, however, account for about 25% of the total algal flo-

Table 3. Some examples of pantropical freshwater algae.

Cyanobacteria
Anabaena sphaerica Born. & Flah. var. *tenuis* G. S. West
Oscillatoria articulata Gardner
Trichodesmium iwanoffianum Nygaard
Bacillariophyta
Aulacoseira agassizii (Ostenfeld) Simonsen
Aulacoseira ikapoensis (O. Müller) Simonsen
Caloneis incognita Hustedt
Eunotia didyma Grunow
Gomphonema brasiliense Grunow
Navicula perrottetii Grunow
Pinnularia schweinfurthii (A. Schmidt) Hustedt
Pinnularia tropica Hustedt
Euglenophyta
Lepocinclis nayali Conrad
Strombomonas ensifera (Daday) Defl.
Trachelomonas armata (Ehr.) Stein var. *duplex* Playf.
Trachelomonas conica Playf. f. *punctata* Defl.
Chlorophyta
Ankistrodesmus bernardii Komarek
Coelastrum indicum Turner
Schroederia indica Philipose
Cosmarium ordinatum (Borges.) West & West
Euastrum praemorsum (Nordst.) Schmidle
Micrasterias foliacea Bailey ex Ralfs var. *ornata* Nordst.
Micrasterias radians Turn.
Pleurotaenium ovatum (Nordst.) Nordst.
Chrysophyta
Mallomonas bangladeshica (Takah. & Hayak.) Nicholls
Mallomonas bronchartiana Compère
Mallomonas tropica Dürrschmidt & Croome
Synura australiensis Playf.
Dinophyta
Ceratium brachyceros Daday
Peridinium gutwinskii Wolosz.

ra in these lakes. Uherkovich (1984) estimates that about 390 taxa have been described from the Amazonas region which have never been found elsewhere. Equally, there are several well-defined species of algae which are endemic to Australia (Tyler, this volume).

In the following section I will give some examples of biogeographical distribution types from the INMAR. It is as yet not possible to present a complete analysis of the biogeographical relationships to other tropical regions. Furthermore, as I consider our current knowledge of most major algal groups to be too scant, I will mainly focus on the desmids and diatoms. There is however little doubt that further studies will demonstrate an Indo-Malaysian North Australian element in the other algal groups, such as *Peridinium baliense* Lindeman, *P. lingii* Thomasson and *Trachelomonas mediospina* Vyverman.

Examples of geographic distribution patterns of algae in the Indo-Malaysian North Australian region.

The Tertiary lakes in the Indo-Malaysian North Australian region have a number of endemic diatoms, which can be considered as local endemics and are either confined to one lake or to a number of lakes in the same area. The majority of them belong to the genus *Surirella* and many of them were described from lakes in Sulawesi and Sumatra (Hustedt, 1937–1939, 1942) (Figure 2). As with the African endemic *Surirella* species, they are often found in the plankton, but little more is known of their autecology. Stratigraphic records show that the genus *Surirella* is relatively young, which might be a reason for the large degree of endemism in this genus (Krammer & Lange-Bertalot, 1988). However, as they add, their planctonic life form in large lakes with a characteristic ecology, may prevent rapid dispersal as similar habitats are rare. Although local endemism among freshwater algae certainly exists, it is a rare phenomenon and has to be interpreted with care.

A much larger number of desmids and diatoms can be regarded as regional endemics. These include species which are confined to the INMAR, species which have their main distribution in the INMAR but also occur further to the north and/or south; and species which also occur on the African continent. The latter species are considered as palaeotropical elements.

We prepared a data-base of the desmids from the Indo-Malaysian North Australian region, resulting in about 2700 species, varieties and forms (Vyverman, in prep.). More than 800 of them (31%) have never been found elsewhere (Table 4). A data-base of the diatoms is in preparation, but at present it is taxonomically too incomplete to be used in a similar analysis. Even among the desmids, many taxa, especially the numerous infraspecific taxa, require further taxonomic investigations. Some typical Indo-Malaysian North Australian desmid species are shown in Figure 3, the geographic distribution of a number of them is shown in Figure 4a,b. Although the number of localities where these taxa have been found is rather small, there is little doubt that they also occur in many other suitable habitats in this region. Typically, they are rare in samples although in some lakes they may be abundant, such as e.g., *Staurastrum freemanii* West & West in Lake Birip (Brook & Hine, 1966; Vyverman, 1991b), and *Cosmarium mikron* Skuja in Lake Kopawunum (Vyverman, 1996).

Table 4. Numbers of desmid taxa per major genus and their number of regional endemic taxa recorded from the Indo-Malaysian North Australian region.

Genus	Total number	% total	IMNAR	% genus
Actinotaenium	41	1.5	8	19.5
Arthrodesmus	18	0.7	7	38.9
Bambusina	14	0.5	--	--
Closterium	321	11.9	45	14.0
Cosmarium	1041	38.8	310	29.8
Cylindrocystis	11	0.4	4	36.4
Desmidium	45	1.7	9	20.0
Docidium	22	0.8	8	36.4
Euastrum	413	15.4	146	35.4
Hyalotheca	11	0.4	4	36.4
Micrasterias	89	3.3	32	35.9
Penium	39	1.5	13	33.3
Pleurotaenium	84	3.1	31	36.9
Spondylosium	12	0.4	4	33.3
Staurastrum	310	11.5	147	47.4
Staurodesmus	86	3.2	15	17.4
Xanthidium	66	2.5	29	43.9
Other genera	71	2.6	11	15.5
Total	2684	100	823	30.7

Why is the distribution of these species restricted to the Indo-Malaysian North Australian region? Like the pantropical algae, the Indo-Malaysian North Australian flora element seems to be restricted in its distribution by temperature limits, both along latitudinal and altitudinal gradients. A study of the altitudinal distribution of non-cosmoplitan desmids and diatoms

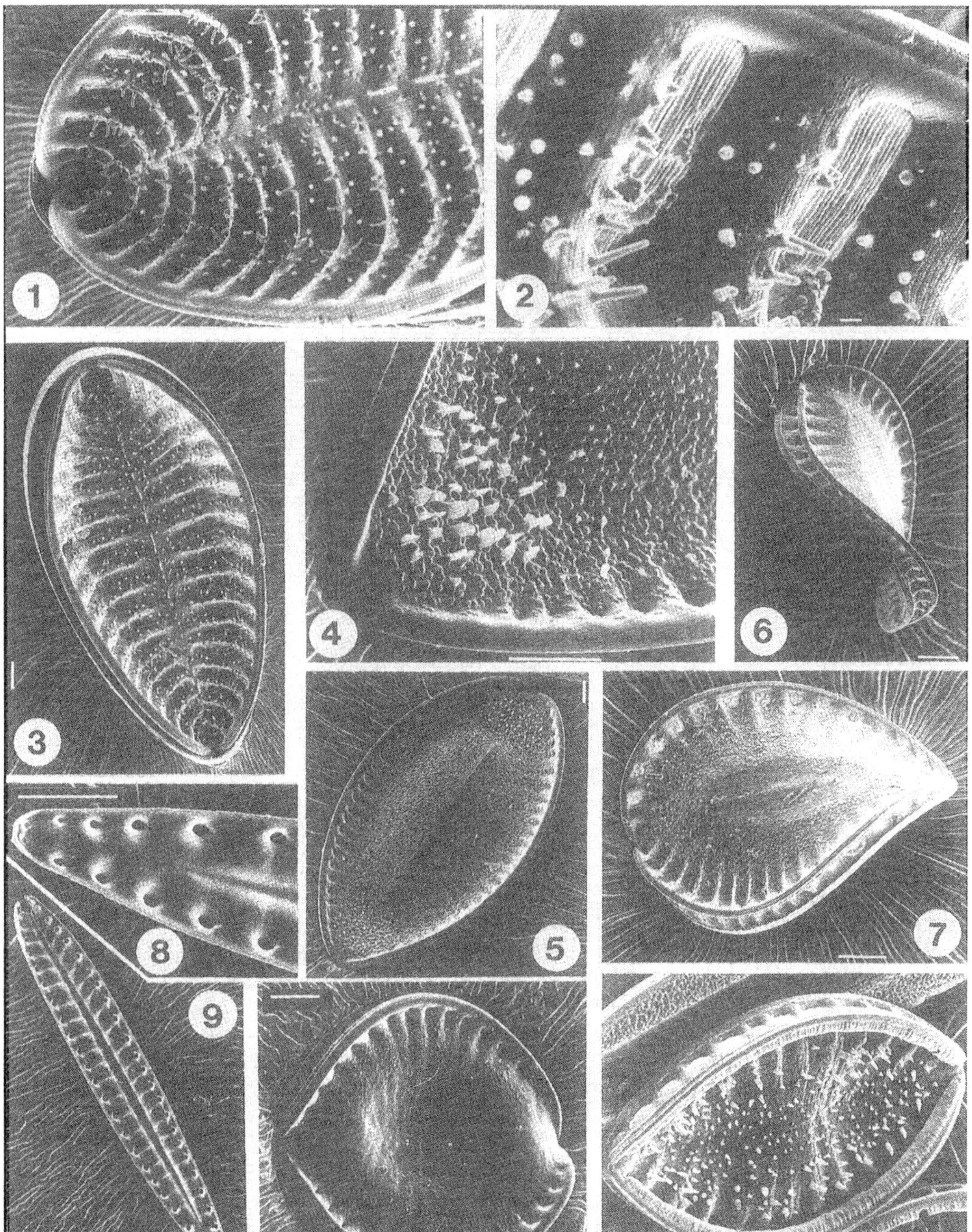

Figure 2. Endemic *Surirella* species from the Indo-Malaysian North Australian region. Figure 2.1–3. *Surirella robusta* Ehr. var. *armata* Hustedt, regional endemic, needs revision; Figure 2.4–5. *Surirella cataractarum* Hustedt; Figure 2.6–7. *Surirella pseudovalis* Hustedt, local endemic in lakes of Sulawesi; Figure 2.8–9. *Surirella excellens* Hustedt, local endemic in Lake Matano, Sulawesi; Figure 2.10. *Surirella woltereckii* Hustedt, local endemic in lakes of Sulawesi; Figure 2.11. *Surirella spinosa* Hustedt. Scale bar = 10 μm.

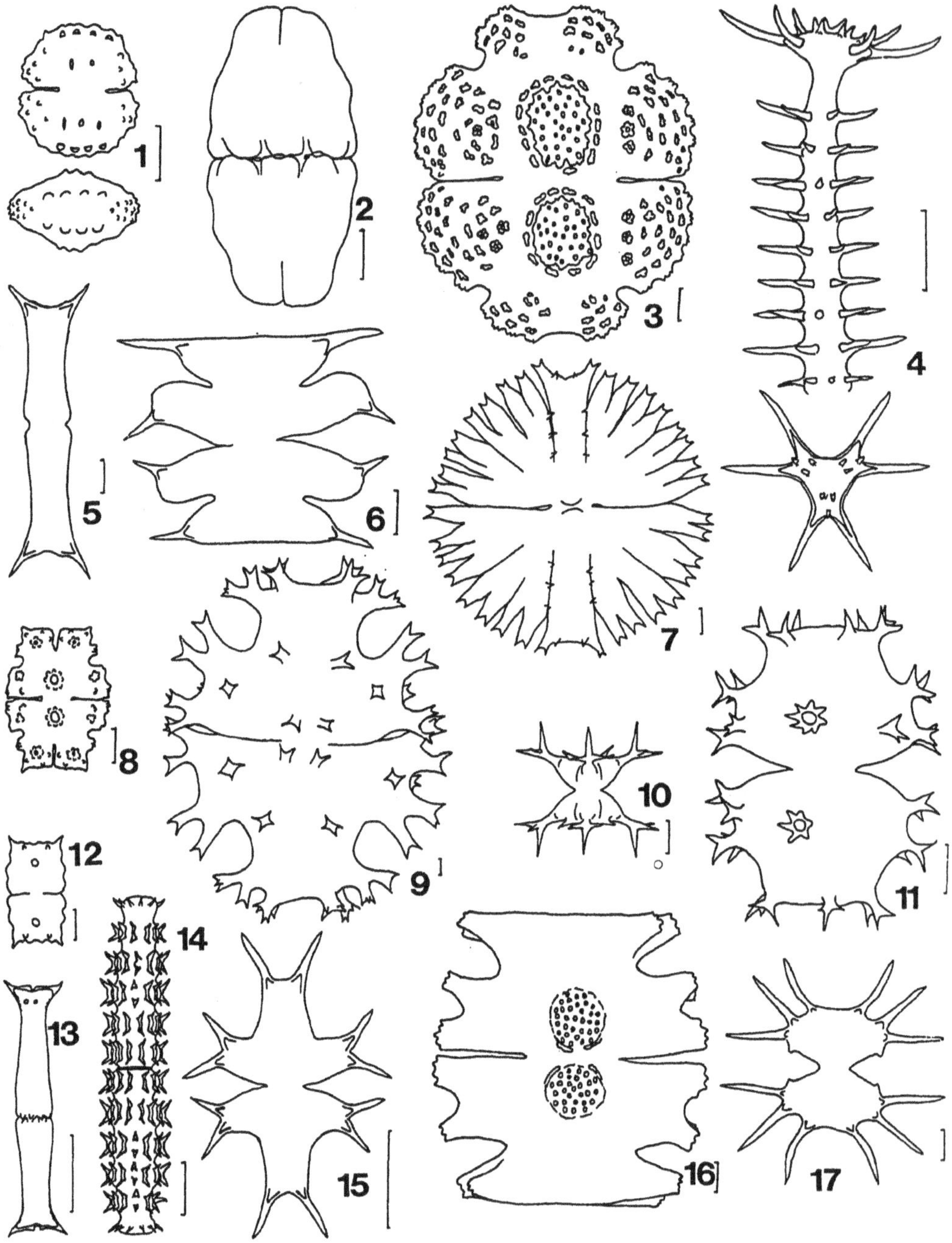

Figure 3. Some characteristic desmids from the Indo-Malaysian North Australian region. Figure 3.1. *Cosmarium ceylanicum* West & West; Figure 3.2. *Euastrum gnatophorum* West & West; Figure 3.3. *Euastrum horikawae* Hinode; Figure 3.4. *Triploceras splendens* Prowse; Figure 3.5. *Ichthyocercus longispinus* (Borge) Krieger; Figure 3.6. *Micrasterias zeylanica* Fritsch; Figure 3.7. *Micrasterias lux* Joshua; Figure 3.8. *Euastrum coralloides* Joshua var. *trigibberum* Lagerh.; Figure 3.9. *Micrasterias anomala* Turner var. *reducta* Scott & Prescott; Figure 3.10. *Staurastrum freemanii* West & West; Figure 3.11. *Xanthidium armatum* (Breb.) Rabenh. var. *anguliferum* Krieger; Figure 3.12. *Euastrum exile* Joshua; Figure 3.13. *Ichthyodontum sachlanii* Scott & Prescott; Figure 3.14. *Pleurotaenium kayei* (Archer) Rabenh.; Figure 3.15. *Micrasterias ceratofera* Joshua; Figure 3.16. *Euastrum moebii* (Borge) Scott & Prescott var. *burmense* West & West; Figure 3.17. *Xanthidium sexmamillatum* West & West var. *pulneyense* Iyengar & Bai. Figures 3.8, 12, 14: redrawn after Scott & Prescott, 1961; Figures 3.4, 10, 13: redrawn after Ling & Tyler, 1986; Figures 3.1–3, 5–7, 9, 11, 15–17: original. Scale bar = 10 μm in all figures except for Figures 3.4, 14 where scale bar = 50 μm.

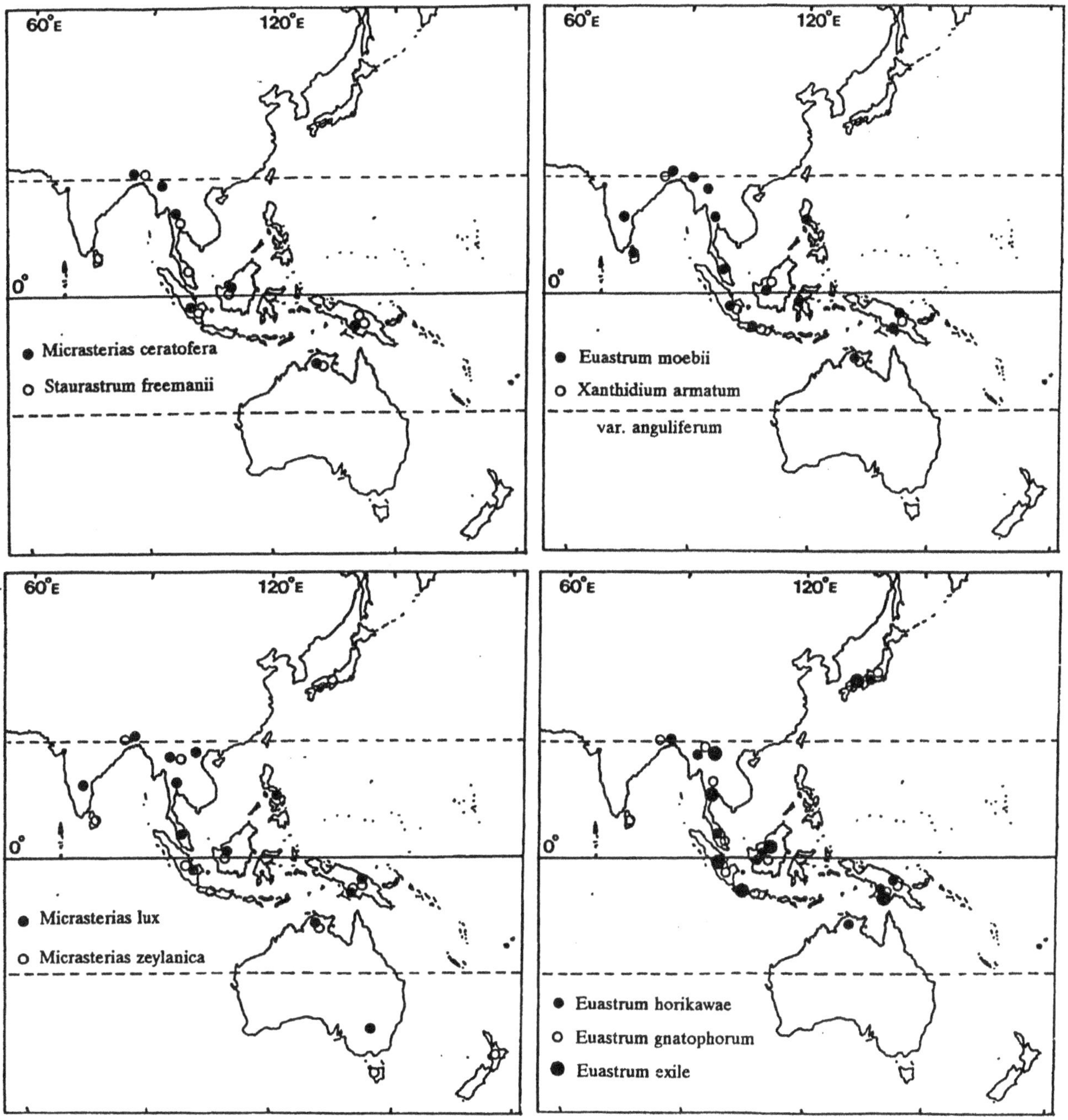

Figure 4. Some typical distribution patterns of desmids endemic to the Indo-Malaysian North Australian region. Figures 4.1–2: species restricted to the IMNAR; Figures 4.3–4: Species with a main distribution in the IMNAR, but also occurring further to the north and south. *Micrasterias zeylanica* also occurs in Africa.

(Vyverman, 1992b) has shown a gradual decrease in the proportion of pantropical, palaeotropical and Indo-Malaysian North Australian taxa along an altitudinal gradient in Papua New Guinea. Similarly, temperate-montane taxa replace the tropical element along higher altitudes. About 20% of the diatoms and 27% of the desmids in this study have a non-cosmopolitan distribution. The zone between 1700 and 2500 m seems to be the transition zone between typical lowland and highland algal floras, which is in agreement with the distribution patterns observed among aquatic macrophytes (Chambers et al., 1987).

There are some aparent differences between the composition of the non-cosmopolitan element in the desmids and diatoms. Among the diatoms, the pantropical element is considerably higher than among the desmids, which have a larger number of typical Indo-Malaysian North Australian species. It was postulated (Vyverman, 1992b) that this may be related to different dispersal capacities and/or speciation rates of diatoms and desmids.

Interestingly, several taxa which have their main distribution in the Indo-Malaysian North Australian region, also occur further to the north and to the south. Examples are *Euastrum horikawae* Hinode, *E. gnatophorum* West & West and *E. exile* Joshua, which occur to the north as far as Japan; and *Micrasterias zeylanica* Fritsch (which is also found in Africa) and *M. lux* Joshua which have been found in south-eastern Australia, Tasmania and New Zealand (Figure 4c,d). Similar distribution patterns were observed in the neotropics (Coesel et al., 1988). A possible explanation for these distribution patterns is passive dispersal by birds along the main north-south orientated migrating routes in the American and Asian/Australian regions. Although this is a plausible hypothesis, no research has yet been undertaken to obtain further evidence.

Palaeotropical species include the above-mentioned *Micrasterias zeylanica* but there are others as well such as *Micrasterias doveri* Biswas var. *africana* Bourrelly and *M. apiculata* (Ehr.) Meneghini ex Ralfs var. *lacerata* Turner. About 3% of the diatoms and 4% of the desmids of Papua New Guinea can be considered as palaeotropical taxa (Vyverman, 1992b). We estimate that for the complete desmid and diatom flora of the IMNAR, comparable or even higher figures will be found. Similar palaeotropical distribution patterns also exist for rotifers (Dumont, 1980) and have been explained by Pleistocene climatic events, allowing dispersal along the near East. In the case of the algae, such distribution patterns need to be further studied, but they indicate that dispersal routes for freshwater algae and rotifers follow similar patterns.

Sometimes, related species show distinct or partly overlapping geographic distributions. *Micrasterias crux-melitensis* (Ehr.) Hassall ex Ralfs and *M. radians* Turner can be used as a good example. Both species are closely related, the former has its main distribution in temperate regions, while the latter has a pantropical distribution. In New Guinea, *M. crux-melitensis* is distributed in some oligotrophic highland lakes, while *M. radians* is widespread in dystrophic lowland lakes and swamps. At medium altitudes, intermediate forms between the two species can be observed (Vyverman & Viane, 1995). *Eunotia serra* Ehr. is another example of a widespread diatom in oligotrophic waters in temperate regions, while related tropical species are *E. serrata* Hustedt, *E. tropica* Hustedt, *E. muelleri* Hustedt, and *E. subrobusta* Hustedt.

Conclusions

The majority of the freshwater lakes on Earth are of relatively young geological age. Most of them are of Pleistocene origin, and have since then been subject to climatic and ecological changes. Variations in water temperature are much smaller than in terrestrial habitats in a similar climatological and topographical setting. From an evolutionary viewpoint, a great dispersal capacity would be of prime importance to freshwater algae, which would contribute to their prevailing cosmopolitan distribution.

The example of the Indo-Malaysian North Australian region shows that, despite the considerable amount of research done, there are still large gaps in our knowledge of species diversity, taxonomy, ecology, and geographic distribution of tropical freshwater algae. This is a serious obstacle for any detailed biogeographical studies. World-wide monographic revisions, supported by modern taxonomic tools such as isozyme characterisation and genetic compatibility studies, combined with autecological studies focussing e.g. on dispersal capacities and on the effect of temperature-dependent processes on growth and survival, would be a first step towards a better understanding of distribution patterns of algae.

There can however be little doubt that freshwater algae do have a biogeography. In addition to a large cluster of pantropical and temperate-montane species with representatives in all major taxonomic groups, there are large numbers of taxa endemic to each of the different tropical regions. This shows that, although the composition and distribution of algal communities within a given region to a large extent is controlled by ecological factors, historical factors must play a role in the origin of regional endemism. At present however, any statement on the origin of such regional centres of endemism would be highly speculative as there is no information on speciation rate nor on the origin of these taxa.

Acknowledgements

The author is a postdoctoral researcher at the National Fund for Scientific Research (Belgium). I am greatly indebted to the National Fund for financial assistance; the King Leopold III Fund and the Laing Island Biological Station are thanked for supporting field work in New Guinea. The Organizing Committee of the 5th International Phycological Congress is thanked for financial support for participation. Many thanks are due to F. Vyverman and R. Vyverman for the assistance with the preparation of the manuscript. The School of Aquatic Sciences, Deakin University, Australia, is thanked for logistic support.

References

Anton, A., 1994. Long-term phytoplankton changes in a tropical reservoir. Mitt. Int. Ver. Limnol. 24: 243–250.

Ball, I. R., 1975. Nature and formulation of biogeographical hypotheses. Syst. Zool. 24: 407–430.

Bayly, I. A. E. & D. W. Morton, 1978. Aspects of the zoogeography of Australian microcrustaceans. Verh. Int. Ver. Limnol. 20: 2537–2540.

Bayly, I. A. E., J. Peterson & V. P. St. John, 1970. Notes on Lake Kutubu, southern highlands of the territory of Papua and New Guinea. Aust. Soc. Limnol. 3: 40–47.

Behre, K., 1956. Die Süsswasseralgen der Wallacea-expedition. Arch. Hydrobiol. Suppl. 23: 73–84.

Brook, A. J. & A. E. Hine, 1966. A population of *Staurastrum freemanii* from the central highlands of New Guinea. J. Phycol. 2: 66–73.

Chambers, M. R., 1987. The freshwater lakes of Papua New Guinea: an inventory and limnological review. J. trop. Ecol. 3: 1–23.

Chambers, M. R., J. H. Kyle, G. J. Leach, P. L. Osborne & D. Leach, 1987. A limnological study of seven highlands lakes in Papua New Guinea. Science in New Guinea 13: 51–81.

Cocquyt, C., W. Vyverman & P. Compère, 1993. A checklist of the algae of the East African Great Lakes. Scripta Botanica 8: 1–55.

Cocquyt, C. & W. Vyverman, 1994. Composition and diversity of the alga flora in the East African Great Lakes: a comparative survey of lakes Tanganyika, Malawi (Nyasa), and Victoria. Arch. Hydrobiol. Beih. Ergebn. Limnol. 44: 161–172.

Coesel, P. F. M., 1992. Desmid assemblies along altitude gradients in Colombia. Nova Hedwigia 55: 353–366.

Coesel, P. F. M., S. R. Duque & G. Arango, 1988. Distributional patterns in some neotropical desmid species (Algae, Chlorophyta) in relation to migratory bird routes. Rev. Hydrobiol. trop. 21: 197–205.

Compère, P. & A. Iltis, 1983. The phytoplankton (qualitative composition of the algal flora). In Carmouze, J.-P., J.-R. Durand & C. Lévêque (eds), Lake Chad. Ecology and productivity of a shallow tropical ecosystem. Junk, The Hague: 145–152.

Costa, H. H., 1994. The status of limnology in Sri Lanka. Mitt. Int. Ver. Limnol. 24: 73–86.

Cronberg, G., 1989. Scaled chrysophytes from the tropics. Beih. Nov. Hedw. 95: 191–232.

Dumont, H., 1980. Zooplankton and the science of biogeography: The example of Africa. In W. C. Kerfoot (ed.), Evolution and ecology of zooplankton communities. The University Press of New England, Hanover (N.H.); Lond.: 685–696.

Dumont, H., 1983. Biogeography of rotifers. Hydrobiologia 104: 19–30.

Giesen, W., 1994. Indonesia's major freshwater lakes: A review of current knowledge, development processes and threats. Mitt. Int. Ver. Limnol. 24: 115–128.

Hecky, R. E. & H. J. Klin, 1987. Phytoplankton ecology of great lakes in the rift valleys of Central Africa. Arch. Hydrobiol. Beih. 25: 197–228.

Ho, S., 1994. Status of limnological research and training in Malaysia. Mitt. Int. Ver. Limnol. 24: 129–146.

Huber-Pestalozzi, G. Phytoplankton aus Seen und Sumpfen Javas, gesammelt von Prof. C. Schröter, Zürich. Schweiz. Bot. Ges. Ber. 46: 131–168.

Hustedt, F., 1937–1939. Systematische und ökologische Untersuchungen über die Diatomeenflora von Java. Bali und Sumatra nach dem Material der Deutschen Limnologischen Sunda-Expedition. Teil I. Systematischer Teil, Teil II. Allgemeiner Teil. Arch. Hydrobiol. Suppl. 15: 131–506, 638–790.

Hustedt, F., 1942. Süsswasser-Diatomeen des Indomalayischen Archipels und der Hawaii-inseln. Int. Revue ges. Hydrobiol. Hydrogr. 42: 1–152.

Ilmavirta, V., 1988. Phytoflagellates and their ecology in Finnish brown-water lakes. Dev. Hydrobiol. 45: 255–270.

Khondker, M., 1994. The status of limnological research in Bangladesh. Mitt. Int. Ver. Limnol. 24: 147–154.

Krammer, K. & H. Lange-Bertalot, 1988. Süsswasserflora von Mitteleuropa. Bacillariophyceae 2. Gustav Fischer, Stuttgart, 596 pp.

Krieger, W., 1932. Die Desmidiaceen der Deutschen limnologischen Sunda-Expedition. Arch. Hydrobiol., Suppl. 11: 129–230.

Lewis, W. M., 1978. A compositional, phytogeographical and elementary structural analysis of the phytoplankton in a tropical lake: Lake Lanao, Philippines. J. Ecol. 66: 213–226.

Ling, H. U. & P. A. Tyler, 1986. A limnological survey of the Alligators Rivers region. Part II: Freshwater algae, exclusive of diatoms. Austral. Gov. Publ. Serv., Canberra, 173 pp.

Maly, E. J. & I. A. E. Bayly, 1991. Factors influencing biogeographic patterns of Australasian centropagid copepods. J. Biogeogr. 18: 455–461.

Nontji, A., 1994. The status of limnology in Indonesia. Mitt. Int. Ver. Limnol. 24: 95–114.

Nygaard, G., 1926. Plankton from two lakes of the Malaysian region. Dansk. Naturhist. Foren. Vidensk. Meddel. 82: 197–240.

Pearson, R. G., 1994. Limnology in the northeastern tropics of Australia, the wettest part of the driest continent. Mitt. Int. Ver. Limnol. 24: 155–164.

Pollingher, U. & T. Berman, 1991. Phytoplankton composition and activity in lakes of the warm belt. Verh. Int. Ver. Limnol. 24: 1230–1234.

Reynolds, C. S., 1984. The ecology of freshwater phytoplankton. Cambridge University Press, Cambridge, 384 pp.

Rott, E., 1983. A contribution to the phytoplankton species composition of Parakrama Samudra, an ancient man-made lake in Sri Lanka. In F. Schiemer (ed.), Limnology of Parakrama, Sri Lanka, Dr W. Junk Publishers, The Hague: 210–226.

Round, F. E., 1981. The ecology of algae. Cambridge University Press, 653 pp.
Round, F. E., R. M. Crawford & D. G. Mann, 1990. The diatoms. Biology and morphology of the genera. Bath Press, Avon, 747 pp.
Ruttner, F., 1952. Planktonstudien der Deutschen limnologischen Sunda-expedition. Arch. Hydrobiol. 21: 1–174.
Schmidt, G. W. & G. Uherkovich, 1973. Zur Artenfülle des Phytoplanktons in Amazonien. Amazoniana 4: 243–252.
Scott, A. & G. W. Prescott, 1961. Indonesian desmids. Hydrobiologia 17: 1–132.
Servant-Vildary, S. & M. Roux, 1990. Variations de température estimées à partir du déplacement en altitude des associations de diatomées dans une séquence holocène de la Cordillère Orientale de Bolivie. C. R. Acad. Sci. Paris, 311(II): 429–436.
Suxena, M. R. & V. Venkateswarlu, 1966. Desmids of Andhra Pradesh, I from Pakhal Lake, Warangal. Hydrobiol. 28: 49–65.
Talling, J. F., 1987. The phytoplankton of Lake Victoria (East Africa). Arch. Hydrobiol. Beih. 25: 229–256.
Thomasson, K., 1967. Phytoplankton from some lakes on Mount Wilhelm, East New Guinea. Blumea 15: 285–296.
Thomasson, K., 1986. Algal vegetation in Australian billabongs. Nova Hedwigia 42: 301–378.
Uherkovich, G., 1984. Phytoplankton. In H. Sioli (ed.), The Amazon, limnology and landscape ecology of a mighty tropical river basin, Dr W. Junk Publishers, Dordrecht: 295–310.
Vyverman, W., 1991a. Desmids from Papua New Guinea. Bibliotheca Phycologia 87: 1–201, 157 pl.
Vyverman, W., 1991b. Diatoms from Papua New Guinea. Bibliotheca Diatomologica, 22: 1–224, 208 pl.
Vyverman, W. & P. Compère, 1991. Ultrastructure of four dinoflagellates from Papua New Guinea. Bull. Jard. Bot. Nat. Belg. 61: 269–277.
Vyverman, W., 1992a. Multivariate analysis of periphytic and benthic diatom assemblages from Papua New Guinea. Hydrobiologia 234: 175–193.
Vyverman, W., 1992b. Distribution of non-cosmopolitan desmids and diatoms along an altitudinal gradient in Papua New Guinea. Br. phycol. J. 27: 49–63.
Vyverman, W., 1992c. Distribution and ecology of desmid assemblages in Papua New Guinea. Nova Hedwigia 551-552: 257–272.
Vyverman, W. & G. Cronberg, 1993. Silica-scale bearing chrysophytes from Papua New Guinea. Nord. J. Bot. 13: 111–120.
Vyverman, W. & R. Viane, 1995. Morphological variation along an altitudinal gradient in the *Micrasterias crux-melitensis – M. radians* complex Zygnemaphyceae, Desmidiales from Papua New Guinea, Nova Hedwigia.
Vyverman, W. & Sabbe, K., 1995. Diatoms-temperature transfer functions based on the altitudinal zonation of diatom assemblages in Papua New Guinea: a possible tool in the reconstruction of regional palaeoclimatic changes. J. Paleolimnology 13: 65–77.
Vyverman, W., 1996. Phytoplankton diversity and abundance along environmental gradients in tropical floodplain lakes (Sepik-Ramu floodplain, Papua New Guinea). In: Schiemer & Boland (eds), Perspectives in Tropical Limnology. SPB Academ. Publ., Amsterdam, The Netherlands.
West, W. & G. S. West, 1902. A contribution to the freshwater algae of Ceylon. Trans. Linn. Soc. Lond. 2: 123–215.
Yusoff, F. M. & I. Patimah, 1994. A comparative study of phytoplankton populations in two Malaysian lakes. Mitt. Int. Ver. Limnol. 24: 251–258.

Hydrobiologia **336**: 121–126, 1996.
J. Kristiansen (ed.), Biogeography of Freshwater Algae.

12. Bipolarity in the distribution of silica-scaled chrysophytes

J. Kristiansen[1] & M. S. Vigna[2]
[1] *University of Copenhagen, Botanical Institute, Dept. of Phycology, Øster Farimagsgade 2D, 1353 Copenhagen K, Denmark*
[2] *University of Buenos Aires, Facultad de Ciencias Exactas y Naturales, Dept. de Ciencias Biologicas, Pabellon 2, Ciudad Universitaria, Nuñez, 1426 Buenos Aires, Argentina*

Key words: silica-scaled chrysophytes, biogeography, bipolar distributions, Tierra del Fuego

Abstract

An investigation of the chrysophyte flora of Tierra del Fuego (Argentina), 54–55 °S., has shown a high degree of similarity with the flora of climatically comparable regions on the northern hemisphere. All the Fuegian species (except two endemic to South America) also occur on the northern hemisphere – some are more or less cosmopolitan, others have pronounced bipolar distributions. Species in common with other Antarctic regions such as Tasmania are all cosmopolitan, and none of the interesting species originally described from Tasmania occur in Tierra del Fuego. Thus the Fuegian flora appears to be mainly climatically determined and a special Antarctic chrysophyte flora does not exist.

Introduction

The idea behind this study originated during an investigation of the silica-scaled chrysophytes in Tierra del Fuego, Argentina. The southern part of this island, with most of the suitable localities, along the Beagle Canal and around the capital Ushuaia, is situated at a southern latitude of 54–55 °, almost corresponding to the northern latitude of the position of Denmark. Accordingly, our main working hypothesis was the question whether this latitudinal similarity would be reflected in the chrysophyte flora – or if we would meet an Antarctic flora more related to e.g. the Tasmanian. Investigations on the higher plants have shown that several species have Antarctic distributions, and the same is true of many animals (Moore, 1993; compare also Seberg, 1991).

The aim of the investigation was in the first hand to collect material for a chrysophyte volume of the 'Flora Criptogamica de Tierra del Fuego', edited by S. A. Guarrera et al., of which several fascicles already have been published, but so far only two on algae, viz. on Chroococcales and on Chlorococcales (Guarrera, 1986; Tell & Mosto, 1982). The field work took place during a stay in November 1992. Some of the results have already been published or are under publication (Kristiansen & Vigna, 1994; Vigna & Kristiansen, 1995a; Vigna & Kristiansen, 1995b; Vigna & Kristiansen, 1996; Kristiansen & Vigna, 1996).

Previous studies of chrysophytes in Tierra del Fuego include records by Thomasson (1955, 1963) and Vigna (1986, 1988, 1993).

Climate and vegetation of Tierra del Fuego

The southern part of Tierra del Fuego belongs to the antiboreal zone, 52–56 °S (Tuhkanen, 1992). The climate is extremely oceanic, especially in the western parts, with heavy precipitation (400–1000 (–4000) mm per year, around Ushuaia around 500), mild winters and cool summers. The region is mountaneous; the lower regions are covered with dense *Nothofagus* forests and extended *Sphagnum* moors.

Corresponding climates in southern and northern temperate regions have been discussed in detail by Tuhkanen (1992). On the southern hemisphere, there are only few regions with corresponding climate, viz.

small subantarctic islands such as Islas Malvinias (Falkland Islands), Marion Island, Crozet Islands, Kerguelen, Macquarie Island, Campbell Island, etc. Some of these have been investigated for chrysophytes (Croome & Tyler, 1988), but with negative results, at least partly because of lack of suitable habitats. Thus any southern hemisphere comparison of distribution must also include Tasmania and New Zealand, although these islands have different climates, except in some areas in the mountains.

Because of the prevailing distribution of the land masses to the north, the northern hemisphere presents considerably larger areas with climate corresponding to the Fuegian, although they are found at somewhat higher latitudes. In Europe, corresponding climate is found along the Norwegian coast, from c. 62 to 71 °; on the Shetland Islands, Faroe Islands, Iceland, and the southernmost tip of Greenland this climate is found from 60 to 66 °. *Nothofagus* has been transplanted from Tierra del Fuego to the treeless Faroe Islands and proves to grow well, also to southernmost Greenland (Odum, 1991). In N. America, the climate corresponding to the Fuegian is found along the Alaskan south coast, extending out into the Aleutian Islands (54–60 °). Thus the possibilities for relevant comparisons are better, although also here further large areas, both in N. America, Greenland, and Scandinavia must be included, in order to secure enough chrysophyte material for comparisons.

Hypotheses:

(1) The Fuegian chrysophyte flora is historically derived, it is an Antarctic flora, related to the Australian-Tasmanian-New Zealand flora, corresponding to the distribution of the Nothofagus forests and of the marsupial animals. Compare the Gondwanaland theory and other geological theories about continental connections in the circum-pacific area (Seberg, 1991).

(2) The Fuegian chrysophyte flora is primarily determined by the climate, and thus related to the corresponding northern temperate floras, or it may be more or less cosmopolitan.

(3) The Fuegian chrysophyte flora is mainly endemic in its composition.

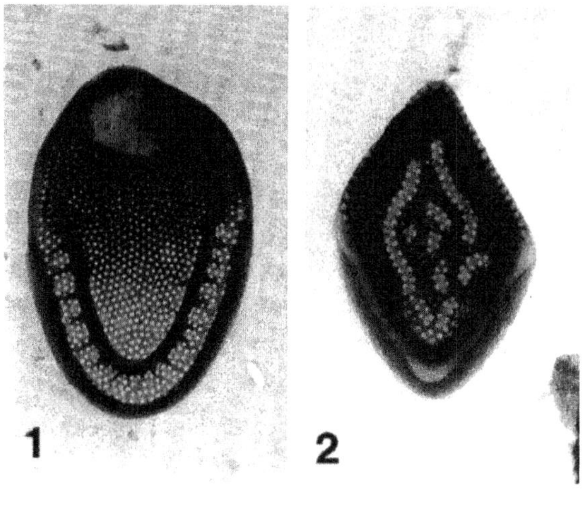

Figures 1, 2. Silica scales of Fuegian chrysophytes endemic to southernmost South America. Figure 1: *Mallomonas americana.* – Figure 2: *M. fuegiana.* EM, 10 000 ×.

Discussion

This discussion of the Fuegian chrysophyte flora will include the genera *Synura, Chrysodidymus*, *Mallomonas*, and *Chrysosphaerella.* The distribution records have been based on Asmund & Kristiansen (1986) for *Mallomonas* and on special papers for the other genera, e.g. Kristiansen & Tong (1989) and Wujek & Bicudo (1993); especially for *Synura curtispina* see Kristiansen & Lind (1995). The species of the genera *Spiniferomonas* and *Paraphysomonas* are still so inadequately investigated as regards their distribution, because of their small size and the special methods required to obtain representative material, that it is premature to include them in any biogeographical speculations.

The composition of the Fuegian flora and its biogeographic characteristics is seen from Table 1. A closer description of the species new to the flora of Tierra del Fuego and their biogeography is given by Vigna & Kristiansen (1996).

In the Fuegian flora there are only two chrysophytes endemic for southern S. America: *Mallomonas americana* Dürrschm.(Figure 1) and *M. fuegiana* Vigna & Kristiansen (Figure 2; Vigna & Kristiansen, 1995a; Vigna & Kristiansen, 1995b). In this connection some remarks are necessary about the many species described by Dürrschmidt (1980, 1981, 1982a, 1982b, 1983) from southern Chile, which at first were thought to be endemic; but most of these have later been found in northern temperate regions also, some of these in

Table 1. Table 1. Distribution types of silica-scaled chrysophytes from Tierra del Fuego. Explanations: *Endemic*: Known distribution restricted to southern South America. *Bipolar*: Only in northern and southern regions, not in the tropics. *Cosmopolitan or widely distributed*: Occurrence also including the tropics. *Tasmania*: Occurrence in Tasmania.

	End.	Bipolar	Cosm. or widely distrib.	Tasman.
Chrysodidymus				
synuroideus			+	+
Chrysosphaerella				
brevispina			+	+
coronacircumspina			+	+
Mallomonas				
matvienkoae f. *matvienk.*			+	+
parvula		+		+
multisetigera			+	+
calceolus			+	+
papillosa			+	+
paxillata		+		
rasilis			+	
guttata			+	+
transsylvanica		+		
americana	+			
heterospina			+	+
akrokomos			+	+
striata			+	+
flora			+	
cristata		+		
areolata			+	+
cyathellata v. *chilensis*			+	+
pillula f. *valvidiana*		+		
annulata			+	+
fuegiana	+			
pumilio		+		
alata		+		
mangofera f. *foveata*			+	
mangofera f. *gracilis*			+	
alveolata		+		
Synura				
curtispina			+	+
echinulata			+	+
petersenii			+	+
sphagnicola			+	
spinosa			+	+

addition to a more or less cosmopolitan distribution: *M. parvula* Dürrschm., *M. multisetigera* Dürrschm., *M. cristata* Dürrschm., and *M. alveolata* Dürrschm. *M. rasilis* Dürrschm and *M. cyathellata* Wujek & Asmund var. *chilensis* Dürrschm. have later been found mainly in tropical regions. Only a few species have still only been found in S. America, as *M. americana.*

Cosmopolitan species, or nearly cosmopolitan species, include most of the species found. All of these have, in addition to their temperate occurrence, also been found in the tropics. Most frequent among these were *M. heterospina* Lund, *M. akrokomos* Ruttner in Pascher, *M. striata* Asmund, *Synura petersenii* Korsh., *S. echinulata* Korsh., *S. spinosa* Korsh. – and there are several others less frequently recorded species with more scattered distributions, such as *M. calceolus* Bradley and *M. guttata* Wujek. However, the distinction between this group and the following is not clear, because many cosmopolitan and widely distributed species, e.g. *M. akrokomos* Ruttner in Pascher and *Chrysosphaerella coronacircumspina* Wujek & Kristiansen, still have their main distribution – so far known – in temperate regions.

Fuegian species with bipolar distributions, occurring in both southern and northern temperate regions, but not in the tropics or subtropics (with few exceptions), are the following: *Mallomonas* parvula Dürrschm. (Figure 4), *M. paxillata* (Bradley) Péterfi & Moméu (Figure 5), *M. transsylvanica* Péterfi & Asmund (Figure 3), *M. cristata* Dürrschm. (Figure 7), *M. pillula* Harris f. *valdiviana* Dürrschm. (Figure 10), *M. pumilio* Harris & Bradley (Figure 8), *M. alveolata* Dürrschm. (Figure 6), and *M. alata* Asmund et al. (Figure 9).

In summary: All the species found in Tierra del Fuego, except the endemic species *M. americana* and *M. fuegiana*, have also been found in northern temperate regions.

Comparisons with Iceland and Southern Greenland where chrysophyte investigations have been made confirm these considerations (Bradley, 1964; Nygaard, 1978; Kristiansen, 1994; Kristiansen, 1995). Out of the 28 species recorded during these investigations, 16 have been found in Tierra del Fuego, further two in Patagonia (Argentina, Thomasson, 1963), and one in Chile (*M. scalaris* Dürrschm., also showing a bipolar distribution). The remaining 9 species are restricted to northern temperate-subarctic areas, a few of them (*M. variabilis* Cronberg and *M. duerrschmidtiae* Siver) with rather restricted occurrences.

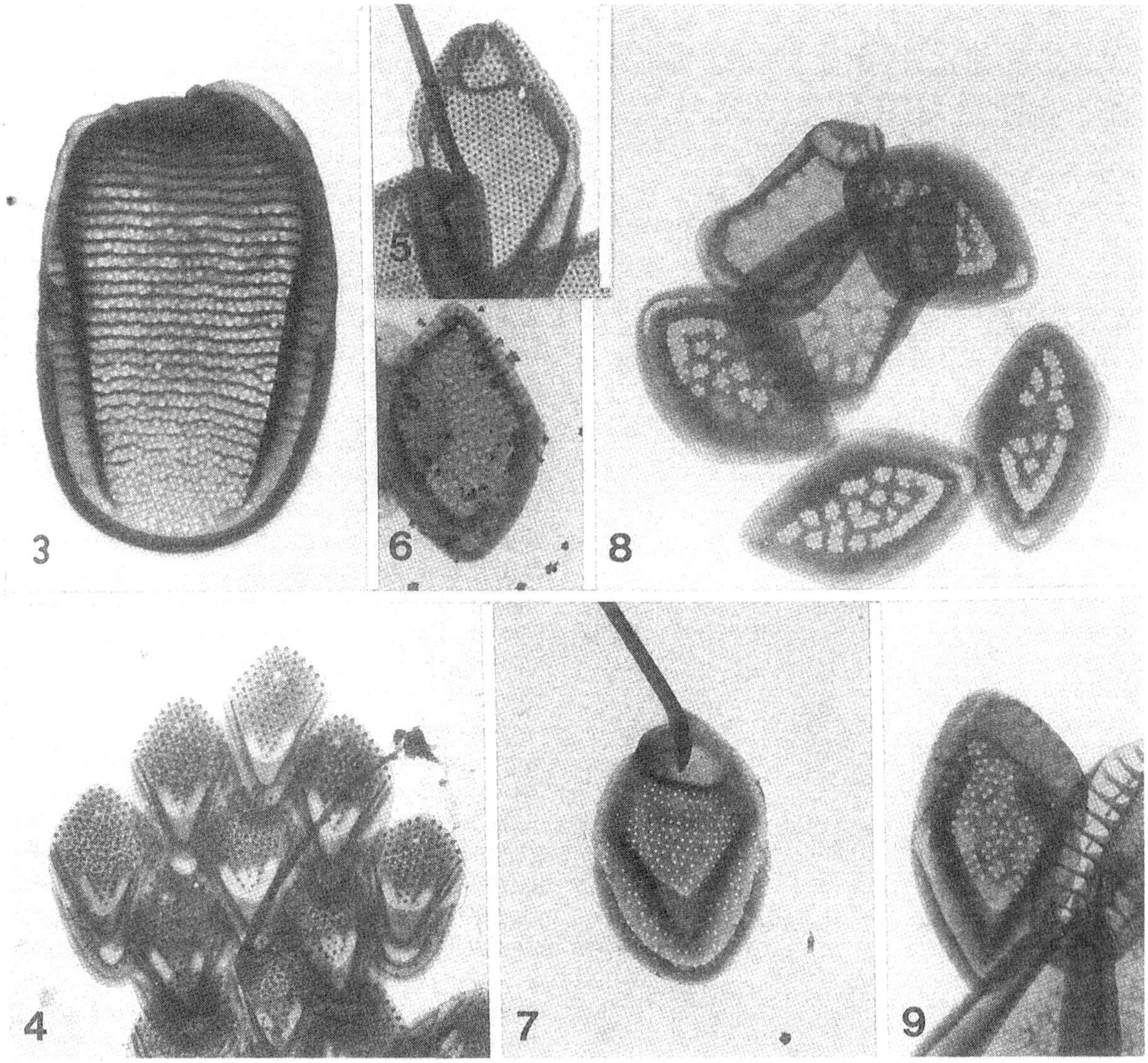

Figures 3–9. Silica scales of Fuegian chrysophytes with bipolar distributions. Figure 3: *Mallomonas transsylvanica.* – Figure 4: *M. parvula.* – Figure 5: *M. axillata.* – Figure 6: *M. alveolata.* Figure 7: *M. cristata.* – Figure 8: *M. pumilio.* – Figure 9: *M. alata.* EM, 10 000 ×. Figure 5 is from Danish material (Asmund & Kristiansen, 1986).

Tasmanian, Australian and New Zealand species: None of the species originally described from Tasmania or described from Australia or New Zealand and occurring in Tasmania have been found in Tierra del Fuego: *M. grossa* Dürrschm., *M. morrisonensis* Croome & Tyler, *M. perpusilla* Dürrschm., *M. plumosa* Croome & Tyler, *M. sabulosa* Croome & Tyler, *M. splendens* (West) Playfair, *M. tasmanica* (Croome & Tyler) Asmund & Krist. (Compare Croome & Tyler, 1985). Thus there are no species found only in Tasmania etc. and in Tierra del Fuego. Species which these regions have in common are all cosmopolitan (except the bipolar *M. parvula* Dürrschm.) And three of the species indicated as cosmopolitan/ widely distributed in Table 1 have not been found in Tasmania/Australia.

There are many species characteristic for the Tropics which occur in tropical and sub-tropical parts of S. America, but do not extend so far south as Tierra del Fuego: e.g. *Synura australiensis* Playfair, *M. matvienkoae* (Matv.) Asmund & Krist. f.*grandis* Dürrschm. & Cronberg, *M. fenestrata* Cronberg & Hickel, *M. bronchartiana* Compère, *M. portae-ferreae*

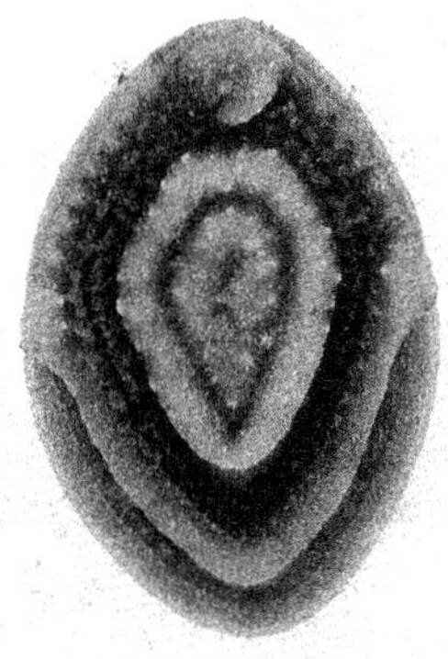

10

Figure 10. Silica scale of a Fuegian chrysophyte with bipolar distribution: *Mallomonas pillula* f.*valdiviana*. EM, 20 000 ×.

(Péterfi & Asmund), *M. rhombica* Cronberg (Cronberg, 1989; Vigna, 1990; Wujek & Bicudo, 1993).

It must also be mentioned that a considerable number of species have been found in northern regions, but not in Tierra del Fuego. *M. caudata* Ivanov em. Krieger, *M. acaroides* Perty em. Ivanov, and *M. multiunca* Asmund are widespread on the northern hemisphere. *M. teilingii* Conrad is mainly found in northern Europe. *M. crassisquama* (Asmund) Fott and *M. alpina* Ruttner in Pascher, which are among the most common species in the northern hemisphere, and almost cosmopolitan, have – though not found in Tierra del Fuego – been recorded from Nahuel Huapi in Argentina (Thomasson, 1963).

It is interesting to compare the distribution of desmids (Lenzenweger, 1993), which shows the same tendency: similarity between the Fuegian flora and the Austrian mountain flora. Also the freshwater algal flora (mainly cyanophytes and chlorophytes) of the Antarctic islands Kerguelen and Crozet is very similar to that of Arctic islands (Therezien & Couté, 1977).

Concluding remarks

It is obvious from the above discussion that the Fuegian chrysophyte flora is very similar to the northern temperate flora. Apart from two endemic species all the Fuegian species also occur in the temperate regions of the northern hemisphere. Most of them are cosmopolitan or widely distributed, several of these with main occurrence in northern temperate areas, and a few are decidedly bipolar, showing only a northern and southern almost exclusively temperate occurrence.

There is little similarity with other southern temperate regions such as Tasmania. Apart from the cosmopolitan species, the regions have no species in common. None of the Fuegian bipolar species occur in Tasmania, and none of the several species characteristic of Tasmania (and described from there or from Australia or New Zealand) have been found in Tierra del Fuego. There are no species with a circumpolar Antarctic distribution, and thus there is no special Antarctic chrysophyte flora.

Then the question remains of the origin of the Fuegian chrysophyte flora. The chrysophytes may have spread from lake to lake southwards along the Andean Range, following the retreat of glaciers after the last, pleistocene glaciation. At that time there were suitable climatic conditions in larger parts of South America than now for these species.

Long distance transport by birds is another, additional possibility to be discussed. There are no migrating routes circumnavigating the southern hemisphere, except for albatrosses and petrels, which cannot be expected to spread freshwater algae. However, there are several north-south routes. Most interesting are the migrations between Alaska and Tierra del Fuego, performed by some species of sandpipers (*Calidris*). And South Polar Skua and Arctic Tern are known to migrate between Antarctic and Arctic islands (Schlichting et al., 1978). There is, however, no evidence of the ability of any of these to act as vectors of chrysophytes, but an investigation of this problem should be highly desired.

Acknowledgments

Lisbeth Haukrogh and Lene Christiansen, Dept. of Phycology, have helped in electron microscopy and in photographic work.

References

Asmund, B. & J. Kristiansen, 1986. The genus *Mallomonas*. Op. Bot. 85: 1–128.

Bradley, D. E., 1964. A study of the *Mallomonas*, *Synura* and *Chrysosphaerella* of Northern Iceland. J. gen. Microbiol. 37: 321–333.

Cronberg, G., 1989. Scaled chrysophytes from the Tropics. Beih. Nov. Hedw. 95: 191–232.

Croome, R. & P. A. Tyler, 1985. Distribution of silica-scaled chrysophytes (Paraphysomonadaceae and Mallomonadaceae) from Australian inland waters. Aust. J. mar. Freshwat. Res. 36: 839–853.

Croome, R. & P. Tyler, 1988. Further observations of silica-scaled Chrysophyceae (Paraphysomonadaceae and Mallomonadaceae) from Australian freshwater. Nov. Hedw. 46: 481–489.

Dürrschmidt, M., 1980. Studies on the Chrysophyceae from Rio Cruces, Prov. Valdivia, South Chile, by scanning and transmission electron microscopy. Nov. Hedw. 33: 353–388.

Dürrschmidt, M., 1981. *Mallomonas cristata* sp. nov. (Chrysophyceae, Synuraceae) from South Chilean inland waters. Phycologia 20: 298–302.

Dürrschmidt, M., 1982a. *Mallomonas parvula* sp. nov. and *Mallomonas retifera* sp. nov. (Chrysophyceae, Synuraceae) from South Chile. Can. J. Bot. 60: 651–656.

Dürrschmidt, M., 1982b. Studies on the Chrysophyceae from South Chilean inland waters by means of scanning and transmission electron microscopy. II. Arch. Hydrobiol. Suppl. 63. Algol. Stud. 31: 121–163.

Dürrschmidt, M., 1983. Three new taxa of *Mallomonas* (Chrysophyceae, Synuraceae) from Lake Lanalhue, Chile. Nord. J. Bot. 3: 423–430.

Guarrera, S. A., 1986. Cyanophyta, Chroococcales. Guarrera, S. A., I. Gamundi de Amos & D. R. de Halperin (eds): Flora Criptogámica de Tierra del Fuego 1: 1–62.

Kristiansen, J., 1994. Preliminary studies on the distribution of silica-scaled chrysophytes in Greenland. Verh. int. Ver. Limnol.25: 2234–2236.

Kristiansen, J., 1995. Silica-scaled chrysophytes from Lake Thingvalla, Iceland. Algol. Stud. 79: 67–76.

Kristiansen, J. & J. F. Lind, 1995. The taxonomic relation between *Synura curtispina* and *S. favus* (Synurophyceae). Nord. J. Bot. 15: 443–448.

Kristiansen, J. & D. Tong, 1989. Studies on silica-scaled chrysophytes from Wuhan, Hangzhou and Beijing, P.R.China. Nov. Hedw. 49: 182–302.

Kristiansen, J. & M. S. Vigna, 1994. Tubular scales in *Synura* and the possible origin of bristles in *Mallomonas* (Synurophyceae). Phycologia 33: 67–70.

Kristiansen, J. & M. S. Vigna, 1996. The chrysophyte flora of Tierra del Fuego (Argentina). Beitr. z. Nov. Hedwigia 114: 109–110.

Lenzenweger, R., 1993. Beitrag zur Kenntnis der Desmidiaceenflora von Feuerland (Argentinien). Arch. f. Protistenk. 143: 143–152.

Moore, D. M., 1983. Flora of Tierra del Fuego. Anthony Nelson, England & Missouri Botanical Garden, U.S.A.

Nygaard, G., 1978. Freshwater phytoplankton from the Narssaq Area, South Greenland. Bot. Tidsskr. 73: 191–238.

Ødum, S., 1991. Choice of species and origins for arboriculture in Greenland and the Faroe Islands. Dansk dendrol. Årsskr. 9: 1–77.

Schlichting, H. E., B. J. Speziale & R. M. Zink, 1978. Dispersal of Algae and Protozoa by Antarctic flying birds. Antarct. J. U.S. 13: 147–149.

Seberg, O., 1991. Biogeographic congruence in the South Pacific. Aust. Syst. Bot.4: 127–136.

Tell, G. & P. Mosto, 1982. Chlorophyceae, Chlorococcales. In Guarrera, S. A., I. Gamundi de Amos & D. R. de Halperin (eds), Flora Criptogámica de Tierra del Fuego 6: 1–147.

Therezien, Y. & A. Couté, 1977. Algues d'eau douce des îles Kerguelen et Crozet (à l'exclusion des diatomacées). CNFRA, 91 pp. Reprint 1982 by Koeltz, Koenigstein.

Thomasson, K., 1955. Studies on South American freshwater plankton. 3. Plankton from Tierra del Fuego and Valdivia. Act. Bot. Gotob. 19: 193–225.

Thomasson, K., 1963. Araucanian Lakes. Act. Phytogeogr. Suec. 47: 1–139.

Tuhkanen, S., 1992. The climate of Tierra del Fuego from a vegetation geographical point of view and its ecoclimatic counterparts elsewhere. Act. Bot. Fennica 145: 1–64.

Vigna, M. S., 1986. Notas Algologicas XI. Dos nuevos generas para Argentina: *Mallomonopsis* Matv. y *Paraphysomonas* de Saed. Darwiniana 27: 299–304.

Vigna, M. S., 1988. Contributions to the knowledge of Argentinian Mallomonadaceae. Nov. Hedw. 47: 19–144.

Vigna, M. S., 1990. Silica-scaled subtropical chrysophytes from the Uruguay River, Argentina. Phycologia 29: 524–529.

Vigna, M. S., 1993. Chrysophytas escamosas (Synurophyceae) de una turbera de Tierra del Fuego (Argentina). Darwiniana 32: 299–302.

Vigna, M. S. & J. Kristiansen, 1995a. *Mallomonas fuegiana* (Synurophyceae) a new species of silica-scaled chrysophytes from Tierra del Fuego, Argentina. Nord. J. Bot. 15: 439–442.

Vigna, M. S. & J. Kristiansen, 1995b. *Mallomonas heterospina* and *M. americana* (Synurophyceae) in Tierra del Fuego, Argentina, with remarks on their taxonomical interrelation. Darwiniana 33: 61–65.

Vigna, M. S. & J. Kristiansen, 1996. Biogeographic implications of new records of scale-bearing chrysophytes from Tierra del Fuego, Argentina. – Arch. f Protislenk. 147: 137–144.

Wujek, D. E. & C. E. de Bicudo, 1993. Scale-bearing chrysophytes from the State of Sao Paulo, Brazil. Nov. Hedw. 56: 247–257.

Hydrobiologia **336:** 127–135, 1996.
J. Kristiansen (ed.), Biogeography of Freshwater Algae.

13. Endemism in freshwater algae

with special reference to the Australian region

Peter Alfred Tyler
School of Aquatic Science and Natural Resources Management, Deakin University, P.O. Box 423, Warrnambool, Victoria, 3280, Australia

Key words: endemics, Australia, flagship taxa, conservation

Abstract

Across the world there is a prevailing view that freshwater algae are cosmopolitan. The notion has seldom been tested and is unlikely to be true in genetic terms. Nonetheless, some morphospecies of several groups of algae do have a worldwide distribution. Others have restricted distributions and may be regarded as endemic to a region. However there is always the possibility that they will be discovered in far away places. Australia has a rather large element of endemicity in its algal flora. From the early days of Australian phycology many new genera and species of freshwater algae have been described. Some are of such distinctive appearance or novelty as to be regarded as 'flagship' taxa. There is little doubt about their endemicity and their existence increases the probability of less-distinguished species also being endemic. The degree of endemicity is probably masked by the 'force-fitting' of European names to Australian species.

Some Australian endemics are robust and are widely distributed in a variety of types of water body. Others, the frail endemics, the ones of greatest novelty and phylogenetic significance, have a very restricted range with their strongholds in dystrophic coastal lagoons where tracts or remnant patches of native vegetation survive. Their survival and the conservation of their biodiversity depends on recognition of the significance of coastal lagoons and swamps.

Most biologists have grown up with the clear notion that animals and plants are discontinuously distributed in the world. There are the latitudinal contrasts between the circum-arctic, coniferous forest and the deciduous woodlands further south, between both of these and the tropical rainforests and between all three of these and the *Eucalyptus* woodlands and evergreen rainforests of Australia. Longitudinally we could contrast the Old World and New World monkeys and butterflies and elements of the two floras. Yet when we turn to freshwater algae we have a prevailing impression of cosmopolitanism (Fritsch, 1953). We have it not because it has been properly investigated and tested but because of the reported world-wide occurrence of many species of cyanophytes, diatoms, chlorophytes and other algae. However, in the literature cosmopolitanism is usually mentioned vicariously, not for its own sake but as counterpoint in advancing the argument that in one group above all, the desmids, there is considerable biogeographic circumscription. From writers like the Wests (West, 1907, 1914; West & West, 1907), Prescott (1948), Krieger (1932) and others we hear of African, Australasian, Asian, etc. elements of the desmid flora and from the sheer volume of floristic works on desmids we must acknowledge the large analytical data base against which these statements are made and therefore admit that endemism is well-established for some desmid species. To this day, there are desmids (and other algae) known only from very restricted parts of the world, such as western Scotland (Brook, 1994a).

Often, however, supposed cosmopolitanism rests on the 'force-fitting' of, for example, southern hemisphere algae into European moulds. Because of the widespread lack of local algal floras such compendia as 'Süsswasserflora von Mitteleuropa' and 'Das Phytoplankton des Süsswassers' are used far away from the areas they most represent. In the case of desmids and diatoms such alliances are probably accurate, because for those groups the art of granule counting and striae measuring is well advanced, but more often it is a matter of getting as close as possible to a European description, a taxonomic analogue of the 'boreal bias' (Platnick, 1992). Floristic works from poorly-studied parts of the world contain many admissions of uncertainty about the credibility of this process, in the form of, for example, cf. *Sphaerocystis schroeteri* or *Cosmarium granatum* fa. Even West (1909) was forced into this, remarking about a species of *Gyrosigma* from Yan Yean 'for the present I have regarded it as possibly a variety of *Gyrosigma elongatum*'. The same has been the case with rotifers (Koste & Shiel, 1986, 1987, 1990; Koste et al., 1988; Shiel & Koste, 1985, 1986; Shiel et al., 1989; Shiel & Sanoamuang, 1992) and for the Cladocera Shiel and Dickson (1995) believe that a 'considerable degree of endemism is hidden by cosmopolitan 'names' in the Australian fauna'. In their opinion 'any cladoceran in Australia described from the Northern Hemisphere should be viewed with suspicion' Duigan (1992) (drawing heavily on the work of Frey) opines that 'force fitting' of chydorid branchipods to European species has reinforced the false and convenient concept that many chydorid species have cosmopolitan distribution. On the other hand one cannot harbour the same misgivings when workers experienced with the European flora find the same species in the far southern hemisphere, as Kristiansen & Vigna (1996) did for scaled chrysophytes, but, among the algae, even with well-recognised species, rarely has the reality of cosmopolitanism been tested in sexual or genetic terms. The desmid *Micrasterias thomasiana* Archer var. *notata* (Nordst.) Grönbl. is considered a cosmopolitan taxon as morphologically-identical populations occur in many parts of the world. However, considerable sexual and genetic incompatibility exists between populations from different areas (Blackburn & Tyler, 1987). Next to that there is also the condition where a cluster of related taxa, spanning a wide range of morphological variation, may all be members of a single worldwide, morphological species (Ling & Tyler, 1995). Because the population biology of most freshwater algae is poorly understood the terms endemic and cosmopolitan almost invariably refer to morphospecies. Even with this simplification recognition of endemic species is not straightforward. It would be easier if all parts of the world had been as well-studied as central Europe.

At present there is reason to believe that certain algae are found only in the High Tatra mountains (e.g. Siemińska, 1964; Starmach, 1968). This can be believed with reasonable authority only because they have not been found in surrounding areas of Europe that have also been comparatively well-studied. However, we cannot say that they do not occur in other, poorly-studied areas of the World. Prescott (1948) lists a number of supposedly endemic desmids which were subsequently located in 'far away stations', on the other side of the globe from the first records, and Croome & Tyler (1983 a–c) described three species of *Mallomonas* as Australian endemics all of which have since been found elsewhere (Dürrschmidt & Croome, 1984; Saha & Wujek, 1990). Zakryś & Walne (1994) give other examples among the euglenophytes and there are many others for which a small number of records have gradually accumulated. Certainly, rarity may be mistaken, initially, for endemism. Even rarity may be no more than circumstantial, the result of insufficient investigation. The Tatra mountains have been better studied than many parts of the world yet new records were still accumulating in recent decades (Mrozińska, 1984; Siemińska, 1958) as they still are for Poland (Szymańska & Zakryś, 1990; Tomascewicz, 1992). For the remoter parts of Britain and Ireland, well-worked by the Wests (West and West, 1904, 1905, 1908, 1911, West & Carter, 1923), new species are still being described (Brook, 1994b,c). Clearly, the smaller and more insignificant an alga, the greater the likelihood of its being overlooked, especially by opportunistic, rather than abiding, studies, the more so if electron microscopy has to be involved in its recognition. Table 1 presents a selection of rarities, dethroned endemics and some which are still thought of as endemics, most of which are easily overlooked.

On the brighter side, the more asiduous among researchers will not overlook even the most insignificant and morphologically-undistinguished of endemics. Brook (1994a) records the eventual rediscovery, on the Isle of Skye, of a desmid not seen since its original description in 1894. Further, publication of a new species today, however lowly and insignificant, may lead to its recognition elsewhere, where perhaps it was known, could not be identified and was given low priority because of its lack of attractive features. Kristiansen (1986) refers to this phenomenon with the comment 'the rules of the game are peculiar' and one could wonder whether new species described today from rel-

Table 1. A selection of some species originally known from one locality only (and hence thought of as endemic) but since located elsewhere, some still considered as endemics and others that are rare but widely distributed

Species	Endemic to	Since located, or known distribution	Reference
CHRYSOPHYTA			
Mallomonas sabulosa	Tasmania	–	Croome & Tyler, 1986
Mallomonas morrisonensis	Tasmania	n. India	Croome & Tyler, 1983a; Saha & Wujek, 1990
Mallomonas tasmanica	Tasmania	n. India	Croome & Tyler, 1983b; Saha & Wujek, 1990
Mallomonas plumosa	Tasmania	New Zealand, Malaya	Croome & Tyler, 1983c; Dürrschmidt & Croome, 1985
Chrysonephele palustris	one swamp, Tasmania	–	Pipes et al., 1989
Catenochrysis hispida	Europe	Tasmania	Bourrelly, 1981
Tetrasporopsis fuscescens	Europe	Europe, Australia	Tschermak-Woess & Kusel-Fetzmann, 1992; Entwisle & Anderson, 1990
EUGLENOPHYTA			
Euglena walnii	one fish pond, Alabama, USA	–	Zakryś, 1994
Euglena jirovecii	Czechoslovakia	N. America	Zakryś & Walne, 1994
CHLOROPHYTA			
Amscottia mira	One locality, Brazil.		Scott et al., 1965; Grönblad 1954a,b
Euastrum pseudopectinatum	Africa	N. America	Prescott, 1948
Staurastrum polonicum	Central Europe	N. America	Prescott, 1948
Euastrum informe	Brazil	N. America	Prescott, 1948
Spinoclosterium curvatum	Java	N. America	Prescott, 1948
Scourfieldia cordiformis	Europe	Tasmania	Croome & Tyler, 1985c; Ettl, 1983
CYANOPHYTA			
Gomphosphaeria aponina var. *multiplex*	Key Islands	Central Australia	Nygaard, 1926 Ling et al., 1989
RHODOPHYTA			
Batrachospermum nodosum (= *Nothocladus*)	SE Australia	–	Entwisle, 1989a; Entwisle & Kraft, 1984
Batrachospermum diatyches	Tasmania	–	Entwisle 1992
Psilosiphon scoparium	SE Australia	–	Entwisle, 1989b
Anfractutofilum umbracolens	Queensland	–	Cribb, 1965

atively well-studied countries (e.g. Zakryś, 1994) are more likely to survive as endemics than those newly-described from more remote parts of the world (e.g. Pipes et al., 1989).

While proof of endemism may be difficult to establish for many putative endemics far greater difficulties are involved in determining the origins of endemism. Some species may have become endemics, via periods of increasing rarity, because of habitat fragmentation or destruction, a process that has been recorded historically (Coesel et al., 1978; Heimans, 1969; Whitton, 1974). Other evidence (Tyler & Wickham, 1988) indicates that, in contrast, an original flora may survive for long periods in stable, protected habitats.

The most pure form of endemic would be species that evolved in a place and in that place only. While we may think of this in terms of long geological time it may in fact happen rather rapidly. Theriot (1992) gives plausible evidence for the evolution of autoapomorphic endemics among diatoms and Brook (1994a) invokes the old notion of clonal populations for narrow endemics among Scottish desmids. Given the generally high level of endemism in the Australian flora and fauna (Tyler, 1992 and see below) it is likely that some Australian freshwater algae are pure endemics. Floristically, the distinction is largely academic.

While 'insignificant' freshwater algae may be repeatedly overlooked, and hence falsely remain on the register as endemics, the same cannot be the case for large algae of striking appearance. A colloquial description in English is showy. Remembering the impression that my first glimpse of *Triploceras* made upon me I cannot imagine anyone overlooking *Amscottia mira* (Grönbl.) Grönbl. in any sample, yet still it is known only from the original Amazonian collection (Scott et al., 1965). The other species of the genus, *A. gulungulana* Ling & Tyler is even more impressive; it occurs sparsely in one area of tropical Australia (Ling & Tyler, 1985). Although the freshwater algal flora of this area of Australia (Ling & Tyler, 1986) has much in common with neighbouring countries to the north (Vyverman, 1995) *A. gulungulana* was not located in Indonesia (Scott & Prescott, 1961) nor in lowland New Guinea by Vyverman (1991) who was aware of it and on the look-out for it (Vyverman, pers. comm.). Such showy algae, especially if they are endemic, a matter of national pride, and intellectually-rewarding by virtue of their beauty or novelty, are known as 'flagship taxa' and it is with these that we have the greatest probability of real endemism. Because they are so showy, or so novel, it is unlikely that such species would be overlooked if indeed they were widely distributed. If *Tessellaria* (Tyler et al., 1989), or freshwater species of *Prorocentrum* (Croome & Tyler, 1987), or perhaps any Australian flagship taxon (Figure 1, Table 2) occurred in Europe or North America then they would have been seen there, long ago. Australia has a fair number of flagship taxa (Table 2) with a degree of novelty about them equivalent to that of marsupality among mammals, and because most of my experience has been with Australia I shall use the continent to stake a claim for pure endemism. An historical viewpoint is relevant to the matter.

The first investigations of freshwater algae in colonial Australia were fitful affairs. Vagrant Europeans paid brief visits and snatched a few samples from near their port of call, or a few samples were sent from Australia to European specialists. Nevertheless, though fragmentary, these early studies immediately revealed new taxa. New species and novelty could reasonably be expected, given the uniqueness of much of Australia's higher flora and fauna. Endemism in the terrestrial biota was well-established and nowhere more so than in Tasmania (Tyler, 1992). Today 11 of the total of 145 resident and regularly-visiting birds are Tasmanian endemics as are 3 of the 39 butterflies. Of the 1400 native Tasmanian vascular plants 320 are endemic, sufficient to fill six volumes of an 'Endemic Flora' (Curtis, 1967–1978). For the Australian freshwater microfauna the degree of endemism reaches extremes (Koste & Shiel, 1986). The first substantial work on Australian freshwater algae was the series of Botany Bulletins issued by Queensland's Colonial Botanist (Bailey, 1893–1913). These were compiled from the publications of Möbius (1892, 1894), Schmidle (1896) and Borge (1896, 1911) to whom Bailey had sent samples. Among about 200 species described from the sample were about 12 new species, but none that I regard as flagship taxa. We do not know how much 'force fitting' they did. The next, an Anglo-Australian alliance between Hardy and West, produced further new species, including flagship taxa (Table 2) and an authoritative taxonomic reference work for south-eastern Australia (West, 1909). The considerable contribution of Hardy in this partnership is summarised by Entwisle (1990). But it was G. I. Playfair, who, during the first two decades of this century, and without European patronage, established both the novelty and the high degree of endemism among Australian freshwater algae. In his series of publications (Playfair, 1907–1923) he described many new species of dinoflagellates, chlorophytes and chrysophytes including such

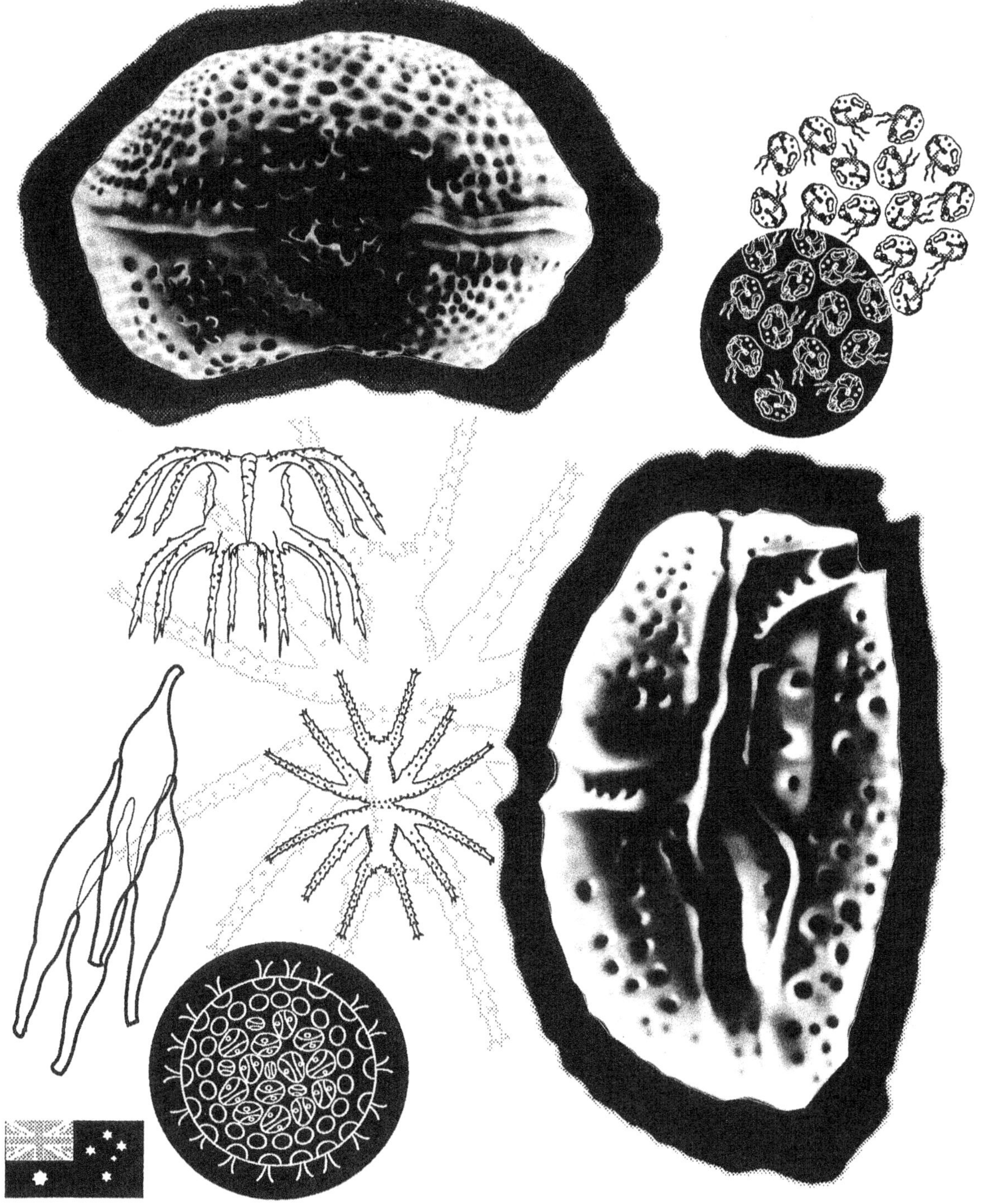

Figure 1. Australian flagship algae. From left to right, top to bottom, not to scale, *Prorocentrum playfairi*, *Chrysonephele palustris*, *Amscottia gulungulana*, *Dinobryon unguentariforme*, *Micrasterias hardyi*, *Thecadiniopsis tasmanica* and *Tessellaria volocina*. References to original descriptions and illustrations are in Table 2. Illustration design by Joshua Fartch, School of Visual Arts, Deakin University, Warrnambool.

Table 2. Some flagship taxa from Australia and elsewhere. The Australian ones are designated as robust (R) or frail (F) endemics, where known (see text)

Taxon	Known Distribution	Reference
CHLOROPHYTA		
Micrasterias hardyi (R)	S-E Australia	Tyler, 1970
Staurastrum victoriense (R)	Australia	West, 1909; Ling & Tyler, 1986
Amscottia gulungulana	Tropical Australia	Ling & Tyler, 1985
Amscottia mira	Amazon	Scott et al., 1965
Ichthyodontum sachlanii	Java	Scott & Prescott, 1956
Allorgeia incredibilis	Sudan	Gauthier-Lièvre, 1958
CHRYSOPHYTA		
Tessellaria volvocina (R)	Australia	Tyler et al., 1989
Dinobryon unguentariforme (F)	Australia	Croome et al., 1988
Chrysonephele palustris (F)	Tasmania	Pipes et al., 1989
DINOPHYTA		
Prorocentrum playfairi (F)	Tasmania	)Croome & Tyler, 1987
Prorocentrum foveolata (R?)	Australia	)
Thecadiniopsis tasmanica (F)	Tasmania	Croome et al., 1987

flagship taxa as *Tessellaria volvocina* (Playfair, 1915, 1918) (Table 2) and the novelty of two species of the marine genus *Prorocentrum* (then *Exuviella*) living in freshwater (Playfair, 1919). Subsequent studies have confirmed many of Playfair's findings (Croome et al., 1985; Croome & Tyler, 1985a, 1987; Tyler et al., 1989) and added further examples of endemism and novelty (Table 2), first *Mallomonas plumosa* Croome & Tyler, whose unusual features (Croome & Tyler, 1983c) soon came to the notice of the bigger world (Crawford & Round, 1989), then new dinoflagellates (Croome et al., 1987) and chrysophytes (Croome & Tyler, 1983a, b, 1986, Croome et al., 1988; Pipes et al., 1989). The Australian catalogue of endemics is greater than the list of flagship taxa; only the showy or novel ones are flagships. However, if flagship taxa are indeed endemics then some doubts about the less showy new species can be neglected and the flagships can 'piggy-back' the lowlier forms onto the list of endemics. Even so of the 3542 species of freshwater Algae recorded from Australia (Day et al., 1995) it is likely that the percentage of endemics is much less than for the aquatic microfauna.

Although we have not sampled every freshwater habitat in Australia we have, during floristic surveys (e.g. Croome & Tyler, 1985b, 1988; Ling et al., 1989), examined many samples from diverse waterbodies across the continent. This gives us a clear impression that many Australian endemics are today restricted to pristine lakes surrounded by undisturbed catchments clothed by native vegetation. Surveys of many lakes in Tasmania's World Heritage Area (Ling et al., 1989; Tyler, 1992) indicate a further restriction in distribution; many flagship taxa are not found in the multitude of pristine, montane lakes of the wilderness. Their stronghold is dystrophic coastal lagoons, particularly in Tasmania's southwest (Bowling et al., 1993), though some also occur on Bass Strait Islands and on Fraser Island, Queensland. We refer to such taxa as fragile endemics (Table 2). Preliminary evidence from King Island, Bass Strait, shows that of the many lowland lagoons only those few that still have a catchment or buffer zone of native vegetation have elements of the endemic flora. Elsewhere, the encroachment of agriculture, particularly access by cattle, appears to have favoured cosmopolitan species of eutrophic preferences. Fragile endemics, however, are capable of colonising new water bodies if conditions are suitable. Coastal lagoons near Strahan, Tasmania (Bowling et al., 1993) are strongholds of fragile endemics. Not far away a dam constructed across Parting Creek, nr. Zeehan, in the 19th Century, created a lake in a valley with a catchment of natural vegetation. The dystrophic water, natural for the region (Tyler, 1992), has been colonised by fragile endemics *Dinobryon ungentariforme* Croome et al. and *Prorocen-*

trum playfairi Croome & Tyler. Others, Lakes Cumberland and Murchison, have been colonised by *Prorocentrum playfairi* and/or *Thecadiniopsis tasmanica* Croome et al. In contrast to the fragile endemics, with their restricted distribution, are the robust endemics (Table 2) which have wider distribution, tolerate a wide range of water quality from ultra-oligotrophic to eutrophic, and are capable of persisting over long time spans. A good example is *Micrasterias hardyi* West, first described from a reservoir (West, 1909) and still found there (Tyler & Wickham, 1988). It is a ready colonist and occurs in numerous reservoirs in south-eastern Australia in heterothallic populations (Tyler, 1970). Another example is *Tessellaria volvocina* Playfair, a species described by Playfair (1918) and not seen again for 70 years (Tyler et al., 1989). Now it is now known colloquially as a Victorian weed. *Prorocentrum foveolata* Croome & Tyler is questionably a robust endemic. It does not occur in the pristine, dystrophic lagoons inhabited by *Pl. payfairi*, but in more saline lagoons (Croome et al., 1987; Tyler, 1995), including some not totally surrounded by native vegetation.

The tentative conclusions that can be drawn from floristic studies to the present day are that the freshwater algal flora of Australia contains a considerable number of endemic species and genera, many of great novelty, some fragile, some robust, some of phylogenetic significance, and that the preservation of much of this unique biodiversity depends on the conservation of coastal lagoons and wilderness lakes surrounded by tracts or remnant patches of native Australian vegetation.

References

Bailey, F. M., 1893–1898. Contributions to the Queensland flora. Botany Bulletins 6, 11 and 15. Queensland Department of Agriculture, Brisbane.

Blackburn, S. I. & P. A. Tyler, 1987. On the nature of eclectic species. A tiered approach to sexual and genetic compability in the desmid *Micrasterias thomasiana* Archer. Br. phycol. J. 22: 277–298.

Borge, O., 1896. Australische Süsswasserchlorophyceen. Bih. kungl. sv. Vet. – Akad. Handl. 22: 1–32.

Borge, O., 1911. Algologische Notizen 6. Süsswasseralgen aus Queensland. Bot. Notiser 1911: 197–207.

Bourrelly, P., 1981. Les Algues d'Eau Douce. Initiation à la systématique. II. Les Algues jaunes et brunes. Soc. Nouvelle des Editions Boubée, Paris. 438 pp.

Bowling, L. C., M. R. Banks, R. L. Croome & P. A. Tyler, 1992. Reconnaissance limnology of Tasmania. II. Limnological features of Tasmania freshwater coastal lagoons. Arch. Hydrobiol. 126: 385–403.

Brook, A. J., 1994a. Algae. In P. S. Maitland, P. J. Boon & O. S. McLusky (eds), The freshwaters of Scotland: A national resource of international significance. John Wiley & Sons Ltd.

Brook, A. J., 1994b. *Cylindrocystis cushleckae*, a new species of saccoderm desmids (Family Mesotaeniaceae). Queckett J. Microsc. 37: 225–231.

Brook, A. J., 1994c. A new species of *Spirotaenia, S. cambrica* (Family Mesotaeniaceae) of the sub-genus Polytaeniae from mid-Wales. Queckett J. Microsc. 37: 222–224.

Coesel, P. F. M., R. Kwakkestein & A. Verschoer, 1978. Oligotrophic and eutrophic tendencies in some Dutch moorland pools as reflected in their desmid flora. Hydrobiologia 61: 21–31.

Crawford, R. M. & F. E. Round, 1989. *Corethron* and *Mallomonas* – some striking morphological similarities. In B. S. C. Leadbeater & W. A. Driver (eds), The Chromophyte Algae, Problems and Perspectives. Oxford University Press, Oxford: 295–305.

Cribb, A. B., 1965. *Anfractutofilum umbracolens* gen. et spec. nov., a freshwater red alga from Queensland. Proc. r. Soc. Qld. 126: 93–95.

Croome, R. L. & P. A. Tyler, 1983a. *Mallomonas morrisonensis* (Chrysophyceae), a new species from Australia. Br. phycol. J. 18: 383–389.

Croome, R. L. & P. A. Tyler, 1983b. *Mallomonopsis tasmanica* spec. nov. (Chrysophyceae) and *Mallomonopsis elliptica* Matvienko from Australia. Br. phycol. J. 18: 357–365.

Croome, R. L. & P. A. Tyler, 1983c. *Mallomonas plumosa* (Chrysophyceae), a new species from Australia. Br. phycol. J. 18: 151–158.

Croome, R. L. & P. A. Tyler, 1985a. *Synura australiensis* (Mallomonadaceae, Chrysophyceae), a light and electron microscopical investigation. Nord. J. Bot. 5: 399–401.

Croome, R. L. & P. A. Tyler, 1985b. Distribution of silica-scaled Chrysophyceae (Paraphysomonadaceae and Mallomonadaceae) in Australian inland waters. Aust. J. mar. freshwat. Res. 36: 839–583.

Croome, R. L. & P. A. Tyler, 1985c. Structure and ecology of the flagellate *Scourfieldia caeca* (Korsh.) Belcher et Swale in two meromictic lakes in Tasmania. Aust. J. mar. freshwat. Res. 36: 413–419.

Croome, R. L. & P. A. Tyler, 1986. *Mallomonas sabulosa* (Chrysophyceae), a new species from Australia. Br. phycol. J. 21: 93–96.

Croome, R. L. & P. A. Tyler, 1987. *Prorocentrum playfairi* and *Prorocentrum foveolata*, two new dinoflagellates from Australian freshwaters. Br. phycol. J. 22: 67–75.

Croome, R. L. & P. A. Tyler, 1988. Further observations of silica-scaled Chrysophyceae (Paraphysomonadaceae and Mallomonadaceae) from Australian freshwaters. Nova Hedwigia 46: 481–489.

Croome, R. L., M. Dürrschmidt & P. A. Tyler, 1985. A light and electron microscopical investigation of *Mallomonas splendens* (West) Playfair. Nova Hedwigia 41: 463–470.

Croome, R. L., G. M. Hallegraeff & P. M. Tyler, 1987. *Thecadiniopsis tasmanica* gen et sp. nova (Dinophyta: Thecadiniaceae), a new peridinoid dinoflagellate from Tasmanian freshwaters. Br. phycol. J. 22: 325–333.

Croome, R. L., H. U. Ling & P. A. Tyler, 1988. *Dinobryon unguentariforme* (Chrysophyceae), an unusual new species from Australia. Br. phycol. J. 23: 129–133.

Curtis, W. M., 1967–1978. The endemic flora of Tasmania. 1–4. Ariel Press, London.

Day, S., R. Wickham, T. Entwisle & P. A. Tyler, 1995. Bibliographic checklist of non-marine Algae in Australia. Flora of Australia, Supplementary Series. Australian biological Resources Survey, Canberra, 276 pp.

Duigan, C. A., 1992. The ceology and distribution of the littoral freshwater Chydoridae (Branchiopoda, Anomopoda) of Ireland, with taxonomic comments on some species. Hydrobiologia 241: 1–70.

Dürrschmidt, M. & R. L. Croome, 1985. Mallomonadaceae (Chrysophyceae) from Malaysia and Australia. Nord. J. Bot. 5: 285–298.

Entwisle, J., 1989a. Macroalgae in the Yarra River basin: flora and distribution. Proc. r. Soc. Vic. 101: 1–76.

Entwisle, J., 1989b. *Psilosiphon scoparium* gen. et spec. nov (Lemaneaceae), a new red alga from south-eastern Australian streams. Phycologia 28: 469–475.

Entwisle, T. J., 1990. The lean legacy of freshwater phycology in Victoria. In P. S. Short (ed), History of systematic Botany in Australia. Australian systematic Botany Society, Melbourne: 239–246.

Entwisle, J., 1992. The setaceous species of *Batrachospermum* (Rhodophyta): a re-evaluation of *B. atrum* (Hudson) Harvey and *B. puiggarianum* Grunow including the description of *B. diatyches* sp. nov. from Tasmania. Muelleria 7: 425–445.

Entwisle, T. J. & R. A. Andersen, 1990. A re-examination of *Tetrasporopsis* (Chrysophyceae) and the description of *Dermatochrysis* gen. nov. (Chrysophyceae): a monostromatic alga lacking cell walls. Phycologia 29: 263–274.

Entwisle, T. J. & G. T. Kraft, 1984. Survey of freshwater red algae (Rhodophyta) of south-eastern Australia. Aust. J. mar. freshwat. Res. 35: 213–259.

Ettl, H., 1983. Chlorophyta 1. Phytomonadina. H. Ettl, J. Gerloff, H. Heynig & D. Mollenhauer (eds), Süsswasserflora von Mitteleuropa 9. Gustav Fischer Verlag, Stuttgart. 807 pp.

Fritsch, F. E., 1953. Comparative studies in a polyphyletic group. The Desmidiaceae. Proc. Linn. Soc. 164: 258–280.

Gauthier-Lièvre, L., 1958. Desmidiacées asymmétriques. Le genre Allorgeia gen. nov. Bull. Soc. hist. nat. Afr. Nord. 49: 93–101.

Grönblad, A., 1954a. A new genus and a new species among desmids. Bot. Notiser 1954: 167–171.

Grönblad, A., 1954b. *Amscottia* Crönbl., nom. nov. Bot. Notiser 1954: 1433.

Heimans, J., 1969. Ecological, phytogeographical and taxonomic problems with desmids. Vegetatio 17: 50–82.

Koste, W. & R. J. Shiel, 1986. New Rotifera (Aschelminthes) from Tasmania. Trans. r. Soc. S. Aust. 110: 93–109.

Koste, W. & R. J. Shiel, 1987. Tasmanian Rotifera: Affinities with the Australian fauna. Hydrobiologia 147: 31–443.

Koste, W. & R. J. Shiel, 1990. Rotifera from Australian waters VI. Proalidae, Lindiidae (Rotifera: Monogonata). Trans. r. Soc. S. Aust. 114: 129–143.

Koste, W., R. J. Shiel & L. W. Tan, 1988. New Rotifers from Tasmania. Trans. r. Soc. S. Aust. 112: 119–131.

Krieger, W., 1932. Die Desmidiaceen der Deutschen Limnologischen Sunda-Expedition. Arch. Hydrobiol. Suppl. 11: 129–230.

Kristiansen, J., 1986. Identification, ecology and distribution of silica-scale bearing Chrysophyceae, a critical approach. In J. Kristiansen & R. A. Andersen (eds), Chrysophytes: Aspects and Problems. Cambridge University Press, Cambridge: 229–239.

Kristiansen, J. & M. S. Vigna, 1996. Bipolarity in the distribution of silica-scaled chrysophytes. In: J. Kristiansen (ed.), Biogeography of Freshwater Algae. Developments in Hydrobiology 118. Kluwer Academic Publishers, Dordrecht: 121–126. Reprinted from Hydrobiologia 336.

Ling, H. U. & P. A. Tyler, 1985. *Amscottia gulungulana* sp. nov. (Desmidiaceae) from tropical Australia. Br. phycol. J. 20: 335–339.

Ling, H. U. & P. A. Tyler, 1986. A limnological survey of the Alligator Rivers Region II. Freshwater Algae, exclusive of diatoms. Research Report 3, Part II. Australian Government Publishing Service, Canberra.

Ling, H. U. & P. A. Tyler, 1995. The *Staurastrum tohopekaligense* species complex. Algol. Stud. 76: 27–60.

Ling, H. U., R. L. Croome & P. A. Tyler, 1989. Freshwater dinoflagellates of Tasmania, a taxonomic and ecological survey. Br. phycol. J. 24: 11–129.

Ling, H. U., D. P. Thomas & P. A. Tyler, 1989. Microalgae. In W. Zeidler & W. F. Ponder (eds), Natural History of Dalhousie Springs. South Australian Museum special Publications, Adelaide: 47–52.

Möbius, M., 1892. Australische Süsswasseralgen. Flora 1892: 309–347.

Möbius, M., 1894. Australische Süsswasseralgen II. Abh. Senckenb. naturf. Ges. 1894: 310–350.

Mrozińska, T., 1984. *Botryosphaerella sudetica* (Lemmermann) Silva (Botryococcaceae, Chlorophyta) a green alga new to Tatra Mountains. Algol. Stud. 37: 45–49.

Nygaard, G., 1926. Plankton from two lakes of the Malayan region. Vidensk. Medd. Dansk naturh. Foren. 82: 197–230.

Pipes, L., P. A. Tyler & G. F. Leedale, 1989. *Chrysonephele palustre* gen. et spec. nov., a new phagotrophic pigmented chrysophyte from Tasmania. Beih. Nova Hedwigia 95: 81–97.

Platnick, N. I., 1992. Patterns of biodiversity. In N. Elderidge (ed.), Systematics, ecology and the biodiversity crisis. Columbia U.P., New York.

Playfair, G. I., 1907a. Some new or less known desmids found in New South Wales. Proc. Linn. Soc. N.S.W. 32: 160–201.

Playfair, G. I., 1907b. Some Sydney desmids. Proc. Linn. Soc. N.S.W. 33: 603–628.

Playfair, G. I., 1910. Polymorphism and life-history in Desmidiaceae. Proc. Linn. Soc. N.S.W. 35: 459–495.

Playfair, G. I., 1912. Plankton of the Sydney water-supply. Proc. Linn. Soc. N.S.W. 37: 512–552.

Playfair, G. I., 1914. Contributions to the knowledge of the biology of the Richmond River. Proc. Linn. Soc. N.S.W. 39: 93–151.

Playfair, G. I., 1915a. The genus *Trachelomonas*. Proc. Linn. Soc. N.S.W. 40: 1–41.

Playfair, G. I., 1915b. Freshwater Algae of the Lismore district with an appendix on the algal Fungi and Schizomycetes. Proc. Linn. Soc. N.S.W. 40: 310–362.

Playfair, G. I., 1916a. Australian freshwater phytoplankton. Proc. Linn. Soc. N.S.W. 41: 823–852.

Playfair, G. I., 1916b. *Oocystis* and *Eremosphaera*. Proc. Linn. Soc. N.S.W. 41: 107–147.

Playfair, G. I., 1917. Rhizopods of Sydney and Lismore. Proc. Linn. Soc. N.S.W. 42: 633–675.

Playfair, G. I., 1918. New and rare freshwater Algae. Proc. Linn. Soc. N.S.W. 43: 497–543.

Playfair, G. I., 1919. Peridineae of New South Wales. Proc. Linn. Soc. N.S.W. 44: 793–818.

Playfair, G. I., 1921. Australian freshwater flagellates. Proc. Linn. Soc. N.S.W. 46: 99–146.

Playfair, G. I., 1923. Notes on freshwater Algae. Proc. Linn. Soc. N.S.W. 48: 206–228.

Prescott, G. W., 1948. Desmids. Bot. Rev. 14: 644–676.

Saha, L. C. & D. E. Wujek, 1990. Scale-bearing chrysophytes from tropical north-east India. Nord. J. Bot. 10: 343–355.

Schmidle, W., 1896. Süsswasseralgen aus Australien. Flora 82: 297–313.

Scott, A. M. & G. W. Prescott, 1956. Notes on Indonesian desmids. II. *Ichthyodontum*, a new desmid from Sumatra. Reinwardia 4: 105–112.

Scott, A. M. & G. W. Prescott, 1961. Indonesian desmids. Hydrobiologia 17: 1–132.

Scott, A. M., R. Grönblad & H. Croasdale, 1965. Desmids from the Amazon basin, Brazil, collected by Dr. H. Sioli. Acta bot. fenn. 69: 1–94.

Shiel, R. J. & J. A. Dickson, 1994. Cladocera recorded from Australia. Trans. r. Soc. Aust. 119 (in press).

Shiel, R. J. & W. Koste, 1985. New species and new records of Rotifera (Aschelminthes) from Australian waters. Trans. r. Soc. S. Aust. 109: 1–15.

Shiel, R. J. & W. Koste, 1986. Australian Rotifera: Ecology and biogeography. In P. de Deckker & W. D. Williams (eds), Limnology in Australia. CSIRO, Melbourne & Dr W. Junk Publishers, Dordrecht.

Shiel, R. J., W. Koste & L. W. Tan, 1989. Tasmania revisited: rotifer communities and habitat heterogeneity. Hydrobiologia 186/187: 239–245.

Shiel, R. J. & L. Sanoamuang, 1993. Trans-Tasman variation in Australasian *Filina* populations. Hydrobiologia 255/256: 455–462.

Sieminska, J., 1958. Nowy dla Tatr i Polski gatunek *Bulbochaete basispora* Wittrock et Lundell. Fragm. florist. geobot. 31: 151–153.

Sieminska, J., 1964. *Rhizochloris tatrica* n. sp. (Xanthophyceae). Acta hydrobiol. 6: 623–625.

Starmach, K., 1968. *Plectonema tatrica* n. sp., a new species of blue-green algae from the Tatra Mountains. Acta hydrobiol. 10: 427–432.

Szymanska, H. & B. Zakrys, 1990. New phycological records from Poland. Algol. Stud. 60: 25–32.

Theriot, E., 1992. Clusters, species concepts, and morphological evolution of diatoms. Syst. Biol. 14: 147–157.

Tomaszewicz, G., 1992. New locality of *Euastrum vrigense* (Chlorophyta, Desmidiaceae) in Poland. Fragm. flor. geobot. 37: 507–510.

Tschermak-Woess, E. & E. Kusel-Fetzmann, 1992. A new find of *Tetrasporopsis fuscescens* (A Braun ex Kützing) Lemmermann (Chrysophyta) in Austria, and some additional observations. Arch. Protistenk. 142: 157–165.

Tyler, P. A., 1970. Taxonomy of Australian freshwater Algae. I. The genus *Micrasterias* in south-eastern Australia. Br. phycol. J. 5: 211–234.

Tyler, P. A., 1992. A Lakeland from the Dreamtime. The Second Founders' Lecture. Br. phycol. J. 27: 353–368.

Tyler, P. A., 1995. The significance of north-east Tasmania for the biogeography of endemic freshwater Algae. Rec. Q. Vic. Mus. (in press).

Tyler, P. A. & R. P. Wickham, 1998. Yan-Yean revisited – a bicentennial window on Australian freshwater phycology. Br. phycol. J. 232: 105–114.

Tyler, P. A., L. Pipes, R. L. Croome & G. F. Leedale, 1989. *Tessellaria volvocina* (Chrysophyta) rediscovered. Br. phycol. J. 24: 329–337.

Vyverman, W., 1991. Desmids from Papua-New Guinea. Bibl. phycol. 87: 1–201.

Vyverman, W., 1996. The Indo-Malayan, North Australian phycogeographical region revised. In: J. Kristiansen (ed.), Biogeography of Freshwater Algae. Developments in Hydrobiology 118. Kluwer Academic Publishers, Dordrecht: 107–120. Reprinted from Hydrobiologia 336.

West, G. S., 1907. Report on the freshwater Algae, including phytoplankton, of the 3rd Tanganyika Expedition conducted by Dr. W. A. Cunnington, 1904–1905. J. Linn. Soc. Bot. 38: 81–197.

West, G. S., 1909. The Algae of the Yan-Yean Reservoir, Victoria: A biological and ecological study.

West, G. S., 1914. A contribution to our knowledge of the freshwater Algae of Columbia. Mem. Soc. Sci. nat. Neuchatel 5: 1013–1050.

West, G. S. & N. Carter, 1923. A monograph of the British Desmidiaceae. 5. Ray Society, London.

West, W. & G. S. West, 1905, 1907, 1908, 1911. A monograph of the British Desmidiaceae. 1–4. Ray Society, London.

Whitton, B. A., 1974. Changes in the British freshwater Algae. In D. L. Hawksworth (ed), The changing flora and fauna of Britain. Academic Press, London.

Zakryś, B., 1994. *Euglena walnei* sp. nova (Euglenophyta) – a new species from Alabama (USA). Algol. Stud. 72: 9–11.

Zakryś, B. & P. L. Walne, 1994. Floristic, taxonomic and phytogeographic studies of green Euglenophyta from the southeastern United States, with emphasis on new and rare species. Algol. Stud. 72: 71–114.

Hydrobiologia **336**: 137–142, 1996.
J. Kristiansen (ed.), Biogeography of Freshwater Algae.

14. Are the impacts of events in the earth's history discernable in the current distributions of freshwater algae?

Annette W. Coleman
Department of Molecular Biology, Cell Biology and Biochemistry; Brown University, Providence, Rhode Island 02912, USA

Key words: algae, flagellates, freshwater algae, historical biogeography, Volvocaceae, waterfowl

Abstract

Freshwater microalgae, lacking a fossil record, have contributed little to the study of historical biogeography. Some of the innate difficulties are discussed, as well as some of the more hopeful possibilities, if distribution records, morphology and DNA sequence analysis are combined with knowledge of the earth's history. Examples of species within the same family showing quite different distributions are given, along with suggested explanations. These include possible examples of the role played by waterfowl in dissemination of freshwater algae.

Freshwater algal genera and species have generally been considered to be geographically widespread. The notion that the current distribution patterns of freshwater microalgae might exhibit differences within and between species and genera which reflect the geological and climatographic history of the continental masses has not been seriously examined, primarily for lack of data. How might one begin to unravel the biogeographical history of a family of microalgae? Here we consider some of the characteristics which might make a group of microalgae useful for historical biogeographic examination. We then present some examples of both very broad and very narrow distributions within one family of microalgae, along with some speculation as to what factors might be responsible in these cases. So few examples prove nothing yet, but may offer a challenge for future work.

Background considerations

First, let us consider some relatively well known distribution pattern examples, ones with a rich fossil record, to see what pitfalls to anticipate. The current Equidae are represented in the wild only by Asian horses and African zebra, and the recently extinct quagga. Yet their origins can be traced to Eocene North America from which they spread across the northern hemisphere and southward. Events of the Pleistocene isolated Asia from North America, and horses became extinct in America until recolonization through human agency. A second example representing an organism of freshwater habitats,is perhaps even more germane. Extant lungfish genera are species of the semitropical southern hemisphere – Africa, South America and Australia, a perfect correlation with a Gondwana origin and maintainance. Yet their fossil record belies this, because fossilized lung fish are scattered across the Northern hemisphere. What such examples illustrate is that present distribution alone may have little informative value concerning the distributional history of a group of genera. Present distribution reflects the sum of origin, mobility, and extinction, and the absence of any fossil record for many freshwater microalgae removes a major source of information necessary to reconstruct evolutionary history.

Except for some of the non-green groups such as diatoms, microalgae not only lack a fossil record but their taxonomy suffers from an additional challenge. Organisms very similar in appearance cannot necessarily be accepted as closely related by descent. The developmental canalization generally assumed to limit the morphs of microalgae are poorly understood as yet, and such may obscure the recognition of deep separations between morphologically similar taxa. When limited to morphological evidence, one can only assume that two disjunct entities, both fitting the same species description, are in fact more closely related by descent to each other than to any other species (morph); each may in fact have arisen by a similar change in the developmental program of two long separated lines of parental organisms.

Nevertheless, the footprints of past events should be found in the genomes of such organisms, since the nucleotide clock ticks inexorably, and past random events affecting DNA sequences are preserved in the genome of current organisms. We have been examing one family of freshwater algae, the Volvocaceae, in the hope that comparative nucleotide sequence data, combined with other characters and distribution records, along with consideration of the major events of historical biogeography, can contribute to reconstructing its evolutionary history. Although lacking fossil record, the family exhibits properties particularly propitious for such a reconstruction, as follows:

1. The individual cells of the Volvocacean colony are essentially identical to those of *Chlamydomonas reinhardtii* in cell structure, and all members of the family undergo a very similar asexual developmental path in forming a new daughter colony, a pattern long recognized as unique among the algae and definitive of the family. There is no reason to doubt *a priori* that this family is monophyletic.

2. Most of the genera and species are readily recognizable even to the inexpert eye, so that reported occurences can generally be accepted and, more important, absences in species flora may represent true absence rather than oversight.

3. All of the genera (with one exception) and many of the species are essentially worldwide in distribution (Ettl, 1983; Iyengar & Desikachary, 1981; many other local floras – list available from the author on request), although records from the tropics are sparse, at least in part for historical lack of attention. There are however some species, particularly of Volvox, which are fairly certainly absent from some continents (Ettl, 1983).

4. All members of the group share remarkably similar habitats, more or less eutrophic, usually ephemeral, bodies of fresh water. Two or more species are often collected in the same sample. There is no suggestion in the literature that differences exist in their physiology which might dictate different ecological niches.

5. There is a comparative wealth of information concerning their life histories, mating types, morphology (Nozaki & Itoh, 1994) and even chromosome numbers, informative characters valuable to compare with sequence data.

6. We can probably rightly presume that we know the dispersal mechansm, the heavy walled zygospore, although bird or wind or both might be the major agents of dispersal.

7. Prior studies, utilizing one or more representatives of all the genera of the family, have begun to outline its evolutionary history based on nucleotide sequence comparisons of small and large subunit (SSU and LSU) nuclear genes for ribosomal RNA (Buchheim & Chapman, 1991; Larson et al., 1992).

A first attempt at estimating the age of the family is the comparison of divergence of SSU rDNA sequences and of histone gene sequences between *Volvox carteri* and *Chlamydomonas reinhardtii* (Rausch et al., 1989). Assuming that the molecular clock ticks in these haploid protistans at a rate similar to that in other organisms, their comparison suggests a divergence time between the presumed unicell ancestry and at least this genus of the family somewhere just prior to the Cretaceous-Tertiary boundary, some 80 mya. The earth's continents and their climates have undergone cataclysmic changes in that period of time. Gondwana fragmented and its components migrated, with the subcontinent of India rejoining Asia ca. 45 mya. Expansion of the North Atlantic ocean continues to distance North America from Europe. More recently, the continents of the northern hemisphere have undergone Pleistocene glaciations which may well have cleansed their higher latitudes of Volvocacean forms repeatedly.

The importance of dispersal and colonization

Such events may have left their mark on current distributions, but only if Volvocaceae are limited in their ability to colonize fresh habitats. The first question then concerns the colonization frequency of Volvocaceae. Since their life cycle includes a heavy walled zygospore stage which is both small and resistant to dessication, long term establishment of a population probably reflects the accumulation of annual zygote crops in the soil surrounding a colonized habitat in numbers sufficient to provide an annual Spring germination (Coleman, 1983). If closely related (by morphology and by sequence analysis) examples of a species are found world wide, the implication is that dispersion and successful colonization are frequent, while if organisms showing the same or greater degree of nucleotide variation are currently only localized in distribution, then colonization may be only rarely successful or long lasting.

Our genetic examination (see Coleman et al., 1994, for methods) thus far has concentrated on the internal transcribed spacer (ITS) region of the nuclear ribosomal DNA, a region which appears to be the most informative at the population, species and genus level of those we surveyed (Coleman & Goff, 1991). This region is composed of both very conserved portions (the 5.8S RNA gene) and the relatively less constrained 5′ ITS-1 and 3′ ITS-2 flanking it (Figure 1). We initially compared four isolates of *Gonium pectorale* Mueller and 12 isolates of *Pandorina morum* Bory de St. Vincent; both species are worldwide in distribution and of very frequent occurence, often together, in nature. For *G. pectorale*, from eastern and western North America, from Japan and from India, pairwise comparisons suggested that all pairs are roughly equally different and there is no significant cline correlating with geographic origin of the collections. The amount of genetic diversity among the isolates is relatively small (a total of 7% variable positions overall, including insertions/deletions), which agrees with both their uniform morphology, chromosome number and their ability to intercross (Coleman et al., 1994). This species has colonized the earth. The evolutionary age of the morph can only be estimated, and probably the least controversial way is by comparison to data from the other species of the family.

For *P. morum* the ITS sequence comparisons suggested a quite different picture (Coleman et al., 1994). *P. morum* has long been known to be subdivided by mating affinity into subspecies which have similar morphology but which differ in chromosome number and in mating affinity. Two of the 'subspecies' or syngens, defined as organisms capable of interbreeding, are worldwide in distribution (the so-called syngen II and syngen XX), and the remaining twenty syngens are restricted in distribution to but one subcontinental area of the earth.

Sequence data correlations with other data

Previously it was not known whether clones originating from different continents but able to mate successfully were indeed closest genetic relatives or whether perhaps the genes for mating type underwent frequent change through a limited array such that a compatible mating type pair might suddenly reappear in a population. The ITS analysis revealed that indeed mating type is a fairly conservative character since, whatever their geographic origin, organisms able to mate (by definition, belonging to the same syngen) are also the most similar in ITS sequence. Thus three North American and two Asian (Thailand and Nepal) isolates of syngen II are clearly most similar inter se; likewise one North American and three Asian (Korea, South China, and India) isolates of syngen XX are most similar inter se and different from syngen II. Finally, two syngens found only in the American midwest, Syngen I and Syngen III, are each different from both II and XX and from each other. However, I and III have very similar ITS sequences as if they derive fairly recently from a common ancester, and both have been found only in the Midwest of North America. With respect to the species as a whole, *P. morum* ITS sequences determined so far resemble each other more than they resemble those of any other Volvocaceae (Coleman, work in progress), which rules out the possibility that the *P. morum* morph might have rearisen repeatedly from various other Volvocaceae.

History or life style?

Have historical biogeographical events contributed to the difference in species structure between *G. pectorale* and *P. morum*? In this case, it is not apparent; instead, a life cycle difference may be responsible. One possible interpretation of the species difference is that *G. pectorale*, exhibiting so little variation in its ITS by comparison with the species *P. morum* as a whole, is a younger species. Alternatively, its apparent ability to spread may make the species panmictic, homogenizing its ITS sequences. In either case, it would appear to be able to colonize with greater ease than *P. morum*. The one characteristic of the *Gonium* life cycle which might confer this ability is the fact that the *G. pectorale* zygotes produce four viable products after meiosis, rather than the single product characteristic of *P. morum* zygotes. Obviously then, it would take only a single zygote of *G. pectorale* germinating in a new habitat to engender a full, sexually reproducing population, while *P. morum* would require two or more zygotes germinating in the same place at the same time to establish a new zygote deposit.

If the zygote character is indeed the culprit, and panmixis the result, one would predict that other widespread *Gonium* species, all of which make four products from the zygote, will show an equivalent degree of sequence similarity among geographical-

```
                    0                   20                  40                  60                  80                  100
                    3'end of SSU
G.soc.UTEX 197      ggaaggatcattgAATCTAT CACAAACCACACCCGCGAAC CATATTTGTTGGCCTCCT-- TGGGCTTCTCTTTTGAGGC- ACCTGGGCCAGGCGCGTGCA
G.soc.UTEX 14       .................... .................... .................... ................R... ....................
G.soc.J-21028-4     .................... ...................M .................... ........A........... ....................
G.soc.UTEX 936      .................... .................... .................... .................... ..T................C
G.pect.UTEX 2570    .................... .....T.....A...T.... .G..C.-.........TCAA ......GC..G........T .GT..--TG..CTTA..-TT

G.soc.UTEX 197      CATACTCGCTATGTGCCGTG CTTGGGTCGTTGGCCC--AT TAATTTATTA.ATGGG---- -CTGGCGTCGGTCTTTTTAC TAACCAACCAACACCAAACC 200
G.soc.UTEX 14       .................... .................... .................... .................... ....................
G.soc.J-21028-4     .................... .................... ..........A...C..... .G.................. ....................
G.soc.UTEX 936      G................... .................... .................... .................... ....................
G.pect.UTEX 2570    GCGCT....------.A... GC.A......-...TAGGGC ATT..C....-...CCTCTA G...-.........--.G.. ....................
                                                                       5.8S start
                                                                            ▽
G.soc.UTEX 197      A-ATCAAATTAA-ATCCGAG TGCTTGGCTTAGAGCCAACA CTCACCAACCAAGACAACTC TCAACAACGGATATCTTGGC TCTCGGATCGATGAAGAACG 300
G.soc.UTEX 14       .................... .................... .................... .................... ....................
G.soc.J-21028-4     .A.................. .................... .................... .................... ....................
G.soc.UTEX 936      .A.................. .................... .................... .................... ....................
G.pect.UTEX 2570    .A.CT...A...A.A..... ...ACA.........TGT.G ....A............... .................... ....A...............

G.soc.UTEX 197      CAGCGAAATGCGATACGTAG TGTGAATTGCAGAAATACGT GAACCATCGAATCTTTGAAC GCAAATTGCGCTCGCGGCTT CGGCCAAGAGCATGTCTGCC 400
G.soc.UTEX 14       .................... .................... .................... .................... ....................
G.soc.J-21028-4     .................... .................... .................... .................... ....................
G.soc.UTEX 936      .................... .................... .................... .................... ....................
G.pect.UTEX 2570    .................... .................... ...T................ ...T..........A..... ..[illegible]..............
                       end of 5.8S
                          ▽
G.soc.UTEX 197      TCAGCGTCGGGTTAATACTC ACCCAACACCCACC---TGT GGGTGTTGTGTGGAACTGGC TTT-CCCGGGTGAACAACTT GTTGTCCCCCCGGGTCTGCT 500
G.soc.UTEX 14       .................... ......C............. .................... .................... ....................
G.soc.J-21028-4     .................... .................... .................... .................... .........-..........
G.soc.UTEX 936      .................... ...AC............... ...................G .C.AT............... .........-..........
G.pect.UTEX 2570    ................-.CT CG.TCT.T...T..CTG... ...ATGA.AAC...T..... .G...T...-CA.T..TA.G A...C-----......A...

G.soc.UTEX 197      GAAGTGCAGAGGTTGATGCA TGGACCCGCTAATGGGCCTC TACTGGGTAGGCAACTTGTT GCTA-TGCTTTAGTGGATGG CTTGGGACTGTGCTTGTTGA 600
G.soc.UTEX 14       ............A....... .................... .................... .................... ....................
G.soc.J-21028-4     .................... .................... .................... .................... ....................
G.soc.UTEX 936      ............A....... .................... .................... .................... ..............-.C...
G.pect.UTEX 2570    .................... .................... ..............T.C... ....A.........A..... .................C..

                                                              5' end of LSU
G.soc.UTEX 197      CCCTAACCAGGAACTCGGGA CTTTGTTCCGAGCAACCCCC CATTTTCtcgacctgagctc aggcaagatt
G.soc.UTEX 14       .................... .................... .................... ..........
G.soc.J-21028-4     .................... .................... .................... ..........
G.soc.UTEX 936      .................... .................... .................... ..........
G.pect.UTEX 2570    ...A............A.-- .C.C..G.T......AA... .................... ..........
```

Figure 1. ITS sequences of*Gonium sociale* and one *Gonium pectorale*. UTEX 197 and 14 are from Europe, UTEX 936 from North America, and J-21028-4 from Japan. UTEX 2570 is from North America.

Table 1. Gonium sociale pairwise sequence comparisons. Percent difference above diagonal, absolute difference below, adjusted for missing data (indels).

	1	2	3	4
1. UTEX 197	–	0.32	0.49	1.46
2. UTEX 14	2	–	0.81	1.62
3. Japan	3	5	–	2.27
4. UTEX 936	9	10	14	–

ly diverse isolates. By contrast, the other species of widespread Volvocaceae, including also species of the genus *Astrephomene* Pocock, would be quite different since they all produce but one product from each zygote. We have now sequenced the ITS region of four isolates of *Gonium sociale* (Dujardin) Warming, two from Europe (UTEX 197 & 14), one from North America (UTEX 936) and one from Japan (J-21028-4, v. Nozaki, 1986). Their sequence uniformity is remarkable, and similar to that found for *G. pectorale*. The four isolates display only 3% nucleotide variation overall (including insertions/deletions) and no indication of geographic substructure (Figure 1, Table 1). Figure 1 also includes one example of *G. pectorale* to illustrate the striking difference between the two species. With *Gonium sociale*, mating type analysis is not helpful since some isolates are heterogamous and some are presumed 'selfers' or have unknown mating affinity; however, two chromosome counts reported in the literature agree on 10 chromosomes (Cave & Pocock, 1951; Rayns, 1961). The results from *G. sociale* fit the prediction of worldwide homogeneity within *Gonium* species, and we are continuing our examination of other widespread species of the family.

Agencies of distribution.

In *P. morum* we see evidence for wide spread dispersal (syngens II and XX) and for endemism (syngens I and III). Perhaps the simplest interpretation compatible with the pairwise differences is that the mating types represented by syngens II and XX have managed to colonize widely (perhaps being older? – at least syngen II displays high levels of sequence variation), while those of syngens I and III are of too recent vintage to be come widespread. Embedded in the *P. morum* results are several further intriguing correlations. Of the 15 known *P. morum* syngens restricted to North America, none is common to east coast, Midwest, and west coast, nor even to two of these three areas. Each of the three areas has its own endemic syngens. This suggests that east/west transport via the prevailing westerlies is considerably less effective than north/south transport. This suggests in turns as first emphasized by Proctor et al. (1967), that transport via migrating birds should receive serious consideration, since waterfowl migration pathways in North America exactly parallel these three divisions of the continent. Perhaps some supporting evidence for this is found in the distribution of *Astrephomene gubernaculifera* Pocock in North America. This species has not been recorded from eastern North America, – only from the Midwest and West. Furthermore, Brooks (1966) found that Midwest collections, while mating *inter se*, were unable to mate with West Coast collections, which were in turn fertile *inter se*. This situation parallels that of the *P. morum* endemic syngens. Recolonization of the North American continent scoured by glaciation may have depended heavily on distribution by waterfowl.

Disjunct distributions

The last example is the rarest Volvocacean of all, the monotypic genus *Platydorina caudata* Kofoid. Colony development in this genus incorporates a secondary flattening of the juvenile colony, a unique developmental step resulting in a readily recognizable colony. The species is generally reported to exist only in North America in the Great Plains (Kofoid, 1899) but also was found after the turn of the century in California (Allen, 1920). While we are vigorously searching for a sample from the reported California site, we have sequenced the ITS of isolates from Iowa and from Kansas and found their ITS sequences to be identical. However, *Platydorina* was also reported from the Volga River by Korshikov in 1938, a report not to be doubted, given Korshikov's expertise and the unique morphology of *Platydorina*. We have for some years now been attempting to obtain samples from the Volga basin in the hopes of recovering the organism. What might one predict if we are successful? Either the Russian *Platydorina* will be very similar genetically to the North American form, and its distribution represents a refugial remnant of a much larger previous coverage (or perhaps a rare vicariant seeded from North America), or the Russian form will have a significantly different ITS sequence, perhaps similar to that of some other Volvocaceae from which it might have arisen by the chance occurrence of an independant mutation similarly affecting development.

Only further work, including that on other freshwater microalgae, can tell whether knowledge of historical biogeography can be satisfactorily combined with evidence in the genomes to help reconstruct phylogenies. Of immense help would be some more solid evidence on rates of nucleotide change in haploid organisms and protistans generally to improve our estimates of age. Footprints of geological past events, such as correlations with continental drift or with routes of dispersal agents in previously glaciated continents, all provide potential explanations of extent distributions and are now subject to potential verification with the help of molecular methods.

Acknowledgments

Dr H. Nozaki kindly provided the Japanese isolate of *Gonium sociale*.

References

Allen, W. E., 1920. A quantitative and statistical study of the plankton of the San Joaquin River and its tributaries in and near Stockton, California, in 1913. U. of Calif. Pub. in Zool. 22: 1–292.

Brooks, A. E., 1966. The sexual cycle and intercrossing in the genus *Astrephomene*. J. Protozool. 13: 367–375.

Buchheim, M. A. & R. L. Chapman, 1991. Phylogeny of the colonial green flagellates: a study of 18S and 26S rRNA sequence data. BioSystems 25: 85–100.

Cave, M. S. & M. A. Pocock, 1951. Karyological studies in the Volvocaceae. Am. J. Bot. 38: 800–811.

Coleman, A. W., 1983. The roles of resting spores and akinetes in Chlorophyte survival. In G. A. Fryxell (ed.), Survival Strategies of Algae. Cambridge U. Press, Cambridge: 1–21.

Coleman, A. W. & L. J. Goff, 1991. DNA analysis of eukaryote algal species. J. Phycol. 27: 463–473.

Coleman, A. W., A. Suarez & L. J. Goff, 1994. Molecular delineation of species and syngens in Volvocacean green algae (Chlorophyta). J. Phycol. 30: 80–90.

Ettl, H., 1983. Susswasserflora von Mitteleuropa. Band 9: Chlorophyta I. Gustav Fischer Verlag, Jena, 807 pp.

Iyengar, M. O. P. & T. V. Desikachary, 1981. Volvocales. Hoe & Co., Madras, 532 pp.

Kofoid, C. A, 1899. Plankton studies III. On *Platydorina*, a new genus of the family Volvocaceae from the plankton of the Illinois River. Bull. of the Ill. St. Lab. 5: 419–440.

Korshikov, A., 1938. Volvocineae. In J. V. Rolla (ed.), Catalog of Freshwater Algae, Ukrainian SSR. Vol. IV. Institute of Botany, Academia of Sciences of Ukrainian Academy of Science, RSR. p. 170. [In Ukrainian].

Larson, A., M. M. Kirk & D. L. Kirk, 1992. Molecular phylogeny of the volvocine flagellates. Mol. Biol. Evol.9: 85–105.

Nozaki, H., 1986. Sexual reproduction in *Gonium sociale* (Chlorophyta, Volvocales). Phycologia 25: 29–35.

Nozaki, H. & M. Itoh, 1994. Phylogenetic relationships within the colonial Volvocales (Chlorophyta) inferred from cladistic analysis based on morphological data. J. Phycol. 30: 353–365.

Proctor, V. W., C. R. Malone & V. L. DeVlaming, 1967. Dispersal of aquatic organisms: viability of disseminules recovered from the intestinal tract of captive killdeer. Ecology 48: 672–676.

Rayns, D. G., 1961. Ph.D. Dissertation, University of London.

Rausch, H., N. Larsen & R. Schmitt, 1989. Phylogenetic relationships of the green alga *Volvox carteri* deduced from small-subunit ribosomal RNA comparisons. J. Mol. Evol. 29: 255–265.

Hydrobiologia **336**: 143–150, 1996.
J. Kristiansen (ed.), Biogeography of Freshwater Algae.

15. Three Rumanian Mallomonas species (Synurophyceae), with special distribution patterns

Leontin Ştefan Péterfi[1] & Laura Momeu[2]
[1]*Department of Plant Biology, Babeş-Bolyai University, Cluj-Napoca, Rumania*
[2]*Institute of Biological Research, Cluj-Napoca, Rumania*

Key words: Mallomonas, Synurophyceae, chrysophytes, distribution

Abstract

Mallomonas is the largest genus of the Synurophyceae (chrysophytes) and includes more than 120 species. Most of them are cosmopolitan, living in freshwater habitats. The Rumanian taxa occur usually in spring or early summer at water temperatures ranging between 4 ° and 10 °C. They are most frequent in neutral or slightly acid, clear-watered, aquatic habitats of various types. However, *Mallomonas paludosa* Fott appears mostly in brown watered peatbog pools of low mineral contents, having a particular geographical area, with its southernmost limit in the Rumanian Eastern Carpathians. Its distribution, like that of other boreal elements, is strictly connected with the occurrence of genuine ombrogenic *Sphagnum*-bogs, which have their southern limit also in Romania. The other two species dealt with – *M. cyathellata* Wujek & Asmund and *M. portae-ferreae* Péterfi & Asmund –, occurring in circumneutral or moderately alkaline waters of eutrophic character, are apparently warm water forms, mostly distributed in tropical, subtropical or mediterranean zones. The Rumanian localities (in the Danube Delta and at Eşelniţa near the Danube, up stream Drobeta Turnu-Severin) might be the northern limits of *M. portae-ferreae* in Eastern Europe. The Danube Delta is the only locality of *M. cyathellata* in Europe, and its northernmost occurrence as well. The occurrence of both species in Rumania is presumably due to migratory waterfowl.

Introduction

The unicellular *Mallomonas*, including more than 120 species, is the largest genus of the Synuraceae (Chrysophytes). Cells are variable in shape, globose to fusiform, ranging between 10 and 50 (100) μm. They are free-swimming, with two parallel and apically inserted flagella. One of the flagella is more elaborated, much longer and bears two rows of tripartite hairs, while the other is smooth, reduced. The external investment of cells, an organized silica armour, consists of imbricated scales and slender, movable bristles arranged in a very precise manner (Siver, 1991). The bristles are tucked under the distal portion of scales. Fine structural patterns of scales and bristles are the most important taxonomic criteria for the identification of taxa (Asmund & Kristiansen, 1986; Siver, 1991).

The inner structure of cells is well known. The cells have one bilobed or two golden or yellow-brown chloroplasts containing a particular pigment complex, fucoxanthin and chlorophylls *a* and c_1 being the most important. The chloroplast exhibits thylakoids grouped in threes, girdle lamella and chloroplast endoplasmic reticulum, connected probably with the external nuclear membrane. The nucleus is centrally located, just below the single dictyosome, chrysolaminarin vacuole large, posterior. The flagellar basal bodies are associated with microtubular roots, connected in a particular manner with the cell cytoskeleton. The photoreceptor consists of a pair of flagellar swellings, but not associated with an eyespot or with the plastid membranes. Production of scales and bristles is endogenous, involving silica deposition vesicles originating from the Golgi apparatus. They are oppressed against- the outer surface of the chloroplast endoplasmic reticulum. Microtubules and microfilaments are involved in the formation and positioning of the cell cover elements, which are cemented then into the cell armour by a glue of organic nature.

Production of siliceous stomatocysts within the cell, seemingly a normal stage of the life cycle, is triggered by sudden changes in environmental conditions. Cyst morphology is of taxonomic significance.

Almost all *Mallomonas* species had long been considered as microthermal, occurring in early spring, in slightly acid waterbodies of low mineral contents. This statement was based on earlier field observations carried out in some limited areas located in the northern or temperate zones (Momeu & Péterfi, 1985). Most recent work pointed out that such generalization cannot be valid for the genus as a whole. The description of eurythermic and especially of warm water forms recorded from subtropical and tropical regions (e.g., Compère, 1975; Cronberg, 1989; Dürrschmidt, 1982; Kristiansen, 1980; Takahashi & Hayakawa, 1979; Wujek & Asmund, 1977) required a change of our earlier opinion.

According to Siver (1995) the individual chrysophytes, including *Mallomonas*, exhibit a particular pH range and can easily be placed into one of the four categories: low acid, mid-acid, neutral and high pH groups. A somewhat similar distribution exhibit the *Mallomonas* taxa towards specific conductance, trophic and temperature gradients. It has often been reported that both species diversity and abundance was higher in spring at temperatures between 10 °C and 15 °C. By the contrary, Siver (1991) reported no significant difference in the number of taxa over temperature gradient ranging between 1 °C and 27 °C. Although, the frequency of occurrence of individual taxa may be controlled by temperature, exhibiting a clear pattern. Siver (1991) defined five temperature groups for *Mallomonas* based on the distribution of taxa (warm-water, cold-water, cold-cool, cool-water and temperature indifferent groups).

It had also been documented that besides the usual, widely distributed, almost cosmopolitan species, there are some having geographically restricted, well-defined areas. Tropical, polar, boreal, bipolar etc. *Mallomonas* elements could be distinguished.

Among the species occurring in Romania, *M. paludosa* Fott, *M. portae-ferreae* Péterfi & Asmund and especially *M. cyathellata* Wujek & Asmund exhibit special distribution patterns. Moreover, the last one is here reported for the first time in Europe.

Observations and conclusions

The species dealt with in this paper have been identified according to the electron microscopical pattern of their scales and bristles (see Figure 1). They are known from 1–2 Rumanian localities and they are rare or lacking in the rest of Europe.

Mallomonas paludosa has been found during summer in two raised peatbogs in Transylvania, the northern part of Rumania, in the Eastern Carpathians (see Figure 2). One of the localities, the Tău lui Dumitru raised peatbog (Péterfi & Momeu, 1976), is located in the Maramureş Mountains; the second locality, Poiana Stampei peatbog, is in the Dorna Valley. Both are ombrogenic, oligotrophic peatbogs, with brown water and pH ranging between 4.0 and 4.5. The bog pools are marked along their margins by populations of *Scheuchzeria palustris, Carex limosa* and *C. rostrata*. They are surrounded by living peat moss cover of oligotrophic character (Pop, 1960).

According to the published records this species is mainly circumboreal or rather bipolar (see Figure 3), being distributed in Europe, Greenland and America: E.g., Čzech Republic (Fott, 1957), Denmark (Kristiansen in Asmund & Kristiansen, 1986), Finland (Eloranta, 1988), Germany (Hartmann & Steinberg, 1989), Sweden (Cronberg & Kristiansen, 1980), Greenland (Kristiansen, 1992), Ontario, Canada (Nicholls, 1982; Kling & Kristiansen, 1983), USA (Wujek & Bland, 1991; Wawrzyniak & Andersen, 1985), Argentina (Vigna, 1990) and Brazil (Wujek & Bicudo, 1993).

This acidobiontic, oligotrophic *M. paludosa* exhibits a distribution pattern very similar to those higher plants, which according to reliable palynological evidence (Pop, 1960, 1976) can be considered remnants of the ancient, extended, late glacial Rumanian flora. During glaciation many boreal elements had extended their distribution areas far to the south of Europe, some became settled in Rumania too, colonizing relatively large zones. The vegetation types (phytocoenoses) in which they had lived here during glaciation could be very near in structure to those in which they are frequently found in boreal regions of the present. The area of such species has been gradually reduced during the Quaternary due to subsequent, postglacial climatic changes, and they could survive competition only in special habitats, such as the ombrogenic peatbogs. Such species can be considered glacial relics (Pop, 1976): E.g., *Scheuchzeria palustris, Carex limosa, C. pauciflora, Rhynchospora alba, Betula nana, Empetrum nigrum, Drosera*

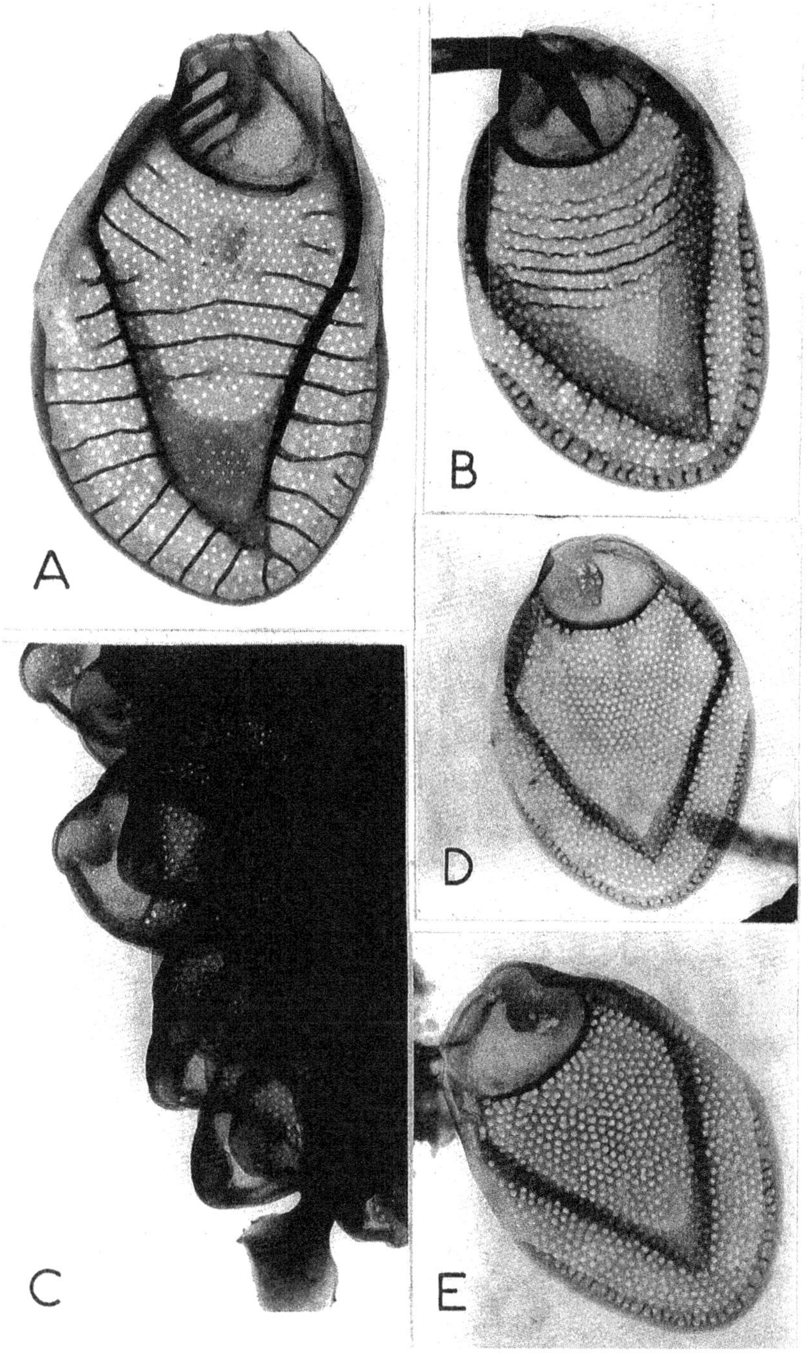

Figure 1. Direct electron micrographs (TEM) of scales of the *Mallomonas* species. A: *M. paludosa*, isolated body scale (× 16 600). B: *M. portae-ferreae*, body scale (× 10 000). C. *M. cyathellata* var. *cyathellata*, part of the cell armour showing the special goblet shaped appendage (× 9000). D,E: *M. cyathellata* var. *cyathellata*, isolated body scales (× 8000 and 10 000).

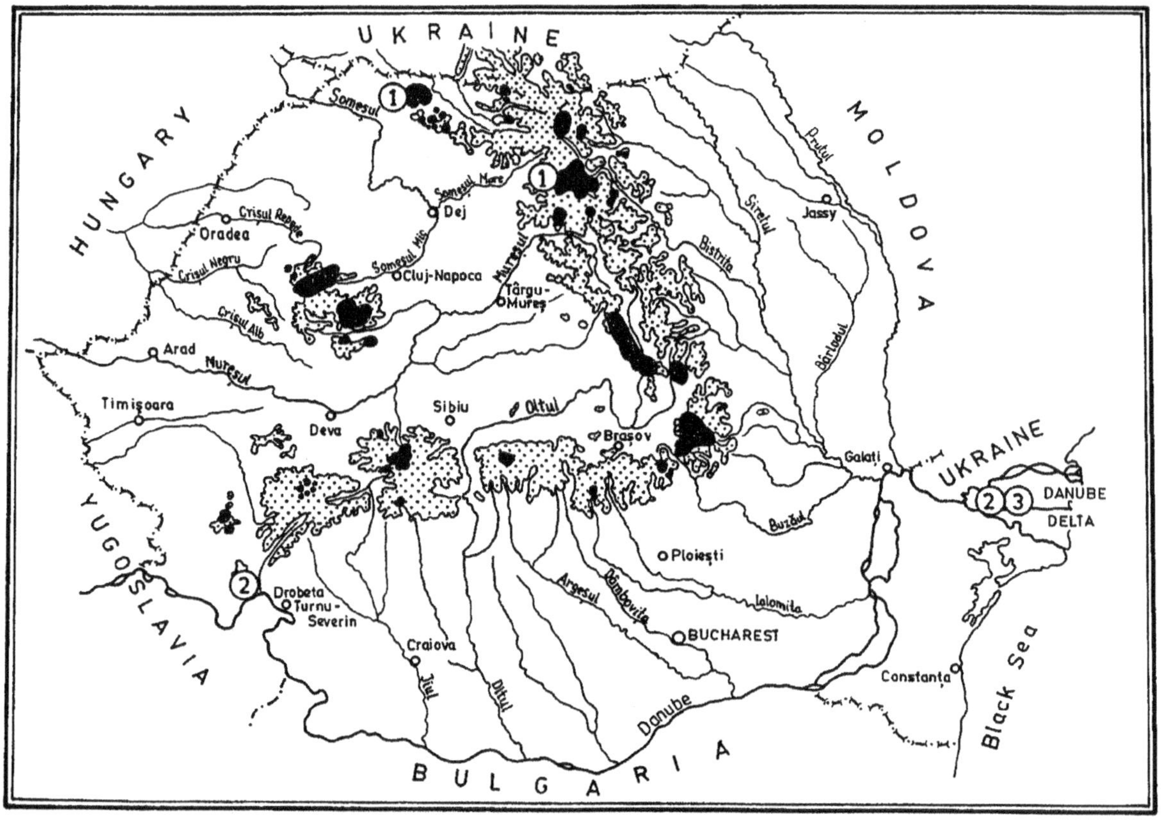

Figure 2. Distribution of *Mallomonas paludosa* (1), *M. portae-ferreae* (2) and *M. cyathellata* (3) in Rumania. Dotted areas: the Carpathians. Solid black spots location of peatbog complexes with genuine ombrogenic, raised peatbogs.

intermedia, Andromeda polyfolia, Vaccinium oxycoccus and others. By analogy, some of the algae, among which *M. paludosa*, can be considered glacial relics. Its ecological behaviour is basically similar to that of the above mentioned associated higher plants. They are stenotopic towards their southern limits of distribution, but exhibit a broader range of ecological tolerance towards the centre of their boreal distribution area. The pH range of *M. paludosa* is 4.1–6.7, the optimum being less than 5.5. In boreal regions the species inhabits not only oligo-dystrophic bog pools, but also other biotopes. The genuine ombrogenic peatbogs, as their inhabitants, have their southernmost limit in the Rumanian Carpathians.

A proof for the continuity of *M. paludosa* from the late glacial to the present would be the evidence of stomatocysts in the ancient lake sediments and in the subsequent peat layers formed.

M. portae-ferreae and *M. cyathellata* are evidently warm water forms, distributed in tropical, subtropical and mediterranean zones (see Figure 4). The Rumanian localities are situated in the southern part of the country with markedly warmer summers (see Figure 2). Both species are alkaliphilic, most frequently found at high water temperatures (up to 28–32 °C). They are preferentially eutrophic, tolerating habitats with pH ranging between 6.5 and 8.7 (pH range of *M. cyathellata* 6.8–8.15; *M. portae-ferreae* – pH range 6.05–8.7) and relatively high amount of organic material.

The type locality of *M. portae-ferreae* is Eşelniţa – near the Danube, up stream Drobeta Turnu-Severin, Rumania. It was found in a small, shallow pit contaminated by waterfowls (Péterfi & Asmund, 1972) and subsequently recorded (in this paper) from the Danube Delta, in a fishpond near Maliuc. The mostly pantropical–subtropical distribution of *M. portae-ferreae* is evident from the following records: Africa: Cameroon (Compère, 1973, 1975); Zimbabwe (Cronberg, 1989); Asia: Bangladesh (Takahashi & Hayakawa, 1979), China (Kristiansen & Tong,

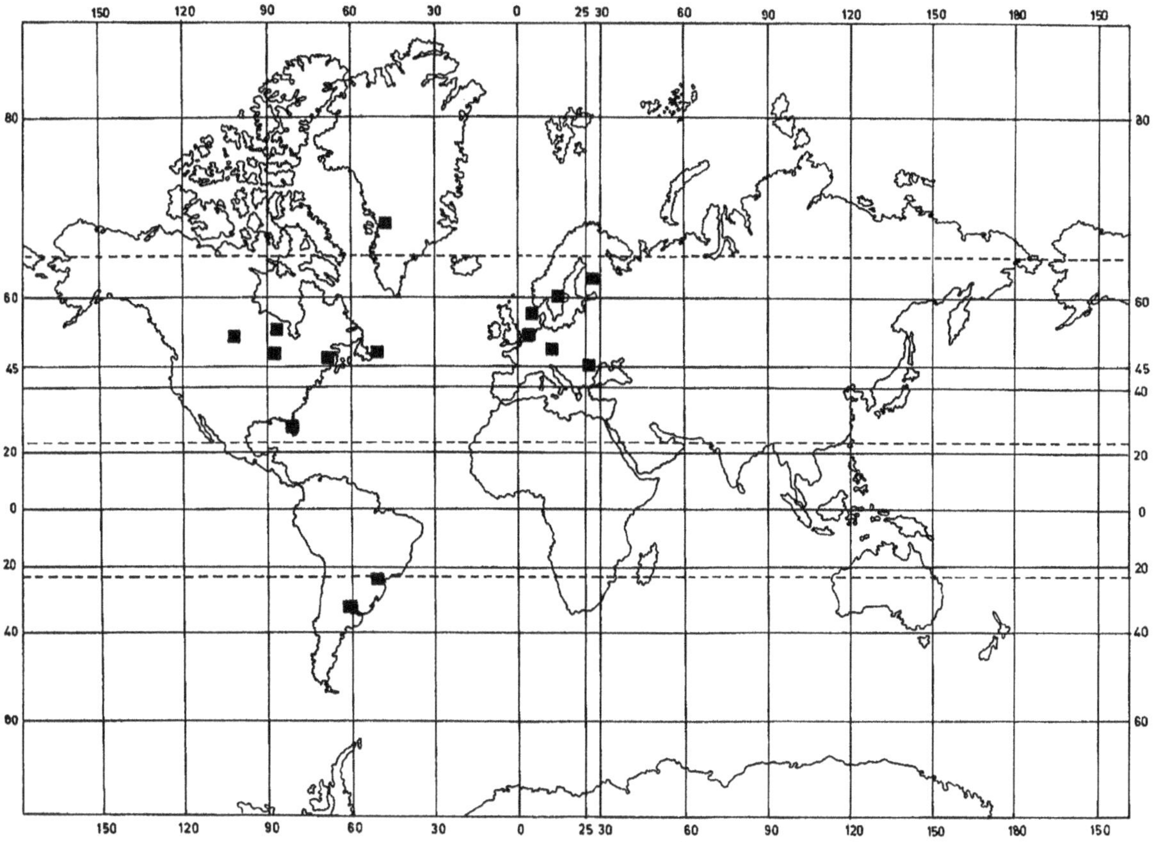

Figure 3. General distribution of *Mallomonas paludosa* (■).

1991; Péterfi & Asmund, 1972), India (Saha & Wujek, 1990), Japan (Ito, 1991), Sri Lanka (Dürrschmidt & Cronberg, 1989); Australia: Tasmania (Croome & Tyler, 1988), Papua-New Guinea (Vyvermann & Cronberg, 1993); North America: Connecticut (Siver, 1991), Florida (Wujek, 1984; Wujek & Bland, 1991), Kansas (Wujek & Weis, 1984), Louisiana (Wee et al., 1993); South America: Amazonas (Cronberg, 1989), Argentina (Vigna, 1988); and Europe: Greece (Kristiansen, 1980), Portugal (Santos & Leedale, 1993). The species has also been recorded from Germany based on the evidence of single scales (Cronberg, 1989; Gutowski, 1993; Hartmann & Steinberg, 1989).

M. cyathellata var. *cyathellata* has a mainly subtropical distribution pattern (Alabama: Wujek & Asmund, 1979; India: Saha & Wujek, 1990; Jamaica: Cronberg, 1989). Its only European locality is Maliuc, the Danube Delta, Rumania, reported here for the first time. It has been found in a fishpond together with *M. portae-ferreae*, occurring in a eutrophic algal community of β-mesosaprobic character. *M. cyathellata* var. *kenyana* Wujek & Asmund is distributed in: Africa: Kenya (Wujek & Asmund, 1979), Zimbabwe (Cronberg, 1989); Asia: India (Saha & Wujek, 1990), Papua New Guinea (Vyvermann & Cronberg, 1993); North America: Florida (Wujek, 1984; Wujek & Bland, 1991), Jamaica (Cronberg, 1989) and South America: Columbia, Guatemala (Cronberg, 1989). *M. cyathellata* var. *chilensis* Dürrschmidt has been recorded from Chile (Dürrschmidt, 1982) and Brazil – Amazonas (Cronberg, 1989), Argentina (Vigna, 1988) and India (Saha & Wujek, 1990). *M. cyathellata* as a whole has a pantropical distribtion.

Evidently both species inhabit mostly tropical and subtropical localities. There is little probability that they are native in Rumania. Their continuity could be tested only by sediment analyses, which are not yet available. At present, one possible explanation for their occurrence is that they are occasionally introduced by migratory waterfowl from tropical Africa. The Danue

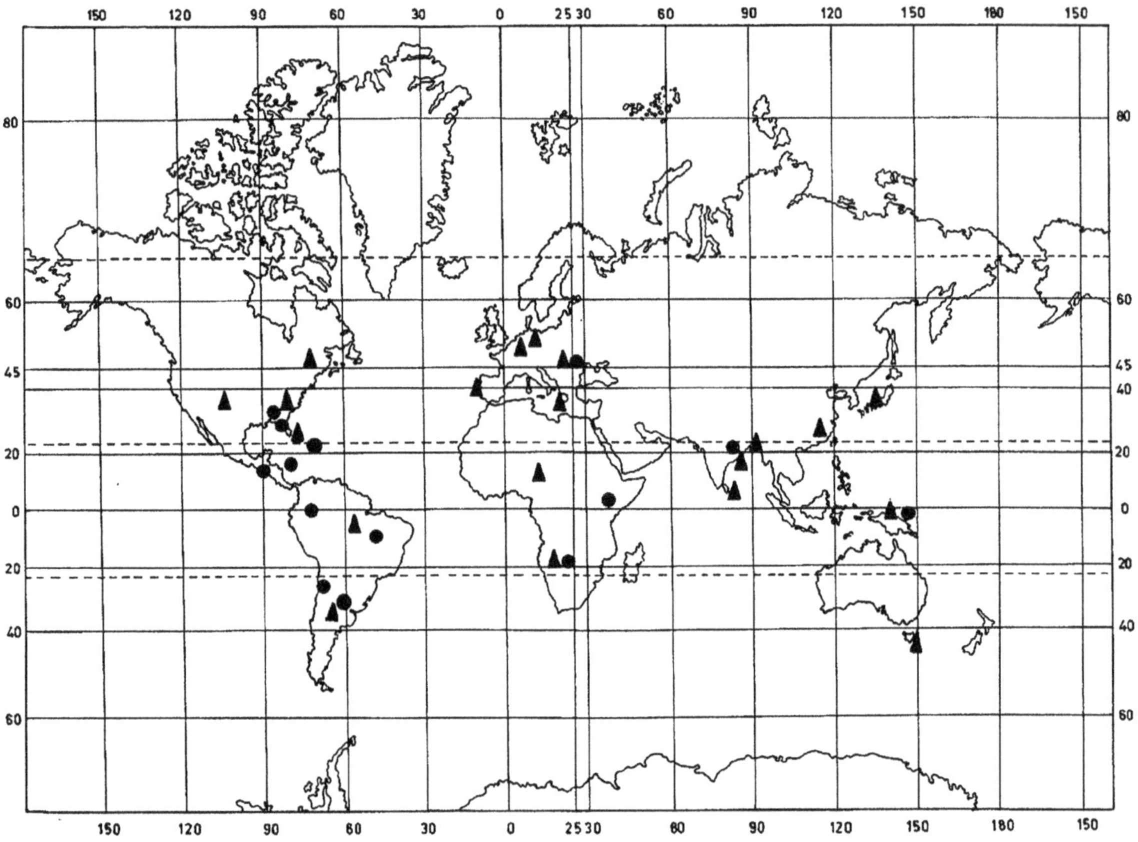

Figure 4. Distribution of *Mallomonas portae-ferreae* (▲) and *M. cyathellata* (●).

Delta, as well as Eşelniţa-Porţile de Fier are located on the main migratory routes of many of our visiting birds (Curry-Lindahl, 1975); Fisher & Peterson, 1964; see also Figure 5). Once introduced they might be able to survive for one or several years as stomatocysts. During summer periods, when the temperature conditions are appropriate, they are able to form rather rich populations. The appearance of *M. portae-ferreae* scales in Germany might have a similar reason.

Acknowledgments

The authors wish to thank Dr Jørgen Kristiansen who helped and encouraged the preparation of the manuscript. We are also indebted to Dr Constantin Crăciun who kindly run the EM and to Dr István Kiss for reading and correcting our English. Finally, we are grateful to Miss A. Stoian and Mr E. Papp for preparing the drawings.

Additional support was provided by the Hungarian Scientific Foundation (Project OTKA No. T 014634).

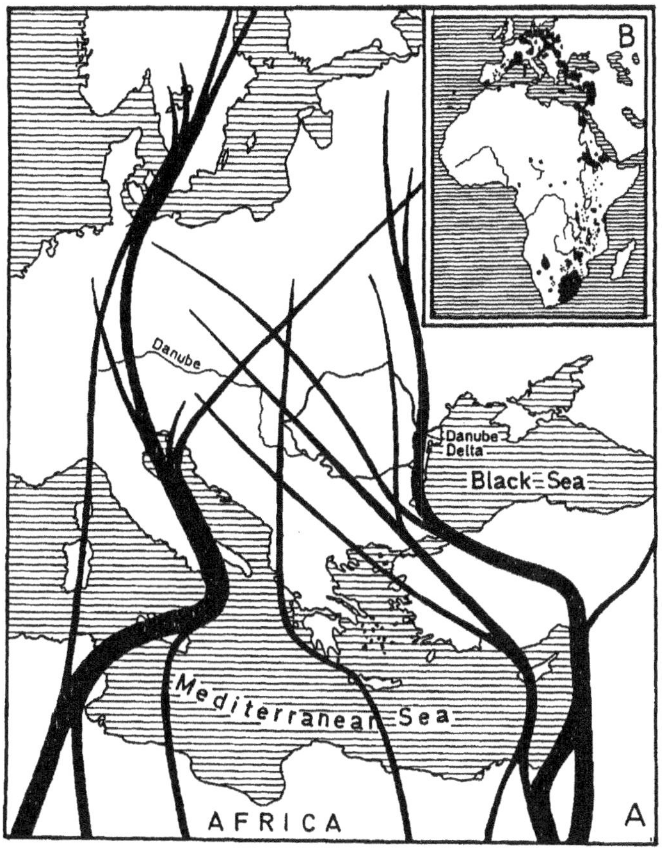

Figure 5. A: The main migration routes of birds in the central and eastern part of Europe. Note the routes crossing the Danube Delta and Eşelniţa – Porţile de Fier (simplified based on the poster: Flying Visitors, 1985). B: Migration routes of white stork (Ciconia ciconia). After N. A. Bobrinschi, from Pop, I. (1977). The dotted areas indicate the main flight routes; black spots indicate the places where birds marked by the Rossiten Ornithological Station have been captured.

References

Asmunds, B. & J. Kristiansen, 1986. The genus Mallomonas (Chrysophyceae). Opera Bot. 85: 5–129.

Compère, P., 1973. *Mallomonas portae-ferreae* Péterfi et Asmund (Chrysophycée) au lac Tchad. Bull. Jard. Bot. Nat. Belg. 43: 235–237.

Compère, P., 1975. Algues de la region du Lac Tchad. Cah. O.R.S T.O.M., Sér. Hydrobiol. 9: 167–192.

Cronberg, G., 1989. Scaled chrysophytes from the tropics. Nova Hedwigia 95: 191–232.

Cronberg, G. & J. Kristiansen, 1980. Synuraceae and other Chrysophyceae from central Småland Sweden. Bot. Not. 133: 595–618.

Croome, R. & P. A. Tyler, 1988. Further observations on silica-scaled Chrysophyceae (Paraphysomonadaceae and Mallomonadaceae) from Australian freshwaters. Nova Hedwigia 46: 481–486.

Curry-Lindahl, K., 1975. Fåglar över land och hav. En global översikt av fåglarnas flyttning. Albert Bonnier Förlag, Stockholm.

Dürrschmidt, M., 1982. Studies on the Chrysophyceae from South Chilean inland waters by means of scanning and transmission electron microscopy II. Arch. Hydrobiol., Suppl. 63/Algol. Stud. 31: 121–163.

Dürrschmidt, M. & G. Cronberg, 1989. Contribution to the knowledge of tropical Chrysophytes: Mallomonadaceae and Paraphysomonadaceae from Sri Lanka. Arch. Hydrobiol., Suppl. 82, 1 (Algol. Stud. 45): 15–37.

Eloranta, P., 1989. Scaled chrysophytes (Chrysophyceae and Synurophyceae) from national park lakes in southern and central Finland. Nord. J. Bot. 8: 671–681.

Fisher, J. & R. T. Peterson, 1964. The world of birds. New York, 288 pp.

Fott, B., 1957. Taxonomie drobnohledné flory našich vod. Preslia 29: 278–319.

Gutowski, A., 1993. Autökologische und taxonomische Studien schuppentragender Chrysophyceen anhand von Untersuchungen im Freiland und in Kulturen. PhD Thesis. Freie Universität Berlin, Berlin Federal Republic of Germany.

Hartmann, H. & C. Steinberg, 1989. The occurrence of silica-scaled chrysophytes in some Central European lakes and their relations to pH. Beih. Nova Hedwigia 95: 131–158.

International Council for Bird Preservation (ed.), 1985. Flying visitors. Poster.

Ito, H., 1991. Chrysophytes in the southern part of Hyogo Prefecture Japan (II). Mallomonas. Jpn. J. Phycol. (Sôrui) 39: 253–262.

Kristiansen, J., 1980. Chrysophyceae from some Greek lakes. Nova Hedwigia 33: 167–194.
Kristiansen, J., 1992. Silica-scaled chrysophytes from West Greenland: Disko Island and the Søndre Strømfjord region. Nord. J. Bot. 12: 527–536.
Kristiansen, J. & D. Tong, 1991. Investigations on silica-scaled chrysophytes in China. Verh. int. Ver. Limnol. 24: 2630–2633.
Momeu, L. & L. S. Péterfi, 1985. Ecological behavior of *Mallomonas* species (Synuraceae, Chrysophyceae) in Romania. Stud. Univ. Babeş-Bolyai, Ser. Biol. 30: 5–10.
Nicholls, K. H., 1982. *Mallomonas* species (Chrysophyceae) from Ontario, Canada including descriptions of two new species. Nova Hedwigia 36: 89–124.
Péterfi, L. S. & B. Asmund, 1972. *Mallomonas portae-ferreae* nova species in the light and electron microscopes. Stud. Univ. Babeş-Bolyai, Ser. Biol. 1: 11–18.
Péterfi, L. S. & L. Momeu, 1976. Romanian *Mallomonas* species studied in light and electron microscopes. Nova Hedwigia 27: 353–392.
Pop, E., 1960. Mlaştinile de turbă din Republica Populară Română. Ed. Acad. Rom., Bucureşti: 512.
Pop, E., 1976. Specii relicte în flora României. In Beldie, A. & I. Morariu (ed.), Flora Republicii Socialiste România. Ed. Acad. Română, Bucureşti: 105–111.
Pop, I., 1977. Biogeografie ecologică. I. Ed. Dacia, Cluj-Napoca, 200 pp.
Saha, L. C. & D. E. Wujek, 1990. Scale-bearing chrysophytes ftom tropical Northeast India. Nord. J. Bot. 10: 343–354.
Santos, L. M. A. & G. F. Leedale, 1993. Silica-scaled chrysophytes from Portugal. Nord. J. Bot. 13: 707–716.
Siver, P. A., 1989. The distribution of scaled chrysophytes along a pH gradient. Can. J. Bot. 67: 2120–2130.
Siver, P. A., 1991. The biology of Mallomonas. Morphology, taxonomy and ecology. Kluwer Academic Publishers, Dordrecht, Boston, London.
Siver, P. A., 1995. The distribution of chrysophytes along environmental gradients: their use as biological indicators. In Sandgren, C. D., J. P. Smol & J. Kristiansen (eds), Chrysophyte algae. Ecology, phylogeny and development. Cambridge Univ. Press: 232–268.
Takahashi, E. & T. Hayakawa, 1979. The Synuraceae (Chrysophyceae) in Bangladesh. Phykos 18: 129–147.
Vigna, M. S., 1988. Contributions to the knowledge of Argentine Mallomonadaceae. Nova Hedwigia 47: 129–144.
Vyvermann, V. & G. Cronberg, 1993. Scale-bearing chrysophytes from Papua – New Guinea. Nord. J. Bot. 13: 111–120.
Wawrzyniak, L. A. & R. Andersen, 1985. Silica-scaled Chrysophyceae from North American boreal forest regions in northern Michigan, USA and Newfoundland, Canada. Nova Hedwigia 41: 127–145.
Wee, J. L., D. J. Booth & M. A. Bossier, 1993. Synurophyceae from the Southern Atlantic Coastal Plain of North America: A preliminary survey in Louisiana, USA. Nord. J. Bot. 13: 95–106.
Wujek, D. E., 1984. Chrysophyceae (Mallomonadaceae) from Florida. Florida Sci. 3: 160–170.
Wujek, D. E. & B. C. Asmund, 1979. Mallomonas cyathellata sp. nov. and Mallomonas cyathellata var. kenyana var. nova (Chrysophyceae) studied by means of scanning and transmission electron microscopy. Phycologia 18: 115–119.
Wujek, D. E. & C. E. Bicudo, 1993. Scale-bearing chrysophytes from the state of São Paulo, Brazil. Nova Hedwigia 56: 247–257.
Wujek, E. D. & R. G. Bland, 1991. Chrysophyceae (Mallomonadaceae and Paraphysomonadaceae) from Florida. II. Additions to the flora. Florida. Sci. 1: 41–48.
Wujek, D. E. & M. M. Weis, 1984. Mallomonadaceae ftom Kansas. Trans. Kans. Acad. Sci. 87: 26–31.

Hydrobiologia **336**: 151–157, 1996.
J. Kristiansen (ed.), Biogeography of Freshwater Algae.

16. Dispersal of freshwater algae – a review

Jørgen Kristiansen
University of Copenhagen, Botanical Institute, Dept. of Phycology, Øster Farimagsgade 2D, 1353 Copenhagen K, Denmark

Key words: dispersal mechanisms, water, fish, insects, mammals, birds, air, distances

Abstract

A survey is given on the various methods for dispersal of freshwater algae. Dispersal factors are either water or air, or organisms - from beetles, dragonflies and mammals to birds, the latter being the most important group. The question of dispersal distances is discussed, in relation to dispersal mechanisms and to the resistance of the algae to transport conditions. Man's recent importance in algal dispersal is emphasized.

Introduction

The many preceding papers in this volume have all in some way and from various aspects treated the distribution of fresh water algae, locally, regionally, and globally. How the algae have arrived there, what dispersal mechanisms have been acting, has only sporadically been touched (e.g. Péterfi & Momeu). Round (1981) has very clearly expressed the dilemma: 'The occurrence of so many common fresh-water algae throughout the world is a reflection of ease of transport – yet for the majority there is no information on transport mechanisms'.

One of the model regions for the study of algal dispersal is Antarctica. This continent has very few endemic species, but a very large number of cosmopolitan ones. All of these must have been transported in some way through the air.

Another model study region is the volcanic island of Surtsey near Iceland. It came into existence by a submarine eruption in 1963. In 1966 the volcanic activity had decreased so much that it is assumed that the immigration of algal species could start that year. In 1968 regular studies of the freshwater – and mainly soil – algal flora were started (Schwabe, 1972). This sterile island was the ideal natural laboratory for the study of immigration mechanisms.

This review will consider four main types of dispersal of freshwater algae, viz. by water, by organisms, by man, and airborne dispersal. All these are passive. Active dispersal is of absolutely minor importance, even though Messikommer (1943) pointed out that the movements of some algae (such as diatoms, blue-green algae, desmids, flagellates) during long periods of time might be notable. Diatoms might even move from one pond to another along wet rock surfaces (Hustedt, 1943).

Dispersal by water

This is the most natural way of dispersal of freshwater algae. Of course it takes place in running water and whenever there is a water connection between water bodies. Atkinson (1988) studied the colonization of a newly constructed reservoir. The source of the inoculum was the inflow. In a short time the same species appeared as in the catchment area. Only after several years did other species appear which were not present in the catchment area. They must have been transported passively overland by other methods, and that had taken a longer time.

Dispersal by organisms

Organisms in the water may enhance the dispersal of plankton algae, such is the case when fish act as vectors. The plankton eating gizzard shad (*Dorosoma cepedianum*) may transport many species within its digestive tract, as shown by Velasques (1940). Many species of plankton algae may pass undamaged and viable through this fish – 30 species of chlorophytes, 12 species of blue-green algae, and some diatoms have been recorded.

Overland transport will in most cases be necessary for the dispersal of algae from one water body to another. This can be effected by animals attached to both water and air. Short distance dispersal from one pond to another can be effected by water beetles, already by Ralfs (1848) postulated as a dispersal mechanism for desmids. They have been observed to carry *Closterium*, chlorophytes and diatoms; in this way algae may be transported between waterbodies too small to be visited by water birds (Migula, 1888).

Some beetles may act as pioneers in new waterbodies, and also in that connection their transport of algae is interesting. Milliger & Schlichting (1968) found about 20 species of viable green algae in the intestine of such beetles, thus the algae may be spread by their faecal pellets.

Odonata may be effective transporters. Messikommer (1943) suspected that dragonflies had brought living *Pandorina morum* into one of his experimental jars. Parsons et al. (1966) investigated many species of dragonflies and damselflies and found viable algae both internally and externally. By washing the animals they obtained 20 genera of blue-green algae, green algae, and euglenoids. Maguire (1963) found dragonflies to be effective even in long distance dispersal.

Water-living mammals may of course also transport algae from one water body to another. In the fur of minks (*Mustela vison*), Irénée-Marie (1938) found 16 species of desmids. Thirteen species of green algae, euglenoids, and blue-green algae were found in cultures from intestinal contents of muskrats (*Ondrata zibeticus*) by Roscher (1967). A raccoon (Maguir, 1963) could probably carry living, non-encysted algae in its fur for many miles during a rainy night, and such an animal would visit small bodies of water seldom or ever visited by waterfowl.

However, the most important transporters of algae are various species of water birds. De Guerne (1888, cited from Zacharias, 1888) was the first to show experimentally that live algae can be transported by birds, especially on their feet. He washed the feet of ducks, and made cultures from the washing water. Several animals such as daphnids and rhizopods appeared, but also some desmids and diatoms.

Seagulls wading from one pool to another may transport algae on their feet (Wille, 1923), desmids may be transported in wet mud on birds' feet for rather long distances (Strøm, 1926). Grønblad (1933) was convinced that birds were the most important transporters of aquatic desmids.

A large-scale investigation on dispersal externally on birds was made by Schlichting (1960). He studied 16 species of waterfowl, 205 specimens were washed, and cultures were established. Mainly green algae appeared, but also blue-green algae, euglenoids, and diatoms. Spores and cysts proved to be favoured, also cells embedded in jelly. He found 86 species on the feet of the birds, 25 on the feathers, and 25 on the bills. These figures were much higher than from cultures based on gullet contents and faeces – 14 species. It is obvious that algae transported externally on birds are in danger of desiccation. Most algae proved to be viable on the birds for up to four hours. Up to eight hours some algae survived, and after more than eight hours there were hardly any algae left. Thus it is proven that birds can disperse algae from lake to lake and certainly even to oceanic islands.

It has been pointed out by Maguire (1963) that not only waterfowl but also nonaquatic birds such as sparrows may transport viable algae attached to their feathers.

The other possible transport mechanism via birds is internally, in the digestive tract. Waterfowl which had been feeding in a *Chara* vegetation proved to have viable *Chara* oospores in their digestive tract. *Chara* can thus be transported from one lake to another (Proctor, 1963). The same is certainly true for *Vaucheria*, found in mute swan faeces (Luther, 1963). However, this observation was based on fixed material. The internal environment may prove more favourable than the outside, on feet or feathers, where there is always the danger of desiccation. Thus Proctor (1966) showed that desmids transported by birds can be carried greater distances in the digestive tract than externally on the birds. And Carausu (1968, after Atkinson, 1970) found many other plankton algae – *Microcystis, Pandorina, Scenedesmus* – alive in cultures based on intestinal contents of several water birds.

Atkinson (1970, 1971, 1972, 1980) has made a long series of experiments where she directly fed various species of ducks with cultures of different algae, with very different results. *Anas platyrryncha* (mallard) was fed with cultures of *Asterionella formosa*. The faeces, collected after 3–17 hours, contained *Asterionella* valves, but no viable *Asterionella* cells (Atkinson, 1970). In an experiment, where faeces did not contain viable *Asterionella, Melosira* could be cultured from the faeces. (Atkinson, 1971). Gut contents from 21 waterbirds showed that on the whole, benthic algae were viable, but plankton algae not (Atkinson, 1972). However, in another experiment, where several species of diatoms were fed to ducks, only *Asterionella formosa* was viable after passage. It was found in faeces 2–20 hours after feeding, two hours corresponding to 220 km of flight (Atkinson, 1980).

In a series of experiments, killdeer (*Charadrius vociferus*), a small wading bird, was fed with cultures of various algae (Proctor et al., 1967). Maximum retention time in the digestive tract for living algae was very different, for desmids it was 3–4 hours, for *Pandorina* 4 hours, and for *Pediastrum* 8 hours. Oospores of *Chara* were viable after up to 24 hours. Accordingly, the thickness and composition of the cell wall may be of importance. The strain of *Gonium* used did not at all tolerate to pass through the birds, contrary to other strains. This strain might have been from an isolated clone, where the lack of resistance against dispersal stress might be the cause of the isolation.

The role of birds in dispersal can also be concluded from direct observations. Compare the observations by Wille (1923), where seagulls were seen flying from one puddle to another. Manguin & Leboime (1948) ascribed the occurrence of 35 species of marine diatoms in artificial ponds in Paris to visiting seagulls from the coast. On the volcanic island of Surtsey, *Synechococcus* and *Oscillatoria* were found together on a place which was the favourite resting place for birds – in contrast to most other algal species found in other places on the island and considered to be airborne (Schwabe & Hickel, 1978).

In Antarctica, all dispersal must be carried out by birds. There is no airborne transport, the air being almost completely sterile. Living algae have been obtained from arctic terns, cape pigeons, giant fulmars, and south polar skuas (Schlichting et al, 1978).

An explanation of the distribution patterns of desmids in South America is that the Andean Chain prevents the dispersal by birds east-west in South America, and that thus birds are essential for the dispersal of desmids (Coesel et al., 1988). Wee et al. (1992) sought an explanation for the very rich chrysophyte flora in Louisiana, U.S.A., in the fact that many birds migrating from Central to North America make their first stops in south-eastern Louisiana, potentially leaving algal propagules there.

A very special short distance transport mechanism was described by Foged (1953). Cormorants were responsible of dispersal of diatoms when collecting water plants for nest material.

Airborne algae

Another important long distance dispersal mechanism is the wind. Already Ehrenberg (1849) found 18 species of diatoms in dust from the air, collected by Darwin (1839) on H. M. S. Beagle, 300 km from the nearest coast, but he did not test their viability. Pouchet (1860) made similar observations. The hazard in airborne dispersal is the danger of desiccation. Thus Strøm (1926) concluded from the immigration routes of desmids to highland lakes, that these algae could not be airborne – except for the zygospores, and these are rarely produced (see also Coesel, 1974).

During his transatlantic flight 1933, Lindberg collected some samples from the air (Gislén, 1948). Later, Overeem (1937) made an exact investigation based on samples taken by filtration from an airplane. Sampling was made from a maximal height of 2000 m and downwards. The filters were then incubated in different culture media, and living algae were found in all samples, with a maximum concentration at 500 m. The samples contained mainly green algae, especially *Chlorococcum*, but also *Stichococcus, Pleurococcus, Hormidium (Klebsormidium)*, and a few blue-green algae. All of them were typical aërophilic algae.

Messikommer (1943) was convinced that the wind was the most important factor for dispersal, and he tried to show it by means of experimental jars exposed to the air. Then he examined what algal species developed in the jars.

Schlichting (1961) found 22 species of viable algae in samples from the air, mainly green algae, but also diatoms, six blue-greens, one chrysophyte and one euglenoid. He found that the species list contained the same algae that give rise to troubles in swimming pools and in the clogging of filters in water purification plants and in air condition units where water is employed. In Colorado and Texas, Maguire (1963) found several living species of algae in rain. Brown et al. (1964) cultivated 62 species from airborne dust, on Bold's Basal Medium. They were mainly soil algae – green algae and blue-green algae, but also some few diatoms and xanthophytes. In dust-filled air he found a maximum of 3000 algae per m^3.

Geissler & Gerloff (1965) found live vegetative algae on air filters. Contents of dust filters, exposed 3 m above ground in southern Jutland, Denmark, contained 59 species of diatoms, most of these from fresh water (Foged, 1975). Rosas et al. (1989) found 16 species of living algae, mainly chlorophytes, in air samples from 4 m above the ground. Filamentous green algae (allegedly *Planktonema lauterbornii*) have been found in samples of coloured water obtained by an aeroplane from clouds above Antarctica (Saxena, 1983). Other investigations (Schlichting & al.1978) show that the Antarctic air is almost sterile.

The importance of the meteorological conditions for the horisontal and vertical distribution of algae in the air has also interested Schlichting (1964). The effect of the prevailing wind directions has been discussed. Vertical 'updrafts' or thermals can contain especially high concentrations of algae. He found 60 species of viable algae, mainly green and blue-green algae.

How can the algae get into the air? Stevenson & Collier (1962) found marine microflagellates and small diatoms in the air when strong winds blew over the sea, so that there would be a particulate interchange between water and air. Along shorelines of lakes fragments of scums and foams, with their contents of algae, can be picked up by the wind and carried aloft (Maynard, 1968). Dried algae can be caught by the wind, just like dust, and carried away, such as e.g. small desmids (Grönblad 1933). Algae can be transported into the air by thermal winds, often after heavy rainfall, when algae had been splashed up by raindrops. Thus concentrations of algae in the air were associated with variations in meteorological conditions (Rosa et al., 1989). Accordingly, they also vary through the year, with highest concentrations in winter and early spring (Schlichting, 1974). However, Tiberg et al. (1984) found maximum concentrations, mainly consisting of *Chlorella, Chlorococcum, Nostoc*, and *Anabaena*, during summer.

As mentioned above, Surtsey has served as a laboratory for dispersal studies (Behre & Schwabe, 1970; Schwabe, 1972; Schwabe & Behre, 1972; Schwabe & Hickel, 1978). Three years after the eruption, 106 algal taxa were observed. The majority were very small, larger forms being decidedly absent, and there were no halophilic species. This indicates long distance wind dispersal. In fact the flora constitutes a selection of the smallest forms found in Iceland, and they have immigrated as viable pioneers. Mainly soil algae, most of them diatoms, and a few blue-greens, almost all of them spread by the wind.

Man as dispersal agent

Above, the importance of wind dispersal for the algal flora of Surtsey has been described, the large majority of the species having arrived in that way. A few have certainly been introduced by birds, and some might even have been introduced inadvertently by man during the scientific investigations on the island (Behre & Schwabe, 1970).

Human activities in Antarctica have given rise to considerable concern regarding the introduction of algal taxa. A study of algae carried on expedition equipment and boots, and in soil on imported vegetables, yielded 50 taxa, most of them soil algae. But these did not establish themselves in the vicinity of the research station, they had apparently a very low survivability under Antarctic conditions. In contrast, airborne algae in the same area could survive the harsh conditions of desiccation and freezing/thawing (Broady & Smith, 1994).

As to the effect of scientific activities for algal dispersal, Talling (1951) gives a relevant citation from Carpenter (1928): 'Even the energetic field naturalist may be unconsciously responsible for extending the range of a species, as he empties the residues of his collections of the day into some convenient puddle or stream'. Talling himself (l.c.) admits that he might have transported two species of desmids to a pond where they suddenly appeared – using the same collecting jars as for a pond where they were abundant.

With the great exchange of goods, animals, and plants worldwide between the countries, there is a great risk of dispersal of algae into hitherto pristine areas. One example is the appearance of *Hydrodictyon* in New Zealand (Cofey & Miller, 1988) where this very distinctive alga had never been seen before. It was found in a pond belonging to a hatchery supplying fish and aquatic plants to aquarists. No doubt *Hydrodictyon* had been imported together with fish or plants from East Asia. It had found optimal conditions, and the authors were very concerned about its possible spreading by fragments or by the resistant cysts to other localities.

Dispersal distances

The possibility for successful dispersal depends on distance and of the tolerance of the algae to the transport conditions. This is clearly seen in an example from the Antarctic coast region (Hirano, 1965; Round, 1981) where the whole flora, consisting exclusively of cosmopolitan species, had immigrated from the north. The species numbers of blue-green algae, diatoms and green algae increase towards the mainland. However, the species numbers of dinophytes, chrysophytes, and charophytes decrease towards the mainland, showing that these algae are less capable for transport.

It is thus probable that the chance of dispersal diminishes rapidly with increasing distance. In the desmids, a correlation can be seen between widespread distribution and the more resistant resting stages (Talling, 1951).

Dispersal distances depend on the dispersal methods. Short distance transport is provided by water beetles from one small pool to another (Migula, 1888). Wille (1923) was convinced that seagulls transport algae between the pools where they are wading, but he questioned whether they could disperse algae from one island to another.

Dragonflies have been shown to be effective dispersal agents. It has been observed that they transport freshwater algae to the Cocos-Keeling Islands, 600 miles from the nearest mainland. These islands do not have a suitable environment for dragonfly naiaids, and therefore dragonfly populations cannot persist there. They appear, however, regularly at certain seasons – during the few days of calm following an easterly or northerly wind (Maguire, 1963).

There have been different opinions about how far viable algae can be transported by birds. Schlichting (1960) found that mud and organic debris seldom remained for more than 30 minutes on ducks suspended in air. Perhaps it would be dislodged in even shorter time from birds in flight. But algae attached to feathers may remain viable for at least four hours.

Internal transport should certainly be more successful (comp. also Proctor (1966) for desmids). In the intestine of a waterfowl, algae could be carried for 100–150 miles, in the caecum many times that distance (Proctor, 1959). Atkinson (1980) found viable *Asterionella* in faeces 2–20 hours after feeding, 2 hours corresponding to 220 km flight. *Chara* oospores may remain alive in the digestive tract of killdeer for up to 24 hours (Proctor et al., 1967). Many ducks in North Europe migrate up to 4800 km, but with many stops. The longest non-stop distance is made by *Anser albifrons* flying from West Greenland to Scotland, 3200 km, in 48 hours (Atkinson, 1980).

The existence of bipolarity in the distribution of chrysophytes also requires possibilities for long distance north-south bird transport (Kristiansen & Vigna, this volume). The same species of blue-green algae have been sampled both from the Arctic and from Antarctica. These algae must have been transported by N-S-migrating birds, e.g. south polar skua and arctic tern. A specimen of south polar skua banded in Antarctica has been recaptured in Greenland (Schlichting et al., 1978).

For airborne transport, it is impossible to give any maximal distances, as even the most remote islands have a flora of freshwater algae; Kerguelen and other Antarctic islands have a flora similar to that of Arctic islands (Therezien & Couté, 1977). Stevenson & Collier (1962) calculated the distances the phytoplankton had been blown over the sea during their observations to be up to at least two miles. Maynard (1968) found viable *Melosira granulata* suspended in the air at a height of 3000 m, a considerable dispersal distance might then be possible. Live concentrations of the freshwater alga *Planktonema lauterbornii* have been found in clouds above Antarctica (Saxena, 1983). Darby et al. (1974) found freshwater diatoms such as *Melosira granulata* and *Stephanodiscus astraea* in snow in the Arctic, on the pack ice 500 km north of Alaska, presumably transported from Alaska or Siberia, just as freshwater diatoms have been found on the Greenland ice cap, certainly transported from North America (Gayley et al., 1989). Brown snow, containing e.g. chrysophyte cysts, was found in Arctic Canada (Saqvaqjuac in the Hudson Bay; according to its chemical composition it must have been derived from northern China (Welch et al., 1991). And it has been shown (Folger, 1970) that dust containing valves of freshwater diatoms has been blown across the Atlantic Ocean.

Concluding remarks

It is evident from this review that our knowledge of algal dispersal mechanisms is fragmentary and that the results of the investigations are inconsistent. Much research remains to be done, especially as regards the relation between migration routes of birds and distribution patterns of fresh water algae.

In general, the evidence from geographical distributions suggests that the dispersal of freshwater organisms is often unusually effective, although the reasons for this are still difficult to define (Talling, 1951). And, as Gislén (1948) wrote: 'Wherever surroundings are suitable, the most astonishing finds can be made of forms which have been described earlier in far away places on the globe... Wherever a biotope suitable for a certain microorganism exists, that organism will appear there as soon as sufficient time has elapsed to allow it to be transported through the air and to settle in that locality'.

Acknowledgements

Much information about literature on airborne algae has been obtained from an unpublished list compiled by Dr Margaret Harper, New Zealand.

References

Atkinson, K. M., 1970. Dispersal of phytoplankton by ducks. Wildfowl 21: 110–111.

Atkinson, K. M., 1971. Further experiments in dispersal of phytoplankton by birds. Wildfowl 22: 98–99.

Atkinson, K. M., 1972. Birds as transporters of algae. Br. phycol. J. 7: 319–321.

Atkinson, K. M., 1980. Experiments in dispersal of phytoplankton by ducks. Br. phycol. J. 15: 49–58.

Atkinson, K. M., 1988. The initial development of net phytoplankton in Cow Green Reservoir (Upper Teesdale), a new impoundment in Northern England. In F. E. Round (ed.), Algae and the aquatic environment. Biopress, Bristol: 30–43.

Behre, K. & G. H. Schwabe, 1970. Auf Surtsey-Island in Sommer 1968 nachgewiesene nicht marine Algen. Schr. naturw. Ver. Schleswig-Holstein 1970: 31–100.

Broady, P. A. & R. A. Smith, 1994. A preliminary investigation of the diversity, survivability and dispersal of algae introduced into Antarctica by human activity. Proc. NIPR Symp. Polar Biol. 7: 185–197.

Brown, R. M., D. A. Larson & H. C. Bold, 1964. Airborne algae; their abundance and heterogeneity. Science 143: 583–585.

Burckle, L. H., R. I. Gayley, M. E. Ram & J. R. Petit. Biogenic particles in Antarctic ice cores and the sources of Antarctic dust. Antarct. J.U.S. 23: 71–71.

Carausu, L., 1968. The ways of populating of the new reservoirs. 1. The rôle of waterfowls. Lucr. Sta. Cerc. biol. geol. geogr. Stejarul 1: 263–265.

Carpenter, K., 1928. Life in inland waters. London.

Coesel, P., 1974. Notes on sexual reproduction in desmids. Acta Bot. Neerl. 23: 361–368.

Coesel, P. F. M., R. Duque & G. Arango, 1988. Distributional patterns in some neotropical desmid species (Algae, Chlorophyta) in relation to migratory bird routes. Revue Hydrobiol. trop. 21: 197–205.

Cofey, B. T. & S. T. Miller, 1988. *Hydrodictyon reticulatum* (L.) Lagerheim (Chlorophyta): A new genus recorded from New Zealand. N. Z. J. Bot. 26: 317–320.

Darby, D. A., L. H. Burckle & D. L. Clark, 1974. Airborne dust on the Arctic pack ice, its composition and fallout rate. Earth and Planet Sci. Lett. 24: 166–172.

Darwin, C., 1839. Journal of researches in the geology and natural history of the various countries visited by H. M. S. Beagle. London, 615 pp.

Ehrenberg, C. G., 1849. Passatstaub un Blutregen. Abh.Königl. Akad. Wiss. Berlin 1847: 269–460.

Foged, N., 1953. Diatomeer indslæbt med mellemskarv, Phalacrocorax carbo sinensis. Bot. Tidsskr. 50: 63–74.

Foged, N., 1975: Luftbårne diatomeeer. Flora & Fauna 81: 51–55.

Folger, D. W., 1970. Wind transport of land–derived mineral, biogenic and industrial matter over the Atlantic. Deep Sea Res. 17: 337–352.

Gayley, R. I., M. Ram & E. F. Stoermer, 1989. Seasonal variation in diatom abundance and provenance in Greenland ice. J. Glaciol. 35: 290–292.

Geissler, U. & J. Gerloff, 1965. Das Vorkommen von Diatomeeen in menschlichen Organen und in der Luft. Nov. Hedw. 10: 565–577.

Gislén, T., 1948. Aerial plankton and its conditions of life. Biol. Rev. 23: 109–126.

Grönblad, R., 1933. A contribution to the knowledge of subaërial desmids. Soc. Sci. Fenn. Comm. Biol. 4: 1–7.

Hirano, M., 1963. Freshwater Algae in the Antarctic regions. In J. van Mieghem & P. van Oye (eds), Biogeography and ecology in Antarctica. Dr W. Junk Publishers, The Hague: 127–193.

Hustedt, F., 1943. Die Diatomeenflora einiger Hochgebirgsseen der Landschaft Davos in der Schweizer Alpen. Int. Revue ges. Hydrobiol. 43: 225–280.

Irénée-Marie, Frère, 1938. Flore desmidiale de la région de Montréal. Laprairie, Canada: 547 pp.

Kristiansen, J. & M. S. Vigna, 1996. Bipolarity in the distribution of silica-scaled chrysophytes. In: J. Kristiansen (ed.), Biogeography of Freshwater Algae. Developments in Hydrobiology 118. Kluwer Academic Publishers, Dordrecht: 121–126. Reprinted from Hydrobiologia 336.

Luther, H., 1963. Botanical analysis of mute swan faeces. Acta Vertebrat. 2: 266–267.

Maguire, B., 1963. The passive dispersal of small aquatic organisms and their colonization of isolated bodies of water. Ecol. Monogr. 33: 161–185.

Manguin, E. & R. Leboime, 1948. Sur la présence anormale de diatomées marines dans les cuvettes d'eau douce de l'Alpinum du Muséum. Bull. Mus. natn. Hist. nat. Paris 2e sér., 20: 311–314.

Maynard, N. G., 1968. Significance of air-borne algae. Z. allg. Mikrobiol. 8: 225–226.

Messikommer, E. L., 1943. Untersuchungen über die passive Verbreitung der Algen. Schweiz. Z. Hydrol. 9: 310–316.

Messikommer, E. L., 1948. Algennachweis in Entenexkrementen. Hydrobiol. 1: 22–27.

Migula, W., 1888. Die Verbreitungsweise der Algen. Biol. Zentr. 8: 514–517.

Milliger, L. E. & H. E. Schlichting, 1968. The passive dispersal of viable algae and protozoa by an aquatic beetle. Trans. am. microsc. Soc.87: 443–448.

Overeem, M. A. van, 1937. On green organisms occurring in the lower troposphere. Trav. Bot. Neerl. 34: 388–442.

Parsons, W. M., H. E. Schlichting & K. W. Stewart, 1966. Inflight transport of Algae and Protozoa by selected Odonata. Trans. am. microsc. Soc. 85: 520–527.

Péterfi, L. S. & L. Momeu, 1996. Three *Mallomonas* species (with special distribution patterns. In: J. Kristiansen (ed.), Biogeography of Freshwater Algae. Developments in Hydrobiology 118. Kluwer Academic Publishers, Dordrecht: 143–150. Reprinted from Hydrobiologia 336.

Pouchet, M. F., 1860. Micrographie atmosphérique. C. r. Acad. Sci. Paris 50: 532–534.

Proctor, V. W., 1959. Dispersal of freshwater algae by migratory waterbirds. Science 130: 623–624.

Proctor, V. W., 1963. Viability of *Chara* oospores taken from migratory water birds. Ecology 43: 528–529.

Proctor, V. W., 1966. Dispersal of desmids by waterbirds. Phycologia 5: 227–232.

Proctor, V. W., C. Malone & V. L. DeVlaming, 1967. Dispersal of aquatic organisms: viability of disseminules from the digestion tract of captive killdeer. Ecology 48: 672–676.

Ralfs, J., 1848. The British Desmidieae. London, 22 + 226 pp.

Rosas, I., G. Roy-Ocotta & P. Mosino, 1989. Meteorological effects on variation of airborne algae in Mexico. Int. J. Biometeorol. 33: 173–179.

Roscher, J. P., 1967. Alga dispersal by muskrat intestinal contents. Trans. am. micr. Soc. 86: 497–498.

Round, F. E., 1981. The ecology of algae. Cambridge, 653 pp.

Saxena, V. K., 1983. Evidence of the biogenic nuclei involvement in Antarctic coastal clouds. J. Phys. Chem. 87: 4130–4134.

Schlichting, H. E., 1960. The role of waterfowl in the dispersal of algae. Trans. am. Micr. Soc. 79: 160–166.

Schlichting, H. E., 1961. Viable species of Algae and Protozoa in the atmosphere. Lloydia 24: 81–88.

Schlichting, H. E., 1964. Meteorological conditions affecting the dispersal of airborn Algae and Protozoa. Lloydia 27: 64–78.

Schlichting, H. E., 1974. Periodicity and seasonality of airborne Algae and Protozoa. In Lieth, H. (ed.), Phenology and seasonality modeling. Springer, Berlin: 407–413.

Schlichting, H. E., B. J. Speziale & R. M. Zink, 1978. Dispersal of Algae and Protozoa by Antarctic flying birds. Antarct. J. U.S. 13: 147–149.

Schwabe, G. H., 1972. Blue-green algae as pioneers on post- volcanic substrate (Surtsey/Iceland). In T. V. Desikachary (ed.), Taxonomy and morphology of the blue-green algae. University of Madras, Madras: 419–424.

Schwabe, G. H. & K. Behre, 1972. Algae on Surtsey in 1969-1970. Surtsey Research Progress Report 6: 129–133.

Schwabe, G. H. & B. Hickel, 1978. The morphological variability of a *Synechococcus* clone from postvolcanic substrate on the island of Surtsey (Iceland). Surtsey Research Progress Report 7: 25–27.

Stevenson, R. E. & A. Collier, 1962. Preliminary observations on the occurrence of air-borne marine phytoplankton. Lloydia 25: 89–93.

Strøm, K. M., 1926: Norwegian mountain algae. Skr. Norske Vid. Akad. Oslo 1: 1–263.

Talling, J. F., 1951. The element of chance in pond populations. The Naturalist 1951: 157–170.

Therezien, Y. & A. Couté, 1977: Algues d'eau douce des îles Kerguelen et Crozet (à l'exclusion des diatomecées). CNFRA, 91 pp. Reprint 1982 by Koeltz, Koenigstein.

Tiberg, E., B. Bergman, B. Wictorin & T. Willén, 1984. Occurrence of microalgae in indoor and outdoor environments in Sweden. In Nilsson, S. (ed.), Nordic Aerobiology. Almqvist & Wiksell, Stockholm: 24–29.

Velasques, G. T., 1940. On the viability of algae obtained from the digestive tract of the Gizzad Shad, *Dorosoma cepedianum*. Am. Midl. Nat. 22: 376–412.

Wee, J. L., D. J. Booth & M. A. Bossier, 1993. Synurophyceae from the Southern Atlantic Coastal Plain of North America: A preliminary survey in Louisiana, USA. Nord. J. Bot. 13: 95–106.

Welch, H., D. C. G. Muir et al., 1991. Brown snow: a long range transport event in the Canadian Arctic. Envir. Sci. Technol. 25: 280–286.

Wille, N., 1923. Karplanter og ferskvandsalger fra øerne Husøy, Ona og Rosholmen paa Nordmør. Nyt Mag. f. Naturv. 61: 53–89.

Zacharias, O., 1888. Über die Verbreitung niederer Wassertiere durch Schwimmvögel. Biol. Zentralbl. 8: 368–369.

Hydrobiologia **336**: 159–161, 1996.
J. Kristiansen (ed.), Biogeography of Freshwater Algae.

17. Biogeography of freshwater algae – conclusions and perspectives

Jørgen Kristiansen
University of Copenhagen, Botanical Institute, Dept. of Phycology, Øster Farimagsgade 2D, 1353 Copenhagen K, Denmark

As indicated by the title, the contributions in this volume in various ways deal with the biogeography of freshwater algae. This problem complex is here approached from many different angles, to give as complete a picture as possible of the state of the art within this field.

The present collection of papers mainly deals with the following subjects. In the first hand, it is of fundamental importance to make clear the units we work with and how stable they are. Next, the distribution of species within the various algal groups must be mapped, this involves problems which can be addressed in various ways, using classical and/or molecular methods. In a comparable way, the algal floras of discrete geographical regions can be treated from various points of view. Distributions can be seen as results of historical and dispersal factors. Finally, the problem of dispersal mechanisms is of fundamental importance.

When discussing distribution patterns of algal species, the species concept is crucial. However, just as important is the stability of the taxa in question, such as discussed by Ichimura. Even if species are static in morphological characters, their genomes may vary and be able to adapt in various ways throughout the occurrence area of the species, in response to environmental factors and to geographical isolation.

The morphological species concept apppears often too broad, and a too coarse classification may conceal species diversity. In the diatoms (Mann & Droop), a very narrow species concept may be necessary, based on minute, often ultrastructural characters. Such a classification makes it possible to demonstrate distribution differences and to clarify possible endemisms. On the other hand, a very narrow species concept will result in enormous numbers of species, for the diatoms an estimate of at least 2×10^5 species on a worldwide basis is given.

Actual geographical and distributional studies have been made for several groups of algae. The bluegreens (Hoffmann) have been treated in a classical way, showing world distribution patterns of selected species. Most species are (almost) cosmopolitan, others have restricted distributions, mainly controlled by temperature, but dispersal possibilities are important, and some distant, isolated areas have decidedly poorer species diversity.

Desmids can be treated in the same way. They are wellknown on a world wide basis, and ten desmid regions can be distinguished. In spite of their wide occurrence it is possible that this group of algae may have a tropical origin (Coesel).

The flora of Euglenophytes shows quantitative and qualitative differences throughout China (Shi). The floras of seven different regions have been statistically compared, and it is shown that there is a progressive change from east towards west, both quantitatively and qualitatively, coinciding with the climatic gradients.

Macroalgae from tundra streams in North America (Sheath et al.) are distributed according to ecological factors, such as stream velocity and freezing periods, and to their ability to survive by means of resistant cells. The importance of grazers for the composition of the flora is emphasized – and attention is drawn to the fact that there are no grazers in antarctic streams.

The uniformity of some *Vaucheria* species throughout their distribution areas can be investigated by means of chloroplast genome analysis (Linne von Berg & Kowallik). The degree of infraspecific sequence variability is strikingly high and can be discussed in relation to the biogeography of the species.

Another approach to algal biogeography is the thorough investigation of the diversity of the algal flora of a single region. In the Primorsky region in the Russian Far East (Gontcharov), altogether 1659 species (2276 taxa) were found, and the diversity in different locality types can be compared. Antarctic islands are quite another region type (Ling). Here the snow algae present special problems. All species have been transported by the wind. Even if they superficially resemble known species of soil algae, they give unexpected difficulties in exact identification. This is a warning against too rash biogeographic speculations and conclusions.

The Indo-Malaysian-North- Australian region contains more than 4700 taxa, among these 2680 species of desmids (Vyverman). Many of the species are restricted to tropical areas, and e.g. 800 desmid species have not been found elsewhere and are probably endemic, favourable localities for such are the old Tertiary lakes. However, along altitudinal gradients tropical species are replaced by taxa characteristic of colder climates.

Such climatic dependence in distribution patterns have been shown in the bipolar distribution of some species of silica-scaled chrysophytes (Kristiansen & Vigna). There is a close resemblance between the chrysophyte flora of Tierra del Fuego and of northern temperate areas, and considerably less similarity with the Australian flora. A special antarctic circumpolar chrysophyte flora does apparently not exist.

In the Australian freshwater algal flora several species are supposed to be endemic (Tyler) – but the problem is difficult, and several of them have recently been found in other regions outside Australia. Some of the Australian endemics are robust, others are found in very special and vulnerable environments, and effective conservation measures are necessary to preserve them (as national 'flagships').

How can the history of an algal flora be explored, when no fossils are available? The recent distribution of various taxa characterized by their DNA sequences can be put in relation to events in Earth's history (Coleman), but also also waterfowl migration routes are important for the understanding of distributions.

The latter explanation is given for the distribution of some silica-scaled chrysophytes in Rumania (Péterfi & Momeu), where the location of the localities can be put in relation to the main routes of migratory birds.

On the whole, distribution patterns are dependent on dispersal mechanisms. The final chapter (Kristiansen) gives a survey of the various agents which have been studied, ranging from water and wind to organisms (insects, birds, mammals). Also the role of man is considerable. Dispersal distances are crucial, in relation to the resistance of the algal cells. However, in contrast to the large amount of work which has been done in this field, the dispersal mechanisms for most algae remain largely unknown.

Many aspects have been addressed in this volume, but important problems are far from solved.

Most algal groups are insufficiently known on a world basis. The desmids, the diatoms and the silica-scaled chrysophytes are in that respect some of the best investigated, and even there gaps are numerous. On the whole, much work remains to be done, especially in the tropics, before algal diversity on a world basis is reasonably well known and before exact distributions can be clarified.

A severe problem in that respect is the species concept. How narrow should it be in order to give a true picture of the diversity? Of similar importance is the correctness of species identifications. In most parts of the world, including the tropics and the southern hemisphere, most identifications have been made by means of European and North American manuals, and it must be feared that many identifications have been biassed by this procedure and many new species have not been recognized as such. All these problems meet serious difficulties because of the increasing lack in taxonomic specialists.

Even if we have a true, morphological species concept and correct identifications, then examinations of gene sequences are necessary to elucidate how a species varies within its distribution area, and perhaps where it has originated. In this volume only two of the papers (Linne von Berg & Kowallik; Coleman) have been based on molecular methods, but it must be expected that gene sequencing will be increasingly used and will soon become a standard tool in biogeographical studies on algae. Similarly, cultivation experiments are necessary to show to what extent species are adapted to their environment, and if cosmopolitan or widely distributed species contain distinct ecotypes.

A field which has not been discussed at all, is the regional distribution of phytoplankton assemblages, in realition to environment and to Earth history. Biogeography of fossil algae, is another field which has not yet been touched. Sediment analyses of mainly diatoms and silica-scaled chrysophytes in different parts of the world, compared with events in Earth history, may shed light on present distribution patterns.

In any biogeographical research, dispersal problems are of paramount interest. The only thing which is known with some certainty is apparently that if a locality is suitable for algae, they may sooner or later arrive. How and in what stage the algae are spread is little known. Neither is it known how often a dispersal "attempt" is successful - it might for many algae be a very rare event, requiring a special combination of favourable circumstances. Much more consistent series of experiments, perhaps with marked algae, must be attempted in order to elucidate these problems, especially as regards long distance dispersal.

We – the authors and the editor – hope that this volume has fulfilled its scope: to give a survey of the state of this branch of phycology, to show gaps and inconsistencies in our knowledge, and to point out some future lines of research.

GPSR Compliance

The European Union's (EU) General Product Safety Regulation (GPSR) is a set of rules that requires consumer products to be safe and our obligations to ensure this.

If you have any concerns about our products, you can contact us on ProductSafety@springernature.com

In case Publisher is established outside the EU, the EU authorized representative is:

Springer Nature Customer Service Center GmbH
Europaplatz 3
69115 Heidelberg, Germany

Batch number: 10868238

Printed by Printforce, the Netherlands